PROTEIN SCIENCE AND ENGINEERING

CASEIN: PRODUCTION, USES AND HEALTH EFFECTS

PROTEIN SCIENCE AND ENGINEERING

Additional books in this series can be found on Nova's website
under the Series tab.

Additional E-books in this series can be found on Nova's website
under the E-book tab.

NUTRITION AND DIET RESEARCH PROGRESS

Additional books in this series can be found on Nova's website
under the Series tab.

Additional E-books in this series can be found on Nova's website
under the E-book tab.

PROTEIN SCIENCE AND ENGINEERING

CASEIN: PRODUCTION, USES AND HEALTH EFFECTS

ANTHONY M. VENTIMIGLIA
AND J.M. BIRKENHÄGER
EDITORS

Nova Science Publishers, Inc.
New York

For permission to use material from this book please contact us:
Telephone 631-231-7269; Fax 631-231-8175
Web Site: http://www.novapublishers.com

Library of Congress Cataloging-in-Publication Data

Casein : production, uses, and health effects / editors, Anthony M. Ventimiglia and J.M. Birkenhdger.
p. cm.
Includes index.
ISBN 978-1-62100-129-4 (hardcover)
1. Casein. 2. Casein--Production control. 3. Casein--Health aspects. I. Ventimiglia, Anthony M. II. Birkenhdger, J. M.
TP248.C4C37 2011
668.4'3--dc23

2011031023

Published by Nova Science Publishers, Inc. † New York

CONTENTS

PREFACE

Casein is the name for a family of related phosphoprotein proteins. These proteins are commonly found in mammalian milk, making up 80% of the proteins in cow milk and between 60% and 65% of the proteins in human milk. Casein has a wide variety of uses, from being a major component of cheese, to use as a food addictive, to a binder for safety matches. In this book, the authors present topical research in the study of the production, uses and health effects of casein, including the technological aspects of caseins of goat and sheep milk; the use of casein and casein-derived proteins as agents to inhibit dental erosion; the genetics of caseins in domestic ruminants; mineral carriers and potential nutraceuticals of casein phosphopeptides; caseins in traditional Portuguese cheeses; using casein protein for microsphere's synthesis, encapsulation drug and textile coatings and the antimicrobial activities of casein and other milk proteins.

Chapter 1 - Goat and sheep dairy farming are a vital part of the national economy in many countries, especially in the Mediterranean and Middle East regions. In the last years, interest concerning small ruminants milk has been increasing also to find a new exploitation for local breeds. For promoting the goat and sheep dairy products there is a clear need to know the quality and technological aspects of milk produced. There are many qualitative differences between goat and sheep in milk composition and both these are strongly different from cow's milk. As a consequence, these milks have unique properties and specific technological destinations. The milk protein composition *in toto* and the peculiar micellar structure determine particular differences in the rheological proprieties of the curds produced after coagulation. Caseins are synthesized in the mammary gland in response to lactogenic hormones and other stimuli, and form colloidal micelles in milk. They are composed of calcium-sensitive caseins (αs- and β-caseins) and of the calcium-insensitive κ-casein, which is responsible for stabilizing the former against precipitation in the presence of calcium. The caseins concentration in ovine milk is higher than bovine and caprine counterparts. Goat and sheep milk have different proportions of the four major caseins (αs1, αs2, β, κ) compared to cow counterparts, and there are great variations, especially between αs1-casein and αs2-casein contents between individuals and breeds of goats and sheep, because of the occurrence of genetic polymorphisms for all milk proteins, which influence greatly their cheesemaking properties. Goat milk has poorer cheesemaking ability compared to cow milk: this difference is mainly due to the lower casein content but also to the differences in casein micelle composition, size and hydration. Conversely, ovine milk contains a higher fat, protein and total solids percentages if compared to cow and goat milk with a greater cheese yield per unit than others. In fact, cheese curd contains primarily the fat and casein from milk. The rennet curd produced from goat milk results more fragile than that obtained with cow milk. The uniqueness of renneting kinetic of goat milk has also been characterized as a shorter coagulation time and greater hardening rate. Moreover, sheep milk is very sensitive to rennet, because it has a higher β/αs-casein ratio, and also coagulation in sheep milk proceeds faster than in cow milk. Casein micelle structure is similar in cow, goat and sheep milk, although sheep milk caseins are richer in calcium than cow casein. Since caseins play a key role in milk coagulation as responsible for primary structure of cheese curd, the clotting of milk would affect the resultant cheese composition, texture and rheology.

Chapter 2 - Dental erosion is an increasing problem in many countries around the world, and research in this field has increased dramatically in recent years. Dental erosion is the dissolution of tooth tissues (enamel and dentine) by acids that are not of bacterial origin; most commonly these originate from the diet. Recently, protein-based technologies including casein and casein phosphopeptide (CPP) have been investigated and found to have erosion-inhibiting properties. The aim of this chapter is to summarize recent studies by the authors' research groups to investigate the use of casein and casein-derived proteins as agents to inhibit dental erosion. A number of *in vitro* techniques and models are employed to investigate erosion, including non-contact optical profilometry, atomic-force microscopy (AFM) nanoindentation and hydroxyapatite dissolution rate experiments (hydroxyapatite is the main mineral component of teeth). Evidence is given for an almost instantaneous 'protective' effect afforded by casein, preventing demineralization of enamel under erosive conditions. Aqueous solutions of casein of 0.5 % w/v applied topically before an erosive challenge are shown to afford between 39 to 45 % reduction in demineralization over a range of clinically relevant timescales. The mechanistic aspects of the 'protective effect' afforded by casein are investigated using AFM and X-ray reflectometry (XRR). Data from both of these techniques show the formation of a thin (~ 6.6 nm) film of casein proteins on mica, a molecularly smooth mineral used as a model substrate.

Chapter 3 - Milk caseins are the major protein source for the newborn, providing amino acids that are essential to sustain growth as well as peptides with immunomodulatory activities. Caseins have also a remarkable impact on the dairy industry, being the main determinant of the efficiency with which milk is transformed into cheese. Casein genes have been thoroughly studied in cattle, sheep and goats, allowing for the stablishment of to causal relationships between their polymorphism and milk composition. In cattle, polymorphism in the β-casein (*CSN2*) and κ-casein (*CSN3*) gene has been reported to have effects on the composition and rheological properties of milk. Similarly, the variability of the goat α_{S1}-casein (*CSN1S1*) gene has been shown to determine the synthesis rate of this protein as well as to have effects, at least in certain breeds, on milk composition and cheese yield and organoleptic properties. Even more, the caprine *CSN3* gene has been associated with milk protein and casein levels. These findings confirm domestic ruminants as a unique model to elucidate the genetic factors that modulate the amount of synthesized caseins and milk composition.

Chapter 4 - The gastrointestinal digestion of casein gives rise to a series of peptides displaying particular features and biological activity, such as the immunostimulant, ACE inhibitory, antihypertensive, opioid agonist/antagonist, antithrombotic and antimicrobial peptides. A last group of peptides, named casein phosphopeptides, are defined mineral carriers and originate from the proteolytic cleavage of αs_1-, αs_2- and β-casein. These casein phosphopeptides are characterized by the presence of phosphate linked to serine residues all over the amino acid sequence but, above all, localized in a particular Ser(P)-Ser(P)-Ser(P)-Glu-Glu sequence, known as the “acidic motif”. Although casein-derived phosphopeptides (CPPs) correspond to different phosphorylated regions of αs_1-, αs_2- and β-caseins, all CPPs present the “acidic motif”, which was highly conserved among species and centuries. The presence of phosphate confer them a high negative charge responsible for their further proteolysis and undiscussed ability to bind minerals such as calcium, iron and zinc. Numerous studies aimed to demonstrate the possibility to use the CPP mineral binding capacity for improving the absorption of calcium and iron at intestinal level, as well as for enhancing remineralization of tooth enamel. Besides the controversial results obtained with calcium transport studies performed using rat intestinal preparations, the use of in vitro human intestinal cell models do confirm that CPPs induce calcium uptake in differentiated intestinal cells and this action is correlated to their ability to form aggregates of defined size with calcium ions. Also in human primary osteoblast like cells CPPs increase the intracellular calcium concentration and the in vitro cell mineralization. In all these events the calcium bound to CPPs represents the fraction of soluble mineral involved in the process. Finally, new and interesting potentialities for CPPs come from recent data displaying a modulation of the intestinal immune system by triggering cytokine secretion and stimulating IgA production, the release of IL-6 cytokine in human epithelial intestinal cell lines, the modulation of cell viability, i.e. proliferation and apoptosis in different human cell cultures. Taken together these findings claimed for a real possibility to use CPPs as functional food, due to their presence in dairy products and their physiological formation in vivo, and for nutraceuticals, due to possibility to add to mouth rinse solutions, toothpaste and chewing gum, but also to confectionary products (breakfast foods, sweets, cakes, ice-cream, milk, powdered milk, yogurt, cheese, sport drinks, mayonnaise..) due to their good solubility in water, which will allow to use these peptides as an easy-to-formulate ingredient in food.

Chapter 5 - Portugal holds a long tradition in cheesemaking, and several unique cheese specialities are indeed available; most such cheeses possess a Protected Denomination of Origin (PDO) status. Each artisanal cheese is intrinsically related to its region of origin, and the corresponding manufacture protocol entails a multicultural history and a complex set of empirical practices passed from generation to generation – which confer unique features to each cheese variety, arising namely from distinct casein degradation patterns. The distinct patterns of casein breakdown are chiefly associated with the specific proteolytic activity of proteases present in the plant (*Cynara cardunculus*) or animal rennet used to bring about milk clotting. However, several other factors also affect the extent and quality of casein hydrolysis in said cheeses, e.g. ripening time, lactation season, proportion of sheep and caprine milk, and mode and amount of salting. Hence, cheesemaking conditions play a critical role upon the profile of casein-derived peptides generated during ripening; they determine not only the final sensory characteristics of Portuguese traditional cheeses, but also their nutritional value. Meanwhile, casein-derived bioactive peptides in cheeses manufactured following traditional procedures have been under extensive scrutiny; peptides possessing angiotensin-converting enzyme-inibitory and antioxidant activities were accordingly found in raw ovine cheese-like systems coagulated with *C. cardunculus*. Finally, adulteration of traditional cheeses bearing PDO labels is a current problem, so methods to monitor the authenticity of said cheeses based on assessment of their casein profiles have been developed – and shown to be accurate and useful in detecting fakes. This chapter reviews the fate of caseins in Portuguese traditional cheeses, from a fundamental point of view – coupled with their being potential precursors of bioactive peptides, or else the applicability of the resulting peptides as markers of authenticity in quality assurance strategies.

Chapter 6 - Milk proteins are natural vehicles, which evolved to deliver essential micronutrients and building blocks, as well as immune system components, from mother to the newborn. Casein is the principal protein found in cow's milk from which it has been extracted commercially for most of the 20th century. It is responsible for the white, opaque appearance of milk in which it is combined with calcium and phosphorus as clusters of casein molecules, called .micelles. The major uses of casein until the 1960s were in technical, non-food applications such as adhesives for wood, in paper coating, leather finishing and in synthetic fibres, as well as plastics for buttons, buckles... During the past 30 years, however, the principal use of casein products has been as an ingredient in foods to enhance their physical (so-called .functional.) properties, such as whipping and foaming, water binding and thickening, emulsification and texture, and to improve their nutrition. The caseins are proline-rich, open-structured rheomorphic proteins which have distinct hydrophobic and hydrophilic domains. The major advances of the past year in harnessing milk proteins for novel health-promoting delivery applications were mainly in nanosizing, conjugation, crosslinking and targeting. Novel milk-protein nanoparticles were introduced for solubilizing and protecting hydrophobic nutraceuticals in clear systems, or for targeting gastric tumors, utilizing the natural digestibility of caseins. Several interesting advancements were reported in the application of milk proteins for drug targeting. Casein like all raw material of animal origin requires sterilization before the processing into consumable products or drugs. Many methods can be used to sterilize casein. Ionizing irradiation can be a very powerful method to sterilize casein and can also reduce the allergenicity of milk proteins by the destruction of the human IgE-binding epitopes in protein allergens. Ionizing irradiation can damage protein. One approach to achieve this goal is the addition of compounds that act a radioprotectors. Milk proteins have a common feature related to their cross-linking affinity, and its effect on texture, as well as nutritional properties. The reactions of milk proteins with some functional groups of other compounds in food have gained increasing attention in food research. It was found that *Camellia sinensis* extract can protect casein against gamma irradiation. Also a decrease of the dose of bacterial decontamination D_{10} was observed after addition of *Camellia sinensis* extract to casein.

Chapter 7 - Casein is a protein that is found in milk and used independently in many foods as a binding agent. Presently, there are numerous applications of casein such as coatings, adhesive, as a paper packaging material, in the leather, textile, and food industries. In this review we describe the use of casein as a building block for the fabrication of proteinaceous microspheres (PM) and film materials. Casein can be used as an interface material holding stable microsphere emulsions made from water and hydrophobic solvents, as well it can be used as matrix material for producing stable films for medical applications. The current manuscript will attempt to review the experiments that have been conducted for the production of microsphere emulsions and the encapsulation of drugs inside the casein spheres. Moreover, the paper will describe the chemical modifications of the casein protein made with the bifunctional compound, glutaraldehyde.

Chapter 8 - The genetic polymorphisms of milk proteins are of significance as they are associated to quantitative and qualitative parameters in milk. Extensive investigation of goat casein polymorphism revealed the presence of high numbers of alleles on the four casein loci.

Goat casein polymorphism is associated with different casein synthesis levels and influence also fat content, micellar size and mineralization, and sensory attributes in milk therefore playing an important role both in milk coagulation performance and cheese quality. Genetic polymorphisms of milk proteins is also associated to different degrees of allergic reaction and to the generation of a wide spectrum of casein derived peptides able to exert multifunctional biological activities. Selection and breeding of particular goat genetic lines could be performed for the production of milk destined to cheesemaking performance and of milk with hypoallergenic and functional properties.

Chapter 9 - Milk proteins and particularly caseins (αs1-, αs2-, β- and κ-) are widely used in the food industry as emulsifiers. Caseins, thanks to their substantial surface activity, adsorb rapidly at the oil-water interface during emulsion formation, and facilitate the formation of a strong, protective film around the oil droplet. The structure and composition of the protein interfacial layer determine to a large extent the physic-chemical properties of the emulsion, which in turn play significant role in the shelf-life and acceptability of the food product. This paper describes the role of the molecular structure of caseins either as monomers or as proteinaceous mixtures in emulsion formation and stabilization. Evidence exists which indicates the conformational changes in the three-dimensional structure of proteins following adsorption at the interface. Furthermore, competitive adsorption of the milk proteins occurs during emulsion formation, which affects the protein composition of the interfacial layer. Additionally, this article reviews recent advances in understanding the effect of processing factors on the interfacial protein coverage, composition and molecular configuration of caseins in the oil-in-water emulsion. These factors include heat treatment, pH, and ionic strength, which seem to affect the protein functionality and thus, emulsion stability. This paper describes the key findings of research undertaken over the last 25 years, which aim to provide some quantitative and qualitative insight on how extrinsic parameters, such as thermal processing, determine the physicochemical behavior of food emulsions. According to experimental evidence, processing conditions can be manipulated by the food scientist depending on the nature of the product, in order to be able to control the mechanisms involved in emulsion destabilization. Nevertheless, the detailed molecular mechanisms involved in emulsion formation and stabilization, which presupposes understanding of the behavior of individual protein molecules, alone or in combination, at interfaces, remains in many cases under investigation.

Chapter 10 - Type 1 Diabetes (T1D) and Celiac Disease (CD) are autoimmune disorders, which share a genetic component associated to HLA-DQ2 and DQ8 haplotypes. However, more than 25% of any given population presents these haplotypes and only a small proportion develops T1D or CD. Therefore, other cofactors of the onset and exacerbation of these diseases has been looked for. Intestinal microbiota changes induced by infections and antibiotic treatments as well as feeding regimes for the first months of life seem to be important. The causal relationship between intake of wheat gluten and related cereal proteins (gliadins) and CD onset in genetically susceptible subjects is well established, and a similar effect has been observed in T1D animal models. Furthermore, there is evidence that bovine milk caseins such as those present in infant formulas correlate to T1D development and to the intestinal inflammatory response typically associated with CD. The autoimmune response cascade in both diseases appears to be related to the chemotactic properties of caseins and gliadins towards human leucocytes. Furthermore, gliadins and caseins are also partially resistant to digestion because of their high proline content, aside of partial similarity in their amino acid sequences. In this chapter we discuss the role of bovine milk caseins as a risk factor for T1D and CD onset and progression on subjects with genetic predisposition. We focus on breastfeeding practices as protective factors and on the gradual and moderate gluten and bovine milk introduction in the infant´s diet while still breastfeeding.

Chapter 11 - Casein, the main milk protein that constitutes approximately 75% of that protein, is in the form of micelles with a radius of 0.1 μm. Most of the methods used for the preparation of casein from milk basically involve a removal of the fat of the milk followed by a destabilisation process of the desired protein by insolubilization. This process is achieved either by an isoelectric precipitation, i.e. by acidification in order to reach the isoelectric point of the casein, or an enzymatic destabilisation of the casein micelle by the addition of the rennin enzyme, which could be of animal, plant or microbial origin. An important and essential use of casein is in the process cheese industry, particularly imitation cheese. Process cheese products are manufactured by blending different types of cheese, or dry casein in the case of imitation cheese, with water, fat, salts and other ingredients like flavour or colour enhancing compounds in the presence of the necessary thermal and mixing energy required. This chapter reviews the different types of casein and their use in the manufacture of process imitation cheese products. These types of protein include mainly rennet casein, acid casein, caseinate, phosphocasein, enriched casein fractions, casein fragments and casein-whey protein aggregates. The most commonly used casein in the process cheese manufacture is the acid- and rennet casein and/or also the sodium caseinate. This is due to the fact that these protein types are the main cause of the physico-chemical stability of the cheese end-product obtained. The current chapter illustrates recent research work done on the use of casein, as a principal source of protein, and its major impact on the physico-chemical (functional) properties, i.e. heat-induced flowability, viscoelastic properties and hardness, of imitation cheese. Traditionally, rennet casein is the casein type used in the manufacture of this type of cheese. Increased casein content resulted in increased viscoelastic properties and hardness, whereas a partial replacement of casein by whey proteins led to a reduced heat- induced flowability due to the interactions between casein and whey proteins. Interestingly, a partial substitution of the casein used by rice starches led to changes in the flowability and the hardness depending on the amylose content of the starch.

Chapter 12 - Casein is an important milk protein. β-casomorphins (β-CM)are opioid peptides produced by the digestion of β-casein. Two major(A^1 and A^2)alleles of β-casein exist in cattle populationsβ-casein A^1 and related minor variants bearing histidine at residue 67 yield bioactive β-CM-7whereas the A^2 variant with proline in that position does not generateβ-CM-7 by gastric and gut enzymes during digestion. A significant relationship was observed between consumption of β-casein A^1 and the incidence of type 1 diabetes. LDL oxidation byβ-CM-7 constitutes an important link between the risk of heart disease and milk protein consumption.It has been postulated that neurological conditions such as autism, schizophrenia, post-partum depression, and and sudden infant death syndrome may be related to β-casein intake since β-CM-7 reaches the brain cells and cause symptoms associated with the diseases Caseinophosphopeptides are produced on an industrial-scale and have been considered for application as components in 'functional foods' and pharmaceutical products randomized clinical trials are needed to reveal if dairy products could be a complementary treatment for oral health.

Chapter 13 - Milk contains a variety of proteins with important biological activities. Many of them including casein, lactoferrin,alpha-lactalbumin and beta-lactoglobulin possess antimicrobial activities either in the native state or after chemical modification. Alpha(S2)-casein and alpha-lactalbumin acquire anti-HIV activity after modification with3-hydroxyphthalic anhydride Casein, alpha-lactalbumin, and beta-lactoglobulin inhibit HIV-1 protease and integrase to an appreciable extent but do not affect HIV-1 reverse transcriptase. Milk proteins (casein and beta-lactoglobulin) other than bovine lactoferrin do not prevent hepatitis C infection, Methylated or ethylated beta-casein alpha-lactalbumin and beta-lactoglobulin exert an antiviral action against bacteriophage M13, proportional to the extent of esterification and, hence, to the increased basicity of the modified proteins The tri-, tetra-, and pentaphosphorylated forms of human beta-casein inhibit *H. influenzae* ,a pathogen implicated in middle ear infection in infants. Kappa-casein inhibits *Helicobacter pylori* adhesion to gastric mucosa.Lactoferrin and its fragments lactoferricin and lactoferrampin manifest inhibitory activity against a large number of viruses, bacteria and fungi.

Chapter 14 - Oral health is a part of general health and hence affects the total well-being of individuals. Dental and oral diseases affect various aspects of quality of life. Dental caries has been associated with numerous adverse effects on children's health including pain, restricted dietary intake, impaired growth, and reduced body weight [Hamissi et al., 2008].Even though the onset of dental caries has decreased through the use of preventive systems (improved oral hygiene, usage of fluoride-containing toothpaste, fluoride content in drinking water, sealing), tooth decay is still one of the most prevalent chronic worldwide diseases [Al-Malik and Rehbini, 2006; de Almeida et al., 2003; Ferrazzano et al., 2006; Ferro et al., 2007; Jacobsen and Young, 2003; Selwitz et al., 2007; Simonsen, 2002].

In: Casein: Production, Uses and Health Effects
Editors: A. M. Ventimiglia, et. al, pp. 1-25
ISBN: 978-1-62100-129-4

Chapter 1

CASEINS OF GOAT AND SHEEP MILK: ANALYTICAL AND TECHNOLOGICAL ASPECTS

Maria Selvaggi and Vincenzo Tufarelli
Department of Animal Production
University of Studies of Bari "Aldo Moro"
70010 Valenzano (BA) Italy.

ABSTRACT

Goat and sheep dairy farming are a vital part of the national economy in many countries, especially in the Mediterranean and Middle East regions. In the last years, interest concerning small ruminants milk has been increasing also to find a new exploitation for local breeds. For promoting the goat and sheep dairy products there is a clear need to know the quality and technological aspects of milk produced. There are many qualitative differences between goat and sheep in milk composition and both these are strongly different from cow's milk. As a consequence, these milks have unique properties and specific technological destinations. The milk protein composition *in toto* and the peculiar micellar structure determine particular differences in the rheological proprieties of the curds produced after coagulation.

Caseins are synthesized in the mammary gland in response to lactogenic hormones and other stimuli, and form colloidal micelles in milk. They are composed of calcium-sensitive caseins (αs- and β-caseins) and of the calcium-insensitive κ-casein, which is responsible for stabilizing the former against precipitation in the presence of calcium.

The caseins concentration in ovine milk is higher than bovine and caprine counterparts. Goat and sheep milk have different proportions of the four major caseins (αs1, αs2, β, κ) compared to cow counterparts, and there are great variations, especially between αs1-casein and αs2-casein contents between individuals and breeds of goats and sheep, because of the occurrence of genetic polymorphisms for all milk proteins, which influence greatly their cheesemaking properties. Goat milk has poorer cheesemaking ability compared to cow milk: this difference is mainly due to the lower casein content but also to the differences in casein micelle composition, size and hydration. Conversely, ovine milk contains a higher fat, protein and total solids percentages if compared to cow and goat milk with a greater cheese yield per unit than others. In fact, cheese curd contains primarily the fat and casein from milk. The rennet curd produced from goat milk results more fragile than that obtained with cow milk. The uniqueness of renneting kinetic of goat milk has also been characterized as a shorter coagulation time and greater hardening rate. Moreover, sheep milk is very sensitive to rennet, because it has a higher β/αs-casein ratio, and also coagulation in sheep milk proceeds faster than in cow milk. Casein micelle structure is similar in cow, goat and sheep milk, although sheep milk caseins are richer in calcium than cow casein. Since caseins play a key role in milk coagulation as responsible for primary structure of cheese curd, the clotting of milk would affect the resultant cheese composition, texture and rheology.

INTRODUCTION

Milk and dairy products are part of a healthy Mediterranean diet which, besides cow's milk, also consists of sheep's, goat's and buffalo's milk as raw material. Goat and sheep dairy farming are a vital part of the national economy in many countries, especially in the Mediterranean and Middle East regions. The economic importance of milk production is largely linked to suitability of the area for grass production along with government policy and the availability of export markets. In the last years, interest concerning small ruminants milk has been increasing also to find a new exploitation for local breeds. For promoting the goat and sheep dairy products there is a clear need to know the quality and technological aspects of milk produced. There are many qualitative differences between goat and sheep in milk composition and both these are strongly different from cow's milk. As a consequence, these milks have unique properties and specific technological destinations. The milk protein composition *in toto* and the peculiar micellar structure determine particular differences in the rheological proprieties of the curds produced after coagulation.

Milk is a secretion of the mammalian gland: it is a complex oil-in-water emulsion containing fat, protein, lactose, minerals as well as enzymes, cells, hormones and immunoglobulins. The composition and the physical characteristics of which vary from species to species (Tables 1, 2, 3). Milk proteins include αs1, αs2, β, and κ-caseins, β-lactoglobulin, α-lactalbumin, serum albumin, lactotransferrin, immunoglobulins and β2-microglobulin (Farrell *et al.*, 2004).

The caseins are a family of phosphoproteins synthesized in the mammary gland in response to lactogenic hormones and other stimuli and secreted as large colloidal aggregates termed micelles, which are responsible for many of the unique physical properties of milk. Their abundance and importance to the dairy industry have made them a popular target for study by biochemists leading to them becoming perhaps the most widely studied food proteins (Swaisgood, 1992). In addition, the caseins are regarded as markers of differentiated mammary function, and have been used as model to study mammary development and regulation (Rosen *et al.*, 1992).

Table 1. Composition of goat, sheep and cow milk

Component	Goat	Sheep	Cow
Fat (%)	3.8	7.9	3.6
Protein (%)	3.5	6.2	3.3
Lactose (%)	4.1	4.9	4.6
Total ash (%)	0.8	0.9	0.7
Ca (mg)	134	193	122
P (mg)	141	158	119
Energy (cal/100 ml)	70	105	69

Park (2007); Park *et al.* (2007).

Table 2. Micelle characteristics of goat, sheep and cow milk

Species	Casein micelle size (nm)	Hydration of micelle (g/g)	Mineralization of micelle (g/Ca/100 casein)
Goat	260	1.77	3.60
Sheep	193	n.a.	3.70
Cow	180	1.90	2.90

Remeuf and Lenoir 1986.

Table 3. Concentrations of caseins and whey proteins in goat, sheep and cow milk

	Goat	Sheep	Cow
Total casein (g/100 ml)	2.70	5.14	2.70
αs1-casein (% of total casein)	5.60	6.66	38.00
αs2-casein (% of total casein)	19.20	22.84	12.00
β-casein (% of total casein)	54.80	61.60	36.00
κ-casein (% of total casein)	20.40	8.90	14.00
Whey proteins (%)	0.80	1.06	0.60

Adapted from: Ovinicoltura/UNAPOC (1992); Assenat (1985); Jenness (1974); Raynal-Ljutovac *et al.*(2008); Park *et al.* (2006).

Initially, the caseins were considered as a homogenous substance. Subsequently, Lindstrom-Lang and Kodama (1925) demonstrated that they were composed of two fractions: one which precipitated in the presence of calcium (calcium-sensitive caseins) and the other which was responsible for stabilizing the former against precipitation (calcium-insensitive caseins). Later, Mellander (1939) demonstrated the presence of three different components designated as α-, β-, and γ-casein. Further studies revealed that an even greater number of components were present in the casein fraction (Wake and Baldwin, 1961) and much effort was dedicated to clarify these minor components. However, with the expansion of molecular biology it is now possible to affirm that only four types of caseins exist and the earlier heterogeneity that was recognized by electrophoresis is due to effects of post-translational processing, alternative splicing of the gene product or genetic polymorphisms (Ng-Kwai Hang and Grosclaude, 1992). In 1984, the American Dairy Science Association Committee on Nomenclature and Classification proposed that the nomenclature developed for the bovine caseins, namely the αs1, αs2, β and κ-caseins, be adopted for investigations of milk proteins in other species.

The caseins comprise the major protein fraction of the milk from most species. Their function is to transport calcium phosphate in milk and hence to provide the suckling infant with a source of calcium and phosphorus for bone formation as well as to contribute to the requirement for amino acids (Stewart *et al.*, 1987). The caseins, being secreted proteins, possess N-terminal signal peptides that direct the passage of the newly synthesized polypeptides into the lumen of the endoplasmic reticulum, whereupon the signal peptides are removed from the primary translation products to yield the mature caseins. The four calcium-sensitive caseins show similar molecular weights (around 24 kDa), promoter regions, leader peptide sequences and locations of the major phosphorylation site. These data support the hypothesis of a common evolutionary origin of these genes from the duplications and modifications of a unique ancestral gene (Rijnkels, 2002). The remainder of the sequences appear to be subject to few functional constraints and have accumulated many nucleotide substitutions, as revealed by comparative sequence analysis (Stewart *et al.*, 1987).

αs1- and αs2-Caseins

The αs1-casein, the primary protein in cow milk, is a structural component of the casein micelle and plays a functional role in cheese curd formation (Walstra *et al.*, 1984). Of the calcium-sensitive caseins, the αs1-caseins are characterized by their greater solubility in the presence of calcium. It is a highly phosphorylated protein (Farrell *et al.*, 2004). The nucleotide sequence of the genes (CSN1S1) encoding for the αs1-caseins in many eutherian species shows the presence of a highly conserved signal sequence of 15 residues but quite divergent sequences for the mature proteins suggesting that they are encoded by a rapidly evolving gene family. The αs1-casein consists of 214 amino acids in each species (Masoodi and Shafi, 2010). Bovine αs1-casein exists in two phosphorylated forms containing 8 and 9 phosphates/mol (Mercier *et al.*, 1971; Mercier, 1981). Naturally occurring variants of the αs1-caseins within individual species have been described. In cattle, eight protein variants of the αs1-casein are known and classified from A to H. The two most common variants are B and C which differ in the amino acid exchange Glu→Gly at amino acid position 192 of the mature protein (Farrell *et al.*, 2004).

The αs2-like caseins represent a more disparate group than the αs1-caseins and have been identified from a number of eutherian species such as ovine, caprine and bovine (Boisnard and Pétrissant, 1985; Bouniol, 1992; Stewart *et al.*, 1987). Because *α*s2-casein is probably the most difficult of all caseins to isolate and purify, it has been the least investigated. However, αs2-casein is found throughout mammalian milks and, in some rare instances, contributes up to 29% of the composition of whole casein. It also is found in higher concentrations in larger caprine casein micelles (Tziboula and Horne, 1999). As in the αs1-caseins, the signal sequence is highly conserved but the mature proteins are quite divergent. Like the αs1-caseins, the αs2-like caseins are highly phosphorylated peptides. The complete amino acid sequence of αs2-casein consists 207 residues (Stewart *et al.*, 1987) with a large number of positively charged side chains, especially in the C-terminal segment. Like κ-casein, it contains two Cys residues. However, it has a low Pro content. There are no reports on the presence of αs2-casein in human milk (Martin *et al.*, 1996). The role of αs2-casein in casein micelles has not been studied in detail. αs2-casein comprises up to 10% of the casein fraction in bovine milk; it consists of 2 major and several minor components exhibiting varying levels of post-translational phosphorylation (Swaisgood, 1992; Farrell *et al.*, 2004) as well as minor degrees of intermolecular disulfide bonding (Rasmussen *et al.*, 1992). In cattle, many protein variants of the αs2-casein are known (A-D) The protein variant A was originally determined chemically (Brignon *et al.*, 1977) and later by cDNA and genomic DNA sequencing (Stewart *et al.*, 1987; Groenen *et al.*, 1993). Variants B and C are specific to zebu and yaks, respectively (Grosclaude *et al.*, 1976, 1978, 1982), whereas the D variant has been found at low frequencies in some European bovine breeds and in the African Namchi taurine breed (Grosclaude *et al.*, 1979; Erhardt, 1993; Jann *et al.*, 2004a; Ibeagha-Awemu *et al.*, 2005). The original variant A is the most frequent in all breeds so far investigated and also almost fixed in most western breeds. The different forms of the gene (CSN1S2) and protein are caused by single nucleotide polymorphisms that result in amino acid changes.

Goat αs1-Casein

Until the late 1980s, it was believed that goat milk lacked the protein αs1-casein (Jenness, 1980; Grosclaude *et al.*, 1987; Brignon *et al.*, 1989; Mora-Gutierrez *et al.*, 1991). Multiple variants of the goat αs1-caseins have been described and they are characterized by markedly different levels of expression (Brignon *et al.*, 1990).

Polymorphism of αs1-casein is one of the most interesting and extensively studied in goat and cow milk (Grosclaude *et al.*, 1987). The types of pertinent mutations are single nucleotide substitutions, deletions or large insertions. αs1-casein has very large individual quantitative variations due to the occurrence of genetic polymorphisms (Addeo *et al.*, 1987; Grosclaude *et al.*, 1987). Nine of the currently identified alleles in goat milk (A, B_1, B_2, B_3, B_4, C, H, L and M), identified as strong alleles, are associated with a high level of αs1-casein (3.5 g/l milk), two (E and I) with medium levels (1.1–1.7 g/l), and three (D, F and G; weak alleles) with low levels (0.45 g/l). The O_1, O_2 and N are "null" alleles and produce no αs1-casein in goat milk (Grosclaude *et al.* 1987; Moioli *et al.*, 1998; Rando *et al.*, 2000; Bevilacqua *et al.*, 2002; Ramunno *et al.*, 2005; Sacchi *et al.*, 2005; Sztankóová *et al.*, 2007; Chianese *et al.*, 1997; Grosclaude and Martìn, 1997).

As αs1-casein participates in cheese curd formation, the impact of low αs1-casein in milk can be significant. For example, one result of this difference is that cheese curds made from goat milk tend to be softer than those of cow milk even at similar casein levels (Storry *et al.*, 1983; Remeuf and Lenoir, 1986; Ambrosoli *et al.*, 1988; Risch, 1992). Optimization of curd firmness is important to cheese-makers to maximize cheese yield. In addition, shorter coagulation time (the point at which coagulation is first notable, the time when casein micelles have aggregated sufficiently to form visible flocs) and higher coagulation rate (the measure of how quickly the curd firms once coagulation has begun) would shorten the processing time required for cheese production. The potential for improving milk coagulation properties by selecting for high αs1-casein milk producers may be significant. Research on milk from Saanen and Alpine goat breeds showed that αs1-casein is correlated with other milk components and coagulation properties (Ambrosoli *et al.*, 1988). Milks with high levels of αs1-casein had higher total solids, total protein, casein and phosphorus and lower pH than milks with low levels of αs1-casein. In particular, goat milks containing higher αs1-casein had significantly greater levels of all milk components than milks without αs1-casein while milk that contained intermediate amounts of αs1-casein had significantly more total solids, solid not fat, protein and casein than null samples. Samples with high αs1-casein also had significantly higher total solids, solid not fat and protein than samples containing intermediate amounts (Clark and Sherbon, 2000a). Milk with high solids, particularly solid not fat and protein, began coagulating later (long coagulation time) than milk with low solids, suggesting protein delays the onset of coagulation. Milks with high levels of αs1-casein also had long coagulation time and formed firm curds; these maybe due to the presence of higher amounts of αs1-casein and αs2-casein in the milk (Clark and Sherbon, 2000a). These two protein fractions may delay curd formation by binding Ca^2 ions, making fewer available for binding after proteolysis of κ-casein by rennet (Ambrosoli *et al.*, 1988). Moreover, average coagulation rate values of milks containing high amounts of αs1-casein were significantly higher than for milks lacking αs1-casein (Clark and Sherbon, 2000a). Ambrosoli *et al.* (1988) showed that high levels of αs1-casein were associated with long coagulation time.

Sheep αs1-Casein

Over the last few years, several studies have been aimed at genetic polymorphism of sheep caseins. By electrophoretic, immunochemical, and chromatographic analyses, it has been shown that ovine αs1-casein exists as a number of distinct genetic variants both in the heterozygous and homozygous forms (Addeo *et al.*, 1992). Many phenotypes (A, B, C, D, E, F, H and I) of αs1-casein in ovine milk have been identified by protein electrophoresis (Chianese *et al.*, 1996a; Pirisi *et al.*, 1999; Wessels *et al.*, 2004; Giambra *et al.*, 2010a; Giambra *et al.*, 2010b). The H variant was formerly defined as X variant. The Welsh variant, initially identified by King (1966) in the Welsh mountains, was later identified as an αs1-casein variant (Davoli *et al.*, 1990; Mauriello *et al.*, 1990). Subsequently, it has been designated as αs1-casein D. Primary structures have been determined only for variants A, C, D (Ferranti *et al.*, 1995) and E (Chianese *et al.*, 2007). The primary structure of A, C and D variants differed from each other by amino acid substitutions and phosphorylation degree (Ferranti *et al.*, 1995). Each variant consisted of three differently phosphorylated forms with 9, 10 and 11 phosphate groups, respectively. The differences between these three genetic variants are simple silent substitutions which involve the degree of phosphorylation. Variant C differs from variant A in the substitution of Ser for Pro at position 13, which determines the loss of the phosphate group on site 12 of the protein chain. A further substitution, of Ser P for Asn at position 68, caused the disappearance of both phosphate groups in the residues Ser_{64}

and Ser_{66} in variant D. Thus, variant D is the least phosphorylated αs1-casein variant. While αs1-casein C occurs in all sheep breeds with high frequency (0.485-0.890) (Chianese *et al.*, 1996b; López-Gálvez *et al.*, 1999; Amigo *et al.*, 2000), variants E and F were identified at low frequencies in Italian breeds only (Chianese *et al.*, 1996b; Pirisi *et al.*, 1999). The high frequency of C variant is most probably the result of indirect selection in the past, as it showed correlation with higher total protein and casein content (Pirisi *et al.*, 1999; Amigo *et al.*, 2000), smaller micelle diameters and higher clotting capacity (Pirisi *et al.*, 1997, 1999). On the other side the variant D, which showed negative effects on milk composition and cheese yield (Pirisi *et al.*, 1999) could not be demonstrated in East Friesian sheep, a breed highly selected for milk production. Variant H (formerly αs1-casein "X"; Wessels *et al.*, 2004) seems to be specific for East Friesian sheep as in Italian milk sheep breeds it was not identified (Chianese *et al.*, 1996b; Pirisi *et al.*, 1999). Samples of animals with αs1-casein H are also associated with a reduced protein expression level of this protein (Giambra *et al.*, 2010a). αs1-casein I variant was identified for the first time in Gray Horned Heath (Giambra, 2010c).

Goat αs2-Casein

It has been speculated that the differences in αs2-casein content account for the unique physicochemical characteristics of goat caseins and explain a large part of differences in the technological behavior between caprine and bovine milks (Mora-Gutierrez *et al.*, 1996, 1997; Mora-Gutierrez and Farrell, 2001). In this context, the proportions of αs-casein fractions are in large part responsible for the differences observed between the two milks in cheesemaking (Ambrosoli *et al.*, 1988; Grosclaude, 1988; Remeuf *et al.*, 1995). Concerning the polymorphism of αs2-casein in goat, at least eight alleles have been identified, associated with three synthesis levels (Sacchi *et al.*, 2005). Three variants (A–C) were described by Boulanger *et al.* (1984) and Bouniol *et al.* (1994), which differ by amino acid substitution (Martìn and Addeo, 1995). Variants A, B, C, E (Lagonigro *et al.*, 2001), F (Ramunno *et al.*, 2001a) and G (Erhardt *et al.*, 2002) are associated with normal synthesis levels; whereas D and O are associated with lower and null synthesis level respectively (Ramunno *et al.*, 2001b). The amount of αs2-casein was associated with allergenic properties of milk. Variants A, B, C, E and F displayed a higher allergenic potency than did D and O (Marletta *et al.*, 2004). D variant is a rare defective allele characterized by a deletion, involving part of the exon 11 and of the following intron, that should result in the loss of the codons Pro (CCC_{122}), Thr (ACC_{123}) and Val (GTG_{124}) and the substitution Thr→Asn at position 121 (Ramunno *et al.*, 2001a). The null allele shows a G→A transition at the 80th nucleotide in exon 11, creating a premature stop codon at position 110 (Ramunno *et al.*, 2001b). Mass spectrometry studies revealed that in goat's milk each αs2-casein variant can exist in two forms; the main component corresponds to the full length of 207 residues, while the minor component, originated by a skipping event, lacks the internal sequence Glu_{34}-Glu_{42} or Lys_{35}-Glu_{43} (Ferranti *et al.*, 2000). Recently, two new truncated subvariants of goat αs2-casein variants A and E have been identified (Cunsolo *et al.*, 2006; Ferranti *et al.*, 2000). An evolutionary pathway has been proposed (Sacchi *et al.*, 2005) starting from the A variant and leading independently to the B, C, F alleles, each characterized by a different amino acid substitution with respect both to goat variant A and to the bovine and ovine sequences. Thus, A allele may be considered the ancestral variant.

Sheep αs2-Casein

Ovine αs2-casein appears as the most heterogeneous fraction due to the multiphosphorylation level. Boisnard *et al.* (1991) found two αs2-casein variants in sheep which differed at the amino acid level involving an amino acid replacement, Asn_{49} and Lys_{200} for Asp_{49} and Asn_{200}, respectively. These variants exist as non-allelic long and short forms in the ratio 60-70/30-40% (Mercier and Gaye, 1983; Boisnard *et al.*, 1991). The long and short forms of ovine αs2-casein are encoded by four mRNAs, differing in the presence or absence of two stretches of 44 and 27 nucleotides in the 5'untranslated region and the coding frame, respectively. The physicochemical properties of the long and short αs2-casein should be quite different and should confer unique properties to ovine milk. These four different transcripts probably arise from the incorrect processing of a unique pre-mRNA, resulting in the partial skipping of both the putative exons, with frequencies greater than 90% and 30%, respectively (Boisnard *et al.*, 1991). Chianese *et al.* (1993a) identified three phenotypes, differing in electrophoretic behavior, discovering at the protein level an ovine αs2-casein low molecular weight variant in the Manchega breed. The phenotypic frequency of this new variant in the Manchega breed was 5.5%. Chessa *et al.* (2003) called two phenotypes (identified by isoelectric focusing) A and B, which were recently characterized by amino acid exchanges at positions 75 and 105 of mature protein. The structural analysis give evidence that two amino acid exchanges account for the differences between the ancestral ovine αs2-casein A and the genetic variant B found in three Italian breeds: particularly a non-silent mutation, Asp→Tyr at position 75 and a silent one, Ile→Val at position 105 (Picariello *et al.*, 2009). αs2-casein A and B variants were the most frequent as observed Gentile di Puglia, Sarda, and Comisana breeds (Chessa *et al.*, 2003; Picariello *et al.*, 2009) and subsequently confirmed by Giambra *et al.* (2010c) in many German breeds such as Gray Horned Heath, Merinoland, Merino Mutton and Rhön sheep. In addition, the same authors found two new variants (C and D) for the first time and their genetic control was confirmed. Further biochemical and molecular characterization of these variants are in progress. All the four variants were identified in Merinoland sheep, while A, B and C variants were observed in East Friesian, Gray Horned Heath, Merinoland and Rhön sheep breed (Giambra *et al.*, 2010c).

The significance of the protein variation described at the genetic level needs to be further investigated. It is possible to suppose that the modification of the biochemical properties of the protein may affect the interactions among the four casein fractions within the micelle. Thus the milk produced by different breeds may exhibit different technological quality.

Since the casein αs1 and αs2 from goat and sheep share a great similarity so they could be thought of having the same effect and function.

β-Caseins

β–caseins are described as "calcium-sensitive" because they precipitate in the presence of low concentrations of this cation (Waugh, 1971). They are important in determining the surface properties of casein micelles and essential for curd formation when milk is clotted by the proteolytic enzyme chymosin (Pearse *et al.*, 1986). The β-casein family includes several eutherian and two marsupial proteins for which the amino acid and nucleotide sequence is known (Alexander and Beattie, 1992; Bonsing *et al.*, 1988; Collet *et al.*, 1992; Greenberg *et al.*, 1984; Provot *et al.*, 1989; Stewart *et al.*,1987). Eutherian β-caseins sequences are characterized by a highly conserved 15 residue signal sequence although the mature proteins show a degree of divergence across species in terms of amino acid identity similar to that observed for the αs1-caseins (Ginger and Grigor, 1999). The β-caseins are particularly rich in glutamines, the location of which is well conserved, as is the presence of a single major phosphorylation site near the N-terminus. The number of phosphorylation sites and the level of phosphorylation are less that observed for the αs1- and αs2-caseins. Three main bovine variants for β-casein, A1, A2 and B have been reported. The A2 variant predominates in *Bos indicus* cattle, whereas the A1 and A2 forms occur at approximately equal allele frequencies in Holstein–Friesian cattle (Ng-Kwai Hang and Grosclaude, 1992).

Goat βs2-Casein

The β-casein is the most abundant protein fraction in goats' milk (Jenness, 1980). The heterogeneity of PAGE and immunoblotting patterns was thought to be originated only by a discrete phosphorylation rate (Chianese *et al.*, 1993b): in fact, two main phosphorylation levels (6 and 5P) corresponding to β1 and β2 proteins occur with comparable relative abundance (Richardson and Creamer, 1974) but other forms (3 and 4P) have also been reported (Chianese *et al.*, 1993b).

In the past, the β-casein was considered monomorphic, until several studies showed the presence of a null phenotype in different goat breeds (Dall'Olio *et al.*, 1989; Mahé and Grosclaude, 1993). Subsequently, three variants A, B and C (Mahé and Grosclaude, 1993; Neveu *et al.*, 2002) differing by a single amino acid substitution, were associated with normal β-casein content. The C variant was found to be predominant in Italian goat breeds (Chessa *et al.*, 2005). Moreover, two null alleles were characterized (Ramunno *et al.*, 1995; Persuy *et al.*, 1999). Currently, genetic polymorphism of the goat β-casein gene (CSN2) is characterized by extensive variability, which is under the control of at least 8 autosomal alleles: A, A1, B, C, D, E, and 2 null alleles (O and O′; Mahé and Grosclaude, 1993; Neveu *et al.*, 2002; Galliano *et al.*, 2004; Cosenza *et al.*, 2005; Caroli *et al.*, 2006). The genetic variants A, A1, B, C, D, and E were found to be associated with normal β-casein content. Two null alleles (O and O′) of the CSN2 gene were identified, both characterized by mutations responsible for premature stop codons in exon 7 (Ramunno *et al.*, 1995; Persuy *et al.*, 1999). CSN2 O originates from a single deletion of an A at position 16 of the 7th exon of the gene creating a premature stop codon at position 73. The null allele CSN2 O′ is characterized by a transition (C→T) at position 373 of the 7th exon of the gene, responsible for a premature stop codon at position 182. These phenomena may be responsible for non-functional messengers and for the absence of β-casein in milk (Martìn and Addeo, 1995). Both the null alleles are associated with a non-detectable amount of this protein in milk. The mRNA analysis revealed that the transcript product amounts were almost 10 (Ramunno *et al.*, 1995) and 100 (Persuy *et al.*, 1999) times smaller for the null alleles than for the A variant. The alignment among the mature β-casein sequence of different species suggests that A is the ancestral allele in comparison with C (Chessa *et al.*, 2005). In fact, the presence of Ala at position 177 of the mature protein coded by goat A allele has been found also in other ruminant species, whereas the C allele codes for Val at the same amino acid position. Studies about the aptitude for coagulation of individual milks with "null" β-casein showed, that longer rennet coagulation times were needed than for normal milk and weaker curd firmness occurred (Park *et al.*, 2007).

Sheep βs2-Casein

Ovine whole casein contains 45% β-casein, represented by two multiphosphorylated forms: β1 and β2-caseins which have similar amino acid compositions to bovine β-casein. Ovine β1-casein has a higher phosphorus content and greater electrophoretic mobility at alkaline pH than ovine β2-casein; they have a common polypeptide chains and differ only in their degree of phosphorylation (Richardson and Mercier, 1979). The occurrence of multiphosphorylated forms of β-casein might affect micelle stability and the availability and distribution of calcium in the milk. This depend on the capacity of single β-casein forms, phosphorylated to different degrees, to bind different amounts of calcium. On the other hand, the stage of lactation, health and age of individuals, as well as an altered availability of phosphate, play an important role in determining the level of phosphorylation of β-casein in individual milks (Amigo *et al.*, 2000).

Chianese (1997), using gel isoelectric focusing in Sarda breed, differentiated between three genetic variants of β-casein designated A, B and, C. The only sequence difference found between A and C was the amino acid substitution of Glu at position 2 in variant A for Gln in variant C but no sequence data for the B variant is available yet. Moreover, an A to G transition leading to the $Met_{183} \rightarrow Val_{183}$ exchange was detected at the DNA level (Ceriotti *et al.*, 2004). Recently, Chessa *et al.* (2010) found two new patterns, called CSN2 X and CSN2 Y; after sequencing the same authors discovered that CSN2 Y pattern, found only in the Garfagnina breed, was due to a silent mutation in the triplet coding for Gln at position 192 (G to A transition), whereas a C to A transversion, responsible for the amino acid exchange $Leu_{196} \rightarrow Ile_{196}$, characterized the CSN2 X pattern. The mutations found do not change the isoelectric point of the coded proteins confirming the monomorphic isoelectrophoretic pattern observed at the protein level (Chessa *et al.*, 2010).

κ-CASEINS

κ-casein constitutes about 15% of the total caseins and contains Cys. The amino acid or nucleotide sequence of κ-casein from several eutherian species is known (Bonsing and Mackinlay, 1987; Brignon *et al.*, 1985; Furet *et al.*, 1990; Gorodetskii and Kaledin, 1987; Mercier *et al.*, 1976; Mercier *et al.*, 1973). κ-casein consists of 169 amino acids residues

(19,023 Da) and occurs in polymeric form via disulfide bonds ranging in sizes from 60 to 600 kDa. κ-casein is the only casein soluble in the presence of calcium ions and has a much smaller phosphate component than any of the other caseins. The phosphorylation sites are confined to the C-terminal region of the molecule and are present as single sites rather than the clusters found in the calcium-sensitive caseins. The signal peptide of κ-casein is 21 residues in length (compared to 15 in calcium-sensitive caseins). κ-casein is the only eutherian casein that has conclusively been shown to contain carbohydrate moieties. These features have led to the conclusion that κ-casein is not related to the calcium-sensitive caseins (Bonsing and Mackinlay, 1987; Jollés *et al.*, 1978; Mercier and Vilotte, 1993). κ-casein is susceptible to cleavage by the aspartate protease, chymosin (rennin). Jollés *et al.* (1978) have shown that cleavage occurs at a specific Phe-Met bond in the C-terminal portion of bovine κ-casein. Nakhasi *et al.* (1984) proposed that κ-casein be classified into two separate groups depending on the species of origin. Group I (cow, sheep, goat, water buffalo) κ-caseins differ from group II (rat, mouse, pig, human) κ-caseins on the basis of hydrophobicity, carbohydrate content, amino acid composition and site of proteolytic cleavage. Group I κ-caseins contain the archetypal Phe-Met bond. In group II κ-caseins however, the cleavage site is specified by Phe-Ile or Phe-Leu. This divergence might reflect differences in the mechanism of clotting in ruminant and non-ruminant mammals (Herskovitis, 1966). The κ-caseins are the only caseins from eutherian milk at least that are glycolsylated. The carbohydrate groups are attached to κ-casein via O-glycosidic linkages to Ser and Thr residues within the C-terminal portion of the molecule. Glycosylation occurs post-translationally and is catalysed by membrane bound O-glycosyl-transferases within the Golgi apparatus of mammary epithelial cells (Jollés and Fiat, 1979; Mepham *et al.*, 1992; Takeuchi *et al.*, 1984). Glycosylation is increased during the colostral period, and in response to infection such as mastitis but decreases with successive periods of lactation (Dziuba and Minkiewicz, 1996). Human κ-casein is more highly glycosylated than that of the sheep or cow and carbohydrate residues may account for up to 55% of the weight of the molecule (Dev *et al.*, 1993). The carbohydrate portion of bovine κ-casein consists of galactose, N-acetyl galactosamine and N-acetyl neuraminic acid. Sheep κ-casein contains N-glycolyl-neuraminic acid in addition to these three sugars (Jollés and Fiat, 1979). Human κ-casein is yet more complex and contains fucose, galactose, N-acetyl glucosamine, N-acetyl galactosamine and N-acetyl neuraminic acid (Dev *et al.*, 1993; Fiat *et al.*, 1980; Yamauchi *et al.*, 1981). Hydrolitic action by chymosin at Phe-Met releases a hydrophobic N-terminal segment (1 to 105) and a hydrophilic soluble C-terminal segment (106 to 169) known as para-κ-casein and macropeptide, respectively (Wong *et al.*, 1996). The macropeptide is glycosylated, with the carbohydrate moieties mostly linked to Thr, containing N-acetylneuraminic acid, galactose and galactosamine. Para-κ-casein possesses no carbohydrates. Macropeptides from cow and sheep contain only one glycosylated Thr, whereas the human one has ten sites of glycosylation (Fiat *et al.*, 1980). The uneven distribution of carbohydrates, plus the high content of Asp and Glu in the C-terminal sequence results in high net negative charges (3.5 at pH 6.8) that are involved in the stabilization of the κ-casein. The overall acidic character of the macropeptide is conserved in κ-caseins for several species, including cow, zebu, buffalo, goat, ewe, sow and human (Mercier and Chobert, 1976).

κ-casein is located mainly on the surface of the casein micelles and is responsible for their stability (Walstra, 1990). The presence of a glycan moiety in the C-terminal region of κ-casein enhances its ability to stabilize the micelle, by electrostatic repulsion and may increase the resistance by the protein to proteolytic enzymes and high temperatures (Dziuba and Minkiewicz, 1996; Minkiewicz *et al.*, 1993). The level of glycosylation does not affect micelle structure but it does affect the susceptibility of κ-casein to hydrolysis by chymosin, with susceptibility decreasing as the level of glycosylation increases (Addeo *et al.*, 1984; Doi *et al.*, 1979; Van Hooydonk, *et al.*, 1984; Vreeman *et al.*, 1986; Zbikowska *et al.*, 1992).

The bovine κ-casein constitutes approximately 12% of the whole milk casein. The genetic polymorphism of the κ-casein gene (CSN3) in ruminant has been well documented in cattle (Pinders *et al.*, 1991; Rijnjkels *et al.*, 1997), buffaloes (Pipalia *et al.*, 2001; Otaviano *et al.*, 2005; Patel *et al.*, 2007; Riaz *et al.*, 2008), goat (Angiolillo *et al.*, 2002; Yahyaoui *et al.*, 2003; Prinzenberg *et al.*, 2005; Kiplagat *et al.*, 2010), sheep (Moioli *et al.*, 2007) and yak (Prinzenberg *et al.*, 2008). Eleven variant alleles of the κ-casein gene have been identified in cattle, namely A, B, C, D, E, F, G, H, Al, and J (Soria *et al.*, 2003). However, only two alleles, namely A and B alleles, are the most commonly identified in *Bos taurus* dairy cattle (Ikonen *et al.*, 1996; Erhadt, 1989). The A and B variants differ in the amino acids 136 and 148 (A: Thr and Asp; B, Ile and Ala respectively) (Grosclaude *et al.*, 1972; Ng-Kwai Hang and Grosclaude, 1992). The bovine protein variants have received considerable research interest in recent years. Several studies on κ-casein variants resulted in its considerable association with lactation performance, milk composition, milk protein contents and rheological properties (Bovenhuis *et al.*, 1992; Verdier-Metz *et al.*, 2001; Boettcher *et al.*, 2004; Kűbarsepp *et al.*, 2005).

Goat κ-Casein

Structural analysis of the goat κ-casein *locus* indicates the existence of a biallelic polymorphism (Di Luccia *et al.*, 1990; Caroli *et al.*, 2001). This casein was analysed at the genomic level in many goat breeds elsewhere. Relevant investigations were carried out by Yahyaoui *et al.* (2001, 2003) and Angiolillo *et al.* (2002), on the following goat breeds: Spanish (Murciano-Granadina and Canaria), French (Alpine and Saanen), Italian (Teramana, Garganica, Girgentana, Sarda, Cilentana Nera, Montefalcone) and by Jann *et al.* (2004) on wild and domestic animals. Further polymorphisms in exon 4 of the κ-casein *locus* were detected. Prinzenberg *et al.* (2005) analyzed the following goat populations: GermanWeisse Edelziege, Saanen, Nigeria Borno goat, Cameroon Red Sokoto, West African Dwarf, Anatolian hair goat and Angora goat at the same *locus*. From these studies, the number of alleles identified in the domesticated goat has increased to 16 (13 protein variants and 3 silent mutations). Both these researches observed that the differences in the occurrence and frequency of the alleles in these goat breeds reflect their geographic origin. The highest number of alleles was found in the goat breeds from the Near East. In particular, Yahyaoui (2003) suggests that the distribution of such alleles has been influenced either by selection pressure for milk production or, more likely, by genetic drift; moreover the same author affirms that A allele appeared later in the evolutionary pathway (Yahyaoui *et al.*, 2003). The distribution of κ-casein variants was also investigated in the Southern Italy native breed (Reale *et al.*, 2005).

Sheep k-Casein

Despite the key-role played by κ-casein in the micellar structure and the clotting process, little is known about ovine κ-casein polymorphism. Ovine κ-casein has been sequenced and was found to be formed by 171 amino acid residues (Jollés *et al.*, 1974). Two variants (A and B) were found through protein analysis differing in glycosylation and phosphorylation degrees (Alais and Jollés, 1967; Soulier *et al.*, 1974; Amigo *et al.*, 2000). Three genomic variants were found with DNA RFLP digestion (Di Gregorio *et al.*, 1991; Lévéziel *et al.*, 1991). Subsequently, the variability in the ovine κ-casein *locus* has been studied by SSCP-PCR. This technique has showed three SSCP patterns anomalies, later confirmed by sequencing. However, these are rare substitutions which do not affect the protein structure, and hence do not result in new variants. In fact, two of them are silent mutations (changes in the third nucleotide of a codon which codes for the same amino acid), while the third is a transversion, the substitution of A for T in an exon segment which in not translated (Barroso *et al.*, 1996; Barroso, 1998). However, no further studies either on the structure of these variants or on their effect on milk constituents were performed. More recently, a single nucleotide polymorphism was found at position 237 of the sheep κ-casein mRNA (Feligini *et al.*, 2005), where a T was substituted by a C, even if the corresponding amino acid remains a Tyr.

Casein Micelle Characteristics and Milk Rheological Properties

The caseins exist predominantly as micelles, which are hydrated spherical structures with dimensions in the sub-micron region. The dry matter of casein micelles consists mainly (>90%) of proteins, with small amounts of inorganic matter, collectively referred to as micellar calcium phosphate (MCP). MCP exists as nanometer-sized clusters of amorphous calcium phosphate which are stabilized by a shell of caseins (αs1-, αs2- and β-caseins). Horne (1998) proposed a model of casein micelle which is dependent on a balance between electrostatic repulsion and hydrophobic interaction, where colloidal calcium phosphate crosslinks the caseins and neutralizes negatively charged phosphoserine groups, allowing the formation of hydrophobic interaction between caseins. Growth of casein micelles is terminated by the solvent-mediated adsorption of κ-casein onto the micellar surface. The hydrophilic C-terminal region of κ-casein or 'brush' protrudes from the surface of the micelles and sterically stabilizes them against aggregation (De Kruif and Zhulina, 1996).

Rennet-induced coagulation of milk proceeds in two phases. The first, or enzymatic phase, involves rapid proteolysis of approximately 90% of the κ-casein (Walstra *et al.*, 1984). When κ- casein is hydrolyzed, the casein micelle becomes unstable and susceptible to precipitation by calcium. Aggregation of the casein micelles occurs during the second, or non-enzymatic phase (Garnot and Olson, 1982; Storry and Ford, 1982; Walstra *et al.*, 1984).

Micelle structures of small ruminants milk differ from cow milk in average diameter, hydration and mineralization (Table 2). Goat casein micelles contain more calcium, inorganic phosphorus, and non-centrifugal casein, are less solvated, less heat stable, and lose β-casein more readily than bovine casein micelles (Jenness, 1980; Remeuf and Lenoir, 1986). Sheep milk is similar to goat milk in micelle characteristics. Average mineralization levels of micelles in goat and sheep milk are higher than in cow milk. There is an inverse relationship between the mineralization of the micelle and its hydration, which also means that goat milk is less hydrated than cow milk (Table 2; Soods *et al.*, 1979; Remeuf and Lenoir, 1986).

During coagulation, destabilized casein micelles and calciumphosphate bonds form a network which entraps fat and other solids. The casein network is formed faster with higher proteins mainly casein content, which results in higher aggregation rate and the development of firmer curd.

Goat milk has poorer cheesemaking ability compared to cow milk counterparts due to lower casein content of caprine milk, and differences in casein micelle composition, size, hydration and αs/β- casein ratio between the two species (Remeuf, 1992). The lower casein content and lower casein number of goat milk compared to cow milk (casein/total nitrogen: 75 *vs.* 78%; Jenness, 1980; Grappin *et al.*, 1981) are responsible for reduced cheese yield (Ricordeau and Mocquot, 1967) and affect cheese texture and rheology. In cheesemaking, a poor coagulum strength can lead to more cheese particle losses in whey, lower cheese yield and likely less textural integrity. The uniqueness of renneting kinetic of goat milk has also been characterized as a shorter coagulation time and greater hardening rate (Puri and Parkash, 1962; Remeuf *et al.*, 1989). Renneting time for goat milk is shorter than for cow milk, and the weak consistency of the gel explains mediocre cheese suitability of goat milk (Parkash and Jenness, 1968; Remeuf and Lenoir, 1986). Storry *et al.* (1983) found that the maximum firmness of the gel of goat milk is usually much less, and even a gel from goat milk with an equal casein content is not as firm as from cow milk. The casein content of milk has a significant influence on rheological properties of the rennet gel, its setting speed and its maximum firmness. Remeuf and Lenoir (1986) observed significant correlations between casein content and the proportion of αs1-casein; between the casein content and levels of colloidal calcium and inorganic phosphorus; between levels of colloidal calcium and inorganic phosphorus and the firmness of the gel and between the degree of hydration of the micelles and their mineralization. Reports have shown that αs1-casein polymorphism has a significant influence on cheesemaking properties of goat milk (Remeuf, 1992), which would in turn have a significant impact on cheese textural quality. The differences in αs1-casein contents observed by comparing milk from different breeds and crosses were accountable for the variations in texture and rheology of their corresponding goat milk cheese products (Clark and Sherbon 2000b). Many authors reported that 'high type' (variants A, B_1, B_2, B_3, B_4, C, H, L and M) were associated with higher amounts of αs1-casein (about 3.5 g/l), 'intermediate type' variants E and I were associated with intermediate amounts (1.1–1.7 g/l), and 'low type' variants D, F and G were associated with low amounts (about 0.45 g/l) of αs1-casein in goat milk (Grosclaude *et al.* 1987; Mojoli *et al.*, 1998; Bevilacqua *et al.*, 2002; Ramunno *et al.*, 2005; Sacchi *et al.*, 2005; Sztankóová *et al.*, 2007; Chianese *et al.*, 1997; Grosclaude and Martìn, 1997). Conversely, Clark and Sherbon (2000a) reported that αs1- casein genetic variants were not highly correlated with coagulation properties.

Sheep milk renneting properties are affected by its physico-chemical properties, including pH, larger casein micelle, more calcium per casein weight, and other mineral concentrations in milk, which cause differences in coagulation time, coagulation rate, curd firmness, and amount of rennet needed (Ramos and Juàrez, 2003). Sheep milk is higher in fat, protein and total solids than cow or goat milk (Haenlein and Wendorff, 2006). The cheese yield is significantly greater per unit of sheep milk than that of cow and goat milk, because cheese curd contains primarily the fat and casein from milk (Anifantakis, 1986). Sheep milk has all major casein fractions of cow milk (Alichanidis and Polychroniadou, 1996). Casein micelle structure is similar in cow, goat and sheep milk, while sheep milk caseins are richer in calcium than cow casein (Alichanidis and Polychroniadou, 1996). Sheep milk is very sensitive to rennet, because it has a higher β/αs-casein ratio, and also coagulation in sheep milk proceeds faster than in cow milk (Muir *et al.*, 1993). This suggests that mode and amount of rennet addition to sheep milk would definitely influence the final texture and rheological characteristics of the cheese. Sheep milk also requires less rennet than cow milk to obtain the same coagulation time (Kalantzopoulos, 1993). Although the rate of curd formation in sheep milk is faster than in cow milk, the rate of syneresis in the former is slower than in the later (Muir *et al.*, 1993); this is due to the higher casein and colloidal calcium content in sheep milk (Storry and Ford, 1982).

References

Addeo F, Martìn P, Ribadeau-Dumas B. 1984 Susceptibility of buffalo and cow κ-casein to chymosin action. Milchwissenschaft 39: 202–205.

Addeo F, Masi P, Rubino R. 1987 The aptitude of caprine milk to cheesemaking Relationship between αs1-casein content and rhelogical properties of curd and cheese. In: L'evaluation des ovis et des caprins mediterraneens, Symposium Philoetios, Fonte-Boa, Portugal.

Addeo F, Mauriello R, Moio L, Laezza P, Di Luccia A. 1992 Ovine casein variant identification using electrophoretic, immunochemical and chromatographic techniques. *Milchwissenshaft* 47: 283-287.

Alais C, Jolles P. 1967 Isolation, purification and analysis of two k-cascins. *Journal of Dairy Science* 50: 1555–1561.

Alexander LJ, Beattie CW. 1992 The sequence of porcine β-casein cDNA. *Animal Genetics* 23: 369–71.

Alichanidis E, Polychroniadou A. 1996 Special features of dairy products from ewe and goat milk from the physico-chemical and organoleptic point of view. In: Proceedings of the IDF-CIRVAL Seminar Production and Utilization of Ewe and Goat milk, Crete, Greece, October 19–21, 1995. International Dairy Federation Publication, Brussels, Belgium, 21–43.

Ambrosoli R, Di Stasio L, Mazzocco P. 1988 Content of alphas1-casein and coagulation properties in goat milk. *Journal of Dairy Science* 71: 24-28.

Amigo L, Recio I, Ramos M. 2000 Genetic polymorphism of ovine milk proteins: its influence on technological properties of milk – A review. *International Dairy Journal* 10: 135–149.

Angiolillo A. Yahyaoui MH, Sanchez A, Pilla F, Folch JM. 2002 Characterization of a newgenetic variant in the caprine kappa casein gene. *Journal of Dairy Science* 85: 2679–2680.

Anifantakis EM. 1986 Comparison of the physico-chemical properties of ewe's and cow's milk. In: Proceedings of the IDF Seminar Production and Utilization of Ewe's and Goat's Milk, Athens, Greece, September 23–25, 1985. International Dairy Federation Publication, Brussels, Belgium, pp. 42–53, Bulletin No. 202.

Assenat L. 1985 Le lait de brebis: composition et propriétés. In: Lait et Produits Laitiers. I. Les laits de la mamelle à la laiterie, Luquet FM (Eds), Technique et Documentation Lavoisier, Schaar, 281-319.

Barroso A, Canon J, Dunner S. 1996 Preliminary results on the search of variability in ovine kappa-casein. *Animal Genetics* 27(Suppl. 2): 40.

Barroso A. 1998 Utilizacion de la tecnica de PCR-SSCP (Polymerase Chain Reaction-Single Strand Conformation Polymorphisms) para el analisis del polimorxsmo genetico de las casennas α y β en las especies bovina y ovina. Ph. D. thesis, Universidad Complutense de Madrid, Spain.

Bevilacqua C, Ferranti P, Garro G, Veltri C, Lagonigro R, Leroux C, Pietrola E, Addeo F, Pilla F, Chianese L, Martin P. 2002. Interallelic recombination is likely responsible for the occurrence of a new rate (αs1-casein variant in the goat species). *European Journal of Biochemistry* 269: 1293–1303.

Boettcher PJ, Caroli A, Stella A, Chessa S, Budelli E, Canavesi F, Ghiroldi S, Pagnacco G. 2004 Effects of casein haplotypes on milk production traits in Italian Holstein and Brown Swiss cattle. *Journal of Dairy Science* 87: 4311-4317.

Boisnard M, Hue D, Bouniol C, Mercier JC, Gaye P. 1991 Multiple mRNA species code for two non-allelic forms of ovine alpha s2-casein. *European Journal of Biochemistry* 201: 633–641.

Boisnard M, Pètrissant G. 1985 Complete sequence of ovine alpha s2-casein messenger *RNA. Biochimie* 67: 1043–51.

Bonsing J, Mackinlay AG. 1987 Recent studies on nucleotide sequences encoding the caseins. *Journal of Dairy Research* 54: 447–61.

Bonsing J, Ring JM, Stewart AF, Mackinlay AG. 1988 Complete nucleotide sequence of the bovine β-casein gene. *Australian Journal of Biological Sciences* 41: 527–37.

Boulanger A, Grosclaude F, Mahé MF. 1984 Polymorphisme des caséines αs1 et αs2 de la chèvre (*Capra hircus*), *Génétics Sélétion Evolution* 16: 157–176.

Bouniol C, Brignon G, Mahé MF, Printz C. 1994 Biochemical and genetic analysis of variant C of caprine αs2-casein (*Capra hircus*). *Animal Genetics* 25: 173–177.

Bouniol C. 1992 Sequence of the goat as2-encoding cDNA. *Gene* 125: 235–236.

Bovenhuis H, van Arendonk JAM, Kerver S. 1992 Associations between milk protein polymorphism and milk production traits. *Journal of Dairy Science* 75: 2549-2559.

Brignon G, Chtourou A, Ribadeau-Dumas B. 1985 Preparation and amino acid sequence of human k-casein. *FEBS Letters* 188: 48.54.

Brignon G, Mahe MF, Ribadeau-Dumas B, Mercier JC, Grosclaude F. 1990 Two of the three genetic variants of goat alpha s1-casein which are synthesized at a reduced level have an internal deletion possibly due to altered RNA splicing. *European Journal of Biochemistry* 193: 237–241.

Brignon G, Ribadeau-Dumas B, Mercier JC, Pélissier JP, Das BC. 1977 Complete amino acid sequence of bovine alpha S2-casein. *FEBS Letters* 76: 274–9.

Brignon, G, Mahé MF, Grosclaude F, Ribadeau-Dumas B. 1989 Sequence of caprine αs1 casein and characterization of those of its genetic variants which are synthesized at a high level, αs1-CN A B and C. *Protein Sequences and Data Analysis* 2: 181–188.

Caroli A, Chiatti F, Chessa S, Rignanese D, Bolla P, Pagnacco G. 2006 Focusing on the goat casein complex. *Journal of Dairy Science* 89: 3178–3187.

Caroli A, Jann O, Budelli E, Bolla P, Jager S, Erhard G. 2001. Genetic polymorphism of goat κ-casein (CSN3) in different breeds and characterization at DNA level. *Animal Genetics* 32: 226–230.

Ceriotti G, Chessa S, Bolla P, Budelli E, Bianchi L, Duranti E, Caroli A. 2004 Single nucleotide polymorphisms in the ovine casein genes detected by Polymerase Chain Reaction-Single Strand Conformation Polymorphism. *Journal of Dairy Science* 87: 2606–2613.

Chessa S, Bolla P, Dario C, Pieragostini E, Caroli A. 2003 Polimorfismi genetici lattoproteici nella razza ovina gentile di puglia: Monitoraggio mediante focalizzazione isoelectrica. *Scienza e Tecnica Lattiero-Casearia* 54: 1–8.

Chessa S, Budelli E, Chiatti F, Cito AM, Bolla P, Caroli A. 2005 Predominance of *β*-casein (*CSN2*) C allele in goat breeds reared in Italy. *Journal of Dairy Science* 88: 1878–1881.

Chessa S, Rignanese D, Berbenni M, Ceriotti G, Martini M, Pagnacco G, Caroli A. 2010 New genetic polymorphisms within ovine β- and αS2-caseins. *Small Ruminant Research* 88: 84–88.

Chianese L, Caira S, Garro G, Addeo F. 2007 Primary structure of ovine deleted variant αs1-CN E. In: Proceedings of the 5th International Symposium on the Challenge to Sheep and Goats Milk Sectors, Alghero/Sardinia, Italy, 70.

Chianese L, D'auria R, Ferranti P, Garro G, Mauriello R, Rubino R, Addeo F. 1996b Occurrence of novel αs1-casein variants in Italian goat breeds. In: Production and Utilization of Ewe and Goat Milk; International Dairy Federation: Brussels, Belgium, 141-147.

Chianese L, Garro G, Addeo F, Lopez-Galvez G, Ramos M, 1993a Discovery of an ovine alpha S2-casein variant. *Journal of Dairy Research* 60: 485–493.

Chianese L, Garro G, Mauriello R, Laezza P, Ferranti P, Addeo F. 1997 Occurrence of three novel αs1 casein in goat's milk. In: International Dairy Federation (Ed.), Milk protein Polymorphism, Special Issue, 9702. *International Dairy Federation*, 316–323.

Chianese L, Garro G, Nicola MA, Mauriello R, Ferranti P, Pizzano R, Cappuccio U, Laezza P, Addeo F, Ramunno L, Rando A, Rubino R. 1993b The nature of β-casein heterogeneity in caprine, *Lait* 73: 533–547.

Chianese L, Garro, Mauriello, R, Laezza P, Ferranti P, Addeo F. 1996a Occurrence of five αs1-casein variants in ovine milk. *Journal of Dairy Research* 63: 49-59.

Chianese L. 1997 The casein variants of ovine milk and the relationships between the αS1-casein variants and milk composition, micellar size and cheese yield. In: Caseins and caseinates: structures, interactions, networks, Hannah Symposium, Scotland, United Kingdom.

Clark S, Sherbon JW. 2000a Genetic variants of alphas1-CN in goat milk: breed distribution and associations with milk composition and coagulation properties. *Small Ruminant Research* 38: 135-143.

Clark S, Sherbon JW. 2000b Alphas1-casein, milk composition and coagulation properties of goat milk. *Small Ruminant Research* 38: 123–134.

Collet C, Joseph R, Nicholas K. 1992 Molecular characterization and in-vitro hormonal requirements for expression of two casein genes from a marsupial. *Journal of Molecular Endocrinology* 8: 13–20.

Cosenza G, Pauciullo A, Gallo D, Di Berardino D, Ramunno L. 2005 A *Ssp*I PCR-RFLP detecting a silent allele at the goat *CSN2 locus*. *Journal of Dairy Research* 72: 456–459.

Cunsolo V, Muccilli V, Saletti R, Marletta D, Foti S. 2006 Detection and characterization by high-performance liquid chromatography and mass spectrometry of two truncated goat αs2-caseins. *Rapid Communications in Mass Spectrometry* 20: 1061–1070.

Dall'Olio S, Davoli R, Russo V. 1989 Una nuova variante di β caseina caprina *Scienza e Tecnologia Lattiero-Casearia* 40: 24-28.

Davoli R, Dall'Olio S, Russo V. 1990 Probe in favore dell'attribuzione all' αs1 -caseina della variante Welsh. *Scienza e Tecnica Lattiero-Casearia* 41: 327-333.

De Kruif CG, Zhulina EB. 1996 k-casein as a polyelectrolyte brush on the surface of casein micelles. Colloids and Surfaces A: *Physicochemical and Engineering Aspects* 117: 151-159.

Dev BC, Sood SM, DeWind S, Slattery CW. 1993 Characterization of human k-casein purified by FPLC. *Preparative Biochemistry and Biotechnology* 23: 389–407.

Di Gregorio P, Rando A, Pieragostini E, Masina P. 1991 DNA polymorphism at the casein *loci* in sheep. *Animal Genetics* 22: 21–30.

Di Luccia A, Mauriello R, Chianese L, Moio L, Addeo F. 1990 k-Casein polymorphism in caprine milk. *Scienza e Tecnica Lattiero-Casearia* 41: 305–314.

Doi H, Ibuki F, Kanamori M. 1979 Heterogeneity of reduced bovine κ-casein. *Journal of Dairy Science* 62: 195–203.

Dziuba J, Minkiewicz P. 1996 Influence of glycosylation on micellestabilizing ability and biological properties of C-terminal fragments of cow's k-casein. *International Dairy Journal* 6: 1017–44.

Erhadt G. 1989 k-casein in Rindermilch- Nachweis eines weiteren Alleles (k-CnE) in Verschiedenen Rassen. *Journal of Animal Breeding and Genetics* l06: 225-231.

Erhardt G, Jaeger S, Budelli E, Caroli A. 2002 Genetic polymorphism of goat αs2-casein and evidence for a further allele. *Milchwissenschaft* 57: 137–140.

Erhardt G. 1993 Allele frequencies of milk proteins in German cattle breeds and demonstration of αS2-casein variants by isoelectric focusing. *Archiv Tierzucht* 36: 145–152.

Farrell HM Jr, Jimenez-Flores R, Bleck GT, Brown EM, Butler JE, Creamer LK, Hicks CL, Hollar CM, Ng-Kwai-Hang KF, Swaisgood HE. 2004 Nomenclature of the proteins of cows' milk--sixth revision. *Journal of Dairy Science* 87: 1641-74.

Feligini M, Vlaco S, Curik VC, Parma P, Greppi G, Enne G. 2005 A single nucleotide polymorphism in the sheep kappa-casein coding region. *Journal of Dairy Research* 72: 317-321.

Ferranti P, Lilla S, Chianese L, Addeo F. 2000 Qualitative and quantitative variation of deleted protein forms in goat αs1 and αs2 casein phenotypes. In: Proceedings 7th International Conference on Goats, Tour, France, 15–18 May 573–575.

Ferranti P, Malorni A, Nitti G, Laezza P, Pizzano R, Chianese L, Addeo, F. 1995 Primary structure of ovine αs1- caseins: localization of phosphorylation sites and characterization of genetic variants A, C, and D. *Journal of Dairy Research* 62: 281-296.

Fiat AM, Jolles J, Aubert JP, Loucheux-Lefebvre M-H, Jolles P. 1980 Localisation and importance of the sugar part of human casein, *European Journal of Biochemistry* 111: 333–339.

Furet JP, Mercier JC, Soulier S, Gaye P, Hue-Delahaie D, Vilotte JL. 1990 Nucleotide sequence of ovine k-casein cDNA. *Nucleic Acids Research* 18: 5286.

Galliano F, Saletti R, Cunsolo V, Foti S, Marletta D, Bordonaro S, D'Urso G. 2004 Identification and characterization of a new beta-casein variant in goat milk by high-performance liquid chromatography with electrospray ionization mass spectrometry and matrix-assisted laser desorption/ionization mass spectrometry. *Rapid Communication in Mass Spectrometry*. 18: 1972–1982.

Garnot P, Olson NF. 1982 Use of oscillatory deformation technique to determine clotting times and rigidities of milk clotted with different concentrations of rennet. *Journal of Food Science* 47: 1912–1915.

Giambra IJ, Chianese L, Ferranti P, Erhardt G. 2010a Molecular genetic characterization of ovine αS1-casein allele H caused by alternative splicing. *Journal of Dairy Science* 93: 792-795.

Giambra IJ, Chianese L, Ferranti P, Erhardt G. 2010b Genomics and proteomics of deleted ovine *CSN1S1*I*. *International Dairy Journal* 20: 195-202.

Giambra IJ, Jäger S, Erhardt G. 2010c Isoelectric focusing reveals additional casein variants in German sheep breeds. *Small Ruminant Research* 90: 11-17.

Ginger MR, Grigor MR. 1999 Comparative aspects of milk caseins. *Comparative Biochemistry and Physiology B* 124: 133–145.

Gorodetskii SI, Kaledin AS. 1987 Analysis of nucleotide sequence of bovine k-casein cDNA. *Genetika* 23: 398–404.

Grappin R, Jeunet R, Pillet R, Toquin A. 1981 Etide des laits de chevre: teneur du lait de chevre en matiere grasse, matiere azotee et fractions azotees (Studies of goat milk: contents of fat, protein and protein fractions). *Le Lait* 61: 117–133.

Greenberg R, Groves ML, Dower HJ. 1984 Human β-casein. Amino acid sequence and identification of phosphorylation sites. *Journal of Biological Chemistry* 259: 5132–8.

Groenen MAM, Dijkhof RJM, Verstege AJM, van der Poel JJ. 1993 The complete sequence of the gene encoding bovine alpha s2-casein. *Gene* 123: 187–193.

Grosclaude F, Joudrier P, Mahé MF. 1978 Polymorphism de la caséine *α*S2 bovine: Etroite liason du *locus* *α*S2-CN avec les *loci* *α*S1-CN, *β*-CN et *κ*-CN; mise en evidence d'une deletion dans le variant *α*S2-CND. *Annales de Genetique et de Selection Animale* 10: 313–327.

Grosclaude F, Joudrier P, Mahé MF. 1979. A genetic and biochemical analysis of a polymorphism of bovine *α*S2-casein. *Journal of Dairy Research* 46: 211–213.

Grosclaude F, Mahé MF, Accolas JP. 1982 Note sur le polymorphisme génétique des lactoprotéines de bovines et de yaks Mongols. *Annales de Genetique et de Selection Animale* 14: 545–550.

Grosclaude F, Mahé MF, Brignon G, Di Stasio L, Jeunet RA. 1987 Mendelian polymorphism underlying quantitative variations of goat αs1-casein, *Génétics Séletion Evolution* 19: 399–412.

Grosclaude F, Mahé MF, Mercier JC, Bonnemaire J, Theissier JH. 1976 Polymorphisme des lactoprotéines de Bovinés Népalais. II. Polymorphisme des caséines "*α*S-mineures": Le *locus* *α*S2-CN est-il lié aux *loci* *α*S1-CN, *β*-CN et *κ*-CN? *Annales de Genetique et de Selection Animale* 8: 481–491.

Grosclaude F, Mahe MF, Mercier JC, Ribadeau-Dumas B. 1972 Localisation des substitutions d'acides aminés différenciant les variants a et b de la caséine × bovine. *Annales de Genetique et de Selection Animale* 4: 515–21.

Grosclaude F, Martìn P. 1997 Casein polymorphism in the goat. In: International Dairy Federation (Ed.), Milk Protein Polymorphism, Special Issue, 9702. International Dairy Federation, Brussels, Belgium 241–253.

Grosclaude F. 1988 Le polymorphisme genetique des principales lactoproteines bovines. Relations avec la quantite', la composition et les aptitudes fromageres du lait. *INRA Productions Animales* 1: 5–17.

Haenlein GFW, Wendorff WL. 2006 Sheep milk - production and utilization of sheep milk. In: Park, YW, Haenlein, GFW (Eds), Handbook of Milk of Non-Bovine Mammals. Blackwell Publishing Professional, Oxford, UK, and Ames, Iowa, USA, 137– 194.

Herskovitis TT. 1966 On the conformation of caseins. Optical rotatory properties. *Biochemistry* 5: 1018–26.

Horne DS. 1998 Casein interactions: casting light on the black boxes, the structure of dairy products. *International Dairy Journal* 8: 171–177.

Ibeagha-Awemu EM, Prinzenberg EM, Erhardt G. 2005 High variability of milk protein genes in *Bos indicus* cattle breeds of Cameroon and Nigeria and characterization of a new αS1 promoter allele. *Journal of Dairy Research* 72: 1–9.

Ikonen T, Ruottinen O, Erhardt G, Ojala M. 1996 Allele frequencies of the major milk proteins in the Finnish Ayrshire and detection of a new k-casein variant. *Animal Genetics* 27: 179–181.

Jann O, Prinzenberg EM, Luikart G, Caroli A, Erhardt G. 2004 High polymorphism in the k-casein (CSN3) gene from wild and domestic caprine species revealed by DNA sequencing. *Journal of Dairy Research* 71: 188–195.

Jann OC, Ibeagha-Awemu EM, Zbeyaz EO, Zaragoza P, Williams JL, Ajmone-Marsan P, Lenstra JA, Moazami-Goudarzi K, and Erhardt G. 2004 Geographic distribution of haplotype diversity at the bovine casein *locus*. *Genetics Seletion Evolution* 36 : 243–257.

Jenness R. 1974 The composition of milk. In: Larson BL, Smith VR (Eds.), Lactation: A comprehensive treatise, Vol 3. NewYork, NY, USA, Academic Press.

Jenness R. 1980. Composition and characteristics of goat milk: review 1968-1979. *Journal of Dairy Science* 63: 1605-1630.

Jollés J, Fiat AM, Schoengten F, Alais C, Jolles P. 1974 The amino-acid sequence of sheep k-casein. *Biochimica et Biophysica Acta* 365: 335–3429.

Jollés P, Fiat A. 1979 The carbohydrate portions of milk glycoproteins. *Journal of Dairy Research* 46: 187–91.

Jollés P, Loucheux-Lefebvre MH, Henschen A. 1978 Structural relatedness of k-casein and fibrinogen g-chain. *Journal of Molecular Evolution* 11: 271–277.

Kalantzopoulos GC. 1993 Cheeses from ewes' and goats' milk. In: Fox, PF (Eds), *Cheese: Chemistry Physics and Microbiology*, Vol. 2, Chapman and Hall, London, 507-543.

King JWB. 1966 The caseins of sheep's milk. In: Polymorphismes Biochimiques des Animaux Paris: *INRA Proceedings of 10th European Conference on Animal Blood Groups and Biochemical Polymorphism*, 427–431.

Kiplagat SK, Agaba M, Kosgey IS, Okeyo M, Indetie D, Hanotte O, Limo MK. 2010. Genetic polymorphism of kappa-casein gene in indigenous Eastern Africa goat populations. *International Journal of Genetics and Molecular Biology* 2: 1-5.

Kűbarsepp I, Henno M, Viinalass H, Sabre D. 2005 Effect of κ-casein and β-lactoglobulin genotypes on the milk rennet coagulation properties. *Agricultural Research* 1: 55–64.

Lagonigro R, Pietrosa E, D'Andrea M, Veltri C, Pilla F. 2001 Molecular genetic characterization of the goat αs2-casein E allele. *Animal Genetics* 32: 391–393.

Lévéziel H, Méténier L, Guérin G, Cullen P, Provot C, Bartaud M, Mercier JC. 1991 Restriction fragment length polymorphisms of ovine casein genes: close linkage between the αs1, αs2, β and κ-casein *loci. Animal Genetics* 22: 1-10.

Linderstrom-Lang K, Kodama SCR. 1925 Studies on casein. I. On the solubility of caseins in hydrochloric acid. *Comptes-Rendus des Travaux du Laboratoire Carlsberg* 16: 1–47.

López-Gálvez G, Amigo L, Ramos M. 1999. Polymorphism of αs-caseins in the milk of two Spanish ovine breeds. *Milchwissenschaft* 54: 17–19.

Mahé MF, Grosclaude F. 1993 Polymorphism of *β*-casein in the Creole goat of Guadeloupe, evidence for a null allele. *Genetics Seletion Evolution* 25: 403–408.

Marletta D, Bordonaro S, Guastella AM, Falagiani P, Crimi N, D'Urso G. 2004 Goat milk with different αs2-casein content: analysis of allergenic potency by REAST inhibition assay. *Small Ruminant Research* 52: 19–24.

Martin P, Addeo F. 1995 Genetic polymorphisms of casein in the milk of goats and sheep. IDF/FIL Publ. No. 9603, 45–58.

Martin P, Brignon G, Furet JP, Leroux C. 1996 The gene encoding αs1-casein is expressed in human mammary epithelial cells during lactation. *Lait* 76: 526-537.

Masoodi TA, Shafi G. 2010 Analysis of casein alpha S1 & S2 proteins from different mammalian species. *Bioinformation* 4: 430-435.

Mauriello R, Addeo F, Pieragostini E, Bufano G. 1990 Polimorfismo delle caseine in pecore di razza Altamurana. *Scienza e Tecnica Lattiero-Casearia* 41: 357-364.

Mellander O. 1939 Elektophoretische untersuchung von casein. *Biochemische Zeitschrift* 300: 240.

Mepham TB, Gaye P, Martin P, Mercier JC. 1992 Biosynthesis of milk protein. In: Fox PF, Editor. Advanced Dairy Chemistry. London: Elsevier, 491–543.

Mercier JC, Addeo F, Pelissier JP. 1976 Primary structure of the casein macropeptide of caprine k-casein. *Biochimie* 58: 1303-1310.

Mercier JC, Brignon G, Ribadeau-Duman B. 1973 Structure primaire de la case´ine kB bovine. Sequence complete. *European Journal of Biochemistry* 35: 222–235

Mercier JC, Chobert JM. 1976 Comparative study of the amino acid sequences of the casein-macropeptides from seven species . *FEBS Letters* 72: 208-214.

Mercier JC, Gaye P. 1983 In: Biochemistry of Lactation (Mepham, T.B., ed.) 177-227, Elsevier, Amsterdam and New York.

Mercier JC, Grosclaude F, Ribadeau-Dumas B. 1971. Structure primaire de la caséine αs1 bovine. Sequénce compléte. *European Journal of Biochemistry* 23: 41–51.

Mercier JC, Vilotte JL. 1993 Structure and function of milk protein genes. *Journal of Dairy Science* 76: 3079-3098.

Mercier JC. 1981 Phosphorylation of caseins, present evidence for an amino acid triplet code post-translationally recognized by specific kinases. *Biochimie* 63: 1–17.

Minkiewicz P, Dziuba J, Muzinska B. 1993 The contribution of N-acetylneuraminic acid in the stabilization of micellar casein. *Polish Journal of Food and Nutrition Sciences* 433: 39–48.

Moioli B, Andrea DM, Pilla F. 2007 Candidate genes affecting sheep and goat milk quality. *Small Ruminant Research* 68: 179-192.

Moioli B, Pilla F, Tripaldi C. 1998 Detection of milk protein genetic polymorphisms in order to improve dairy traits in sheep and goats: A review. *Small Ruminant Research* 27: 185-195.

Mora-Gutierrez A, Farrell HM Jr. 2001 Hydration of native and cold solubilized rennet-treated caprine caseins as determined by oxygen 17 nuclear magnetic resonance. *Journal of Dairy Science* 84(Suppl.): 93–99.

Mora-Gutierrez A, Kumosinski TF, Farrell HM Jr. 1996 Comparison of hydration behavior of bovine and caprine caseins as determined by oxygen-17 nuclear magnetic resonance: pH/pD dependence. *Journal of Agricultural and Food Chemistry* 44: 796–803.

Mora-Gutierrez A, Kumosinski TF, Farrell HM Jr. 1997 Oxygen-17 nuclear magnetic resonance studies of bovine and caprine casein hydration and activity in deuterated sugar solutions. *Journal of Agricultural and Food Chemistry* 45: 4545–4553.

Mora-Gutierrez A, Kumosinski TF, Farrell HM. 1991 Quantification of αs1-casein in goat milk from French-Alpine and Anglo- Nubian breeds using reverse-phase high performance liquid chromatography. *Journal of Dairy Science* 74: 3303.

Muir DD, Horne DS, Law AJR, Sweetsur AWM. 1993 Ovine milk. 2. Seasonal changes in indices of stability. *Milchwissensch* 48: 442–445.

Nakhasi HL, Grantham FH, Gullino PM. 1984 Expression of k-casein in normal and neoplastic rat mammary gland is under the control of prolactin. *Journal of Biological Chemistry* 259: 14894–14898.

Neveu C, Mollé D, Moreno J, Martin P, Léonid J. 2002 Heterogeneity of caprine beta-casein elucidated by RP-HPLC/ MS: Genetic variants and phosphorylation. *Journal of Protein Chemistry* 21: 557–567.

Ng-Kwai Hang KF, Grosclaude F. 1992 Genetic polymorphisms of milk proteins. In: Fox PF, editor. Advanced Dairy Chemistry. London: Elsevier, 405–455.

Otaviano AR, Tonhati H, Sena JAD, Munoz MFC. 2005 Kappa-casein gene study with molecular markers in female buff aloes (*Bubalus bubalis*). *Genetics and Molecular Biology* 28: 237-241.

Ovinicoltura/UNAPOC. 1992 Unione Nazionale Associazioni Produttori Ovi-Caprini. Roma, UNAPOC.

Park YW. 2007 Rheological characteristics of goat and sheep milk. *Small Ruminant Research* 68: 73-87.

Park YW, Juàrez M, Ramos M, Haenlein GFW. 2007 Physicochemical characteristics of goat and sheep milk. *Small Ruminant Research* 68: 88-113.

Park YW, Zhang H, Zhang B, Zhang L. 2006 Goat milk. In YW Park, Haenlein FW (Eds.), Handbook of non-bovine mammals. Oxford, UK: Blackwell Publishing.

Parkash S, Jenness R. 1968 The composition and characteristics of goat's milk: a review. *Journal of Dairy Science* 30: 67–75.

Patel RK, Chauhan JB, Singh KM, Soni KJ. 2007. Genotyping and allelic frequencies of κ-CN and β-LG in Indian river buffalo bulls. *Buffalo Bulletin* 26: 63-66.

Pearse MJ, Linklater PM, Hall RJ, Maciunlay AG. 1986 The effect of casein composition and casein dephosphorylation on the coagulation and syneresis of artificial micelle milk. *Journal of Dairy Research* 53: 381-390.

Persuy MA, Printz C, Medrano JF, Mercier JC. 1999 A single nucleotide deletion resulting in a premature stop codon is associated with marked reduction of transcripts from a goat beta-casein null allele. *Animal Genetics* 30: 444–451.

Picariello G, Rignanese D, Chessa S, Ceriotti G, Trani A, Caroli A, Di Luccia A. 2009 Characterization and Genetic Study of the Ovine αS2- Casein (CSN1S2) *Allele B. Protein Journal* 28: 333–340.

Pinders J, Perryc BN, Skidmoraend J, Sawa D. 1991 Analysis of polymorphism in the bovine casein genes by use of the polymerase chain reaction. *Animal Genetics* 22: 11-20.

Pipalia DL, Ladani DD, Brahmkshtri BP, Rank DN, Joshi CG, Vataliya PH, Solanki JV. 2001 Kappa-casein genotyping of Indian buffalo breed using PCR-RFLP. *Buffalo Journal* 2: 195-202.

Pirisi A, Piredda G, Fraghì A, Papoff CM, Chianese L, 1997 Influence of sheep AA, CC and DD αs1-casein variants on milk composition and cheese yield. In: Proceedings of IDF Seminar on "Milk Protein Polymorphism" Palmerston North, New Zealand, 254–258.

Pirisi A, Piredda G, Papoff CM, Di Salvo R, Pintus S, Garro G, Ferranti P, Chianese L. 1999 Effects of sheep alpha s1-casein CC, CD and DD genotypes on milk composition and cheesemaking properties. *Journal of Dairy Research* 66: 409-419.

Prinzenberg EM, Gutscher K, Chessa S, Caroli A, Erhardt G. 2005 Caprine k-casein (CSN3) polymorphisms: new developments in molecular knowledge. *Journal of Dairy Science* 88: 1490–1498.

Prinzenberg EM, Jianlin H, Erhardt G. 2008 Genetic variation in the -Casein gene (CSN3) of Chinese yak (*Bos grunniens*) and phylogenetic analysis of CSN3 sequences in the Genus Bos. *Journal of Dairy Science* 91: 1198-1203

Provot C, Persuy MA, Mercier JC. 1989 Complete nucleotide sequence of ovine beta-casein cDNA: inter-species comparison. *Biochimie* 71: 827–832.

Puri BR, Parkash S, 1962 Studies in physico-chemical properties of milk. 12. The renneting time of milk of different species. *Indian Journal of Dairy Science* 15: 114–122.

Ramos M, Juàrez M. 2003 Sheep milk. In: Roginski H, Fuquay JW, Fox PF (Eds), Encyclopedia of Dairy Sciences, Vol 4. Academic Press, Amsterdam, The Netherlands, 2539-2545.

Ramunno L, Cosenza G, Pappalardo M, Longobardi E, Gallo D, Pastore N, Di Gregorio P, Rando A. 2001a Characterization of two new alleles at the goat CSN1S2 *locus. Animal Genetics* 32: 264–268.

Ramunno L, Cosenza G, Rando A, Pauciullo A, Illario R, Gallo D, Di Berardino D, Masina P. 2005 Comparative analysis of gene sequence of goat CSN1S1 F and N alleles and characterization of CSN1S1 transcript variants in mammary gland. *Gene* 345: 289–299.

Ramunno L, Longobardi E, Pappalardo M, Rando A, Di Gregorio P, Cosenza G, Mariani P, Pastore N, Masina P. 2010b An allele associated with a non detectable amount of αs2 casein in goat milk. *Animal Genetics* 32: 19–26.

Ramunno L, Mariani P, Pappalardo M, Rando A, Capuano M, Di Gregorio P, Cosenza G. 1995 Un gene ad effetto maggiore sul contenuto di caseina β nel latte di capra. XI Convegno ASPA, 185–186, Grado (GO) Italy.

Rando A, Ramunno L, Masina P. 2000 Mutations in casein genes. *Zootecnia e Nutrizione Animale* 26: 105–114.

Rasmussen L, Hojrup P, Petersen TE. 1992 Localization of two interchain disulfide bridges in dimers of bovine αS2-casein. Parallel and antiparallel alignments of the polypeptide chains. *European Journal of Biochemistry* 203: 381–386.

Raynal-Ljutovac K, Lagriffoul G, Paccard P, Guillet I, Chilliard Y. 2008 Composition of goat and sheep milk products: an update. *Small Ruminant Research*79: 57-72.

Reale S, Yahyaoui MH, Folch YM, Sanchez A, Pilla F, Angiolillo A. 2005 Genetic polymorphism of the k-casein (CSN3) gene in goats reared in Southern Italy. *Italian Journal of Animal Science* 4: 97–101.

Remeuf F, Lenoir J, Duby C. 1989 Etude des relatios entreles caracteristiques physico-chemiques des laits de chevre et leur aptitude a la coagulation par la pressure (Study of the relationships of physico-chemical characteristics of goat milk and aptitude for coagulation under pressure). *Le Lait* 69: 499–518.

Remeuf F, Lenoir J. 1986 Relationship between the physicochemical characteristics of goat's milk and its rennetability. *Bulletin of the International Dairy Federation* 202: 68.

Remeuf F, Verdalet-Guzman I, Lenoir J. 1995 Technological adjustment of goat milk containing low synthesis rate αS1-casein variants. *International Dairy Journal* 5: 381–392.

Remeuf F. 1992 Physico-chemical properties of goat milk in relation to processing characteristics. In: Proceedings of the National Symposium Dairy Goat Production and Marketing, Oklahoma City, OK, 98–110.

Riaz MN, Malik NA, Nasreen F, Qureshi JA. 2008 Moleculer marker assisted study of kappa-casein gene in NILI-RAVI (Buffalo) breed of Pakistan. *Pakistan Veterinary Journal* 28: 103-106.

Richardson BC, Creamer LK. 1974 Comparative micelle structure III. The isolation and chemical characterization of caprine β1-casein and β2-casein, *Biochimica et Biophysica Acta* 365: 133–137.

Richardson BC, Mercier JC. 1979 The primary structure of the ovine β-caseins. *European Journal of Biochemistry* 99: 285–297.

Ricordeau G, Mocquot G. 1967 Influence des variations saisonnieres de la composition du lait de chevre sur le rendement en fromage (Influence of seasonal variations in the composition of goat milk and effect on cheese). *Annales de Zootechnie* 16: 165–181.

Rijnkels M, Kooiman PM, deBoer HA, Pieper FR. 1997 Organization of the bovine casein gene *locus*. *Mammalian Genome* 8: 148-152.

Rijnkels M. 2002 Multispecies comparison of the casein gene *loci* and evolution of casein gene family. *Journal of Mammary Gland Biology and Neoplasia* 7: 327–345.

Risch E. 1992 Rheological properties of goat milk cheese: methods of true axial stress versus apparent axial stress. *Journal of Texture Studies* 23: 403-413.

Rosen JM, Dale T, Gavigan S, Buhler T. 1992 Molecular biology of milk proteins. In: Mechanisms Regulating Lactation and Infant Nutrient Utilization. Houston: Wiley-Liss, 1–19.

Sacchi P, Chessa S, Budelli E, Bolla P, Ceriotti G, Soglia D, Rasero R, Cauvin E, Caroli A. 2005 Casein haplotype structure in five Italian goat breeds. *Journal of Dairy Science* 88: 1561–1568.

Soods SM, Gaind DK, Dewan RK. 1979 Correlation between micelle salvation and calcium content. *New Zealand Journal of Dairy Science Technology* 14: 32–44.

Soria LA, Iglesias GM, Huguet MJ, Mirande SL. 2003 A PCR-RFLP test to detect allelic variants of the bovine kappa-casein gene. *Animal Biotechnology* 14: 1-5.

Soulier S, Ribardeau-Dumas B, Denamur R. 1974 Purification of k-casein from sheep. *European Journal of Biochemistry* 50: 445–452.

Stewart AF, Bonsing J, Beattie CW, Shah F, Willis IM, Mackinlay AG. 1987. Complete nucleotide sequences of bovine αs2- and β-casein cDNAs: Comparisons with related sequences in other species. *Molecular Biology and Evolution* 4: 231–241.

Storry JE, Ford GD. 1982 Some factors affecting the post clotting development of coagulum strength in renneted milk. *Journal of Dairy Research* 49: 469–477.

Storry JE, Grandison AS, Milliard D, Owen AJ, Ford GD. 1983 Chemical composition and coagulating properties of renneted milks from different breeds and species of ruminant. *Journal of Dairy Research* 50: 215–229.

Swaisgood HE. 1992 Chemistry of the caseins. In Advanced Dairy Chemistry. I: Proteins. 63-110 P.F. Fox, ed. Elsevier Applied Science, New York, NY.

Sztankóová Z, Mátlová V, Malá G. 2007 Genetic polymorphism at the CSN1S1 gene in two Czech goat breeds. *Czech Journal of Animal Science* 52: 199-202.

Takeuchi M, Tsuda E, Yoshikawa M, Sasaki R, Chiba H. 1984 Post-translational modification of k-casein in the Golgi apparatus of mid-lactating mammary gland from rat and cow. *Agricultural and Biological Chemistry* 48: 2789–2797.

Tziboula A, Horne DS. 1999 The role of αS1-casein in the structure of caprine casein micelles. *International Dairy Journal* 9: 173–178.

Van Hooydonk ACM., Olieman C, Hagedoorn HG. 1984 Kinetics of the chymosin-catalysed proteolysis of kappa casein in milk. *Netherlands Milk and Dairy Journal* 38: 207-222

Verdier-Metz I, Coulon JB, Pradel P. 2001 Relationship between milk fat and protein contents and cheese yield. *Animal Research* 50: 365–371.

Vreeman HJ, Visser S, Slangen KJ, van Riel JAM. 1986 Characterization of bovine κ-casein fractions and the kinetics of chymosin-induced macropeptide release from carbohydrate-free and carbohydrate-containing fractions determined by highperformance gel-permeation chromatography. *Biochemical Journal* 240: 87–97

Wake RG, Baldwin RL. 1961 Analysis of casein fractions by zone electrophoresis in concentrated urea. *Biochimica et Biophysica Acta* 47: 225–39.

Walstra P, Jenness R, Badings HT. 1984 Dairy Chemistry and Physics. Wiley, New York.

Walstra P. 1990 On the stability of casein micelles. *Journal of Dairy Science* 73: 1965-1979.

Waugh DF. 1971 Formation and structure of casein micelles. In: H.A. McKenzie, ed. Milk proteins: chemistry and molecular biology. Vol. 2. Academic Press, New York.

Wessels G, Hamann H, Erhardt G, Distl O. 2004 Genotypeffekte von Milchprotein-Polymorphismen auf die Milchleistung beim Ostfriesischen Milchschaf. Berliner und Münchenen Tierärztliche Wochenschrift 117: 414–419.

Wong DWS, Camirand WM, Pavlath AE. 1996 Structures and functionalities of milk proteins. *Critical Reviews in Food Science and Nutrition* 36: 807-844.

Yahyaoui MH, Angiolillo A, Pilla F, Sanchez A, Folch JM. 2003 Characterization and genotyping of the caprine kappa casein variants. *Journal of Dairy Science* 86: 2715-2720.

Yahyaoui MH, Coll A, Sanchez A, Folch, JM. 2001 Genetic polymorphism of the caprine kappa casein gene. *Journal of Dairy Research* 78: 209–216.

Yamauchi K, Azuma N, Kobayashi H, Kaminogawa S. 1981 Isolation and properties of human k-casein. *Journal of Biochemistry* 90: 1005–1012.

Zbikowska A, Dziuba J, Jaworska H, Zaborniak A. 1992 The influence of casein micelle size on selected functional properties of bulk milk proteins. *Polish Journal of Food and Nutrition Sciences* 1423: 23–32.

In: Casein: Production, Uses and Health Effects
Editors: A. M. Ventimiglia, et. al, pp. 27-58
ISBN: 978-1-62100-129-4

Chapter 2

CASEIN AND DENTAL EROSION

A.J. White[1], W.H. Briscoe[2] and M.E. Barbour[1]
[1] School of Oral & Dental Sciences, University of Bristol, UK.
[2] School of Chemistry, University of Bristol, UK.

ABSTRACT

Dental erosion is an increasing problem in many countries around the world, and research in this field has increased dramatically in recent years. Dental erosion is the dissolution of tooth tissues (enamel and dentine) by acids that are not of bacterial origin; most commonly these originate from the diet. Recently, protein-based technologies including casein and casein phosphopeptide (CPP) have been investigated and found to have erosion-inhibiting properties.

The aim of this chapter is to summarize recent studies by the authors' research groups to investigate the use of casein and casein-derived proteins as agents to inhibit dental erosion. A number of *in vitro* techniques and models are employed to investigate erosion, including non-contact optical profilometry, atomic-force microscopy (AFM) nanoindentation and hydroxyapatite dissolution rate experiments (hydroxyapatite is the main mineral component of teeth).

Evidence is given for an almost instantaneous 'protective' effect afforded by casein, preventing demineralization of enamel under erosive conditions. Aqueous solutions of casein of 0.5 % w/v applied topically before an erosive challenge are shown to afford between 39 to 45 % reduction in demineralization over a range of clinically relevant timescales.

The mechanistic aspects of the 'protective effect' afforded by casein are investigated using AFM and X-ray reflectometry (XRR). Data from both of these techniques show the formation of a thin (~ 6.6 nm) film of casein proteins on mica, a molecularly smooth mineral used as a model substrate.

1. Introduction

1.1. Dental Erosion

1.1.1. Definitions

During a person's lifetime, teeth are exposed to a variety of physical and chemical conditions that can result in deterioration of the tooth structure. Excepting major physical trauma, this damage can be classified into two main categories: caries and wear.

Dental caries occurs when acidogenic bacteria which reside in the plaque biofilm formed on a tooth surface, metabolize fermentable sugars producing organic acids as a by-product [1, 2]. These acids dissolve the apatite crystals [3], resulting in a dental cavity.

Dental wear can be defined as any loss of material from the tooth – mechanical or chemical – which does not involve bacteria [4, 5]. There are many processes that can cause wear, including abrasion, demastication, attrition, abfraction and erosion. These processes are summarized in table 1 (Adapted from [6]). Within these, *dental erosion* is defined as an irreversible, pathologic, chronic, localized loss of dental hard tissues due to chemical etching and/or chelation without bacterial involvement [6-8].

An eroded tooth surface is softened and therefore more susceptible to abrasive wear, and mechanical insults such as toothbrushing can readily remove the superficially demineralized dental hard tissue [9-12]. This is an example of how erosion and other forms of wear can interact in a synergistic manner.

Table 1. Non- carious chronic destructive processes leading to tooth wear

Terminology	Cause of loss of hard tissue
Abrasion	Mechanical process involving foreign objects or substances (such as toothbrushing)
Demastication	Mechanical interaction between food and teeth
Attrition	Mechanical process involving tooth-to-tooth contact
Abfraction	Mechanical process involving tooth flexure by eccentric occlusal forces (tensile and compressive)
Erosion	Chemical etching and dissolution

Adapted from [6].

1.1.2. Prevalence

Growing emphasis is given today to the prevention of tooth wear [7]. Reports suggest that the prevalence of dental erosion is increasing [13, 14]. Dental erosion affects a wide demographic; the elderly (with an ageing population retaining many of their teeth [4, 11, 15]), but also children; studies in 2001/2 [16-18] showed that 51 % of 14-year-old British children suffered moderate erosion in their permanent teeth, and a 2003 study [14] showed that 53 % of British five-year-olds had evidence of tooth wear.

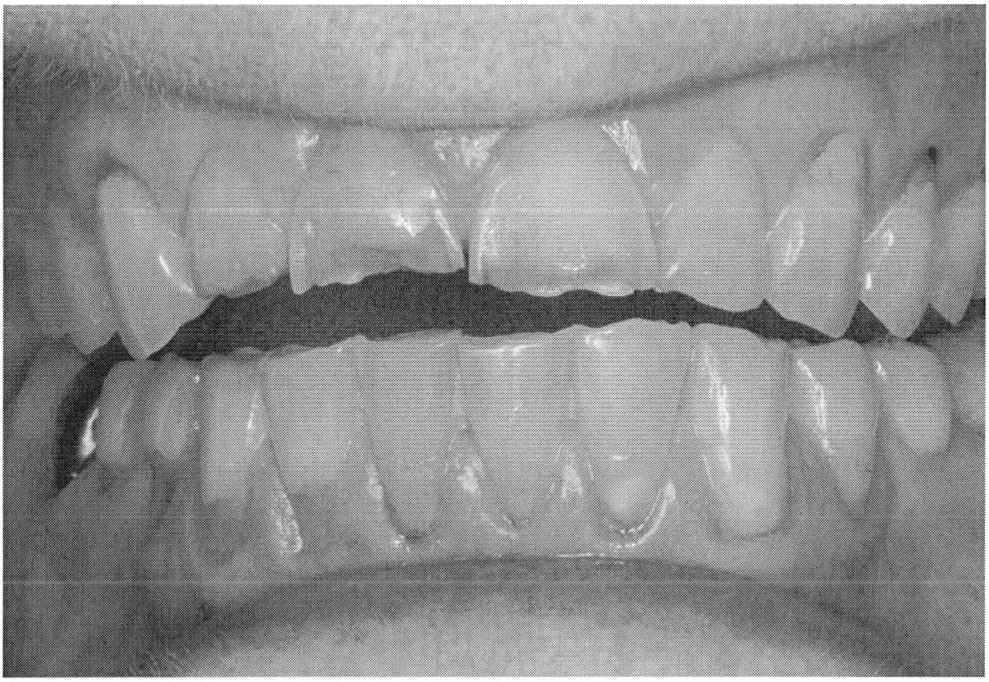

Photograph courtesy of S. M. Hooper, Bristol Dental Hospital.

Figure 1. An example of dental erosion.

1.1.3. Aetiology of Erosion

Erosion may be caused by extrinsic or intrinsic agents. Extrinsic agents include substances such as carbonated drinks, and fruit drinks and juices [8, 19, 20] which contain organic acids such as citric, phosphoric and malic acids. Intrinsic causes of erosion include endogenous acids which can contact the teeth due to, for example, disorders such as bulimia or recurrent vomiting [11, 21].

The major cause of much dental erosion today is thought to be the consumption of acidic foodstuffs and beverages [1, 8, 11, 15, 19, 20, 22-26]. The chemistry of these foodstuffs and beverages is highly important, therefore, in determining the rate of any erosion that occurs. To understand these chemical factors fully, we must first understand the chemistry of the erosion process itself.

1.1.4. Tooth Structure and Chemistry

The functional requirements of tooth enamel are such that, under oral conditions, opposing teeth may experience maximal bite forces of over 800N in men [27] with contact stresses of 0.45-2.5 GPa [28]. Enamel has evolved, therefore, to be the most highly mineralized tissue known, consisting of 96% mineral and 4% organic material and water [29]. Nearly the entire volume of enamel is occupied by densely packed calcium-deficient carbonated hydroxyapatite (HA) crystals:

$$Ca_{10-x}Na_x\,(PO_4)_{6-y}\,(CO_3)_z\,(OH)_{2-u}\,F_u \qquad (1)$$

with a fine, lacy network of organic material between the crystals. Enamel varies in thickness, from a maximum of approximately 2.5 mm [29] over working surfaces to a feather edge at the gingival margin.

Underneath the enamel is a layer of dentine. Dentine is softer and more elastic than the brittle enamel. By weight dentine is 70 % HA, 20 % organic material and 10 % water. It is full of microscopic tubules that run roughly parallel to each other from the inner surface of the dentine towards the dentine surface. If dentine is exposed (due to loss of enamel from erosion, for example), fluid can flow down these tubules, which terminate at odontoblasts. These odontoblasts are connected to nerves, and their stimulation arising from the flow of fluid down the tubules can cause sensitivity in the tooth.

1.1.5. Erosion Chemistry

Erosion occurs when acids come into contact with the enamel and/or dentine. The HA dissolves as illustrated by the simplified equilibrium equation 2 [30].

$$Ca_{10}(PO_4)_6(OH)_2 \leftrightarrow 10Ca^{2+} + 6PO_4^{-3} + 2OH^- \quad (2)$$

The "erosive potential" of a beverage or foodstuff to erode also is related to chemical, biological and behavioural factors.

Common Ions

Erosion rate is reduced if any of the common ions of the HA matrix are present in the dissolving solution. Thus, erosion is inversely related to pH and is reduced in the presence of calcium and/or phosphate.

Buffer Capacity

Many of the acids which cause enamel erosion *in vivo* are weak acids; that is, they are not fully dissociated in aqueous solution. As dissolution proceeds, the H+ ions are consumed and the undissociated molecules dissociate to replenish the supply of H+, thus maintaining a low pH for longer than would be the case for a strong acid. This buffering is an important parameter in determining the erosive potential of a solution.

Fluoride

Fluoride, common in toothpastes and oral rinses and, in some countries, added to water or salt supplies, can substitute for hydroxyl in the HA lattice, producing a salt which is less acid-soluble than the original tissue, thus offering protection against dissolution. For this reason, fluoride is widely used in oral care products such as toothpastes and mouthrinses. Fluoride in acidic solutions significantly reduces the erosion that they cause.

Surrounding Tissues

The position of the teeth in the dental arch may be a factor in the site-specifity of erosion. Different locations on the dental arch are not only exposed to varying amounts of saliva, but also an increased prevalence to abrasive attacks – even caused by the tongue [31].

Saliva and Salivary Pellicle

In terms of biological aspects, the main factors are saliva, the acquired pellicle, surrounding soft tissue and their position relative to the teeth [31]. Saliva and the pellicle are discussed in depth later.

Behavioural Aspects

Human Behavior, Of Course, Profoundly Affects Dental Erosion [32]. This Can include anything from unusual eating and drinking habits, to poor oral hygiene. It is also noteworthy that perceived 'healthier' lifestyles can be high in fruit juices (such as citric acid) which can lead to an increased prevalence of erosion.

In summary, erosion is a multi-factored problem, encompassing not only chemical aspects, but also biological and behavioural components.

1.1.7. Inhibiting Dental Erosion by Food and Beverage Product Modification

As much dental erosion is due to the consumption of acidic foodstuffs and beverages, it is possible to engineer these products to be less erosive especially as the chemistry and biology behind the erosion process is understood. Increasing the pH or adding calcium and phosphate supplements to beverages has been investigated and shown to have some success at modifying their erosive potential [33-42].

Both these techniques of increasing pH and adding calcium and phosphate are felt impractical however, as raising the pH can remove the 'tangy' taste of soft drinks, and the addition of calcium and phosphate can lead to a 'chalky' or 'metallic' taste unless the calcium is well masked [26]. The addition of fluoride to drinks is forbidden in the European Union (EU) and in several non-EU countries [43].

It is therefore desirable to find food-approved, food-derived (neutraceutical) substances that do not detrimentally affect taste and which have anti-erosion properties. These substances, in their very nature as being food-derived, would not require the same regulation that other additives do.

1.2. Saliva and Salivary Pellicle

Saliva is important in protecting teeth against erosion for several reasons. It reduces surface mineral loss [44] and can remineralize slightly eroded enamel, due to its high buffering capacity and ability to act as an ion reservoir [45], although the degree of remineralization varies within the intraoral sites [46]. It is a lubricant aiding speech and mastication and reducing friction between teeth, and its high glycoprotein content makes it mucinous and hence quite sticky lending itself to coating functions [47]. It is also a washing agent, mechanically flushing away bacteria and debris from the mouth, including sugars and acids. With an average unstimulated saliva flow rate of 0.3 mL/minute, clearance half-times are in the region of 2.2 minutes, with stimulated salivary flow rates ranging wildly depending on the stimuli ranging from 3.2 to 7.1 mL/minute [48].

Selective adsorption of salivary proteins form what is known as the salivary pellicle [49, 50], a protective membrane coating the teeth [29]. The term 'pellicle' was first coined by Dawes et al. referring to the cuticular material formed at the enamel surface. Pellicle formation is a selective process; only a fraction of the proteins in saliva form the pellicle (309 proteins discovered in saliva whereas only 130 in the acquired enamel pellicle - as of 2008) [51].

The first stage in pellicle formation is almost instantaneous, starting within seconds and taking a few minutes to complete [47, 48]. After a minute the pellicle is known to be a 10-20 nm electron dense layer [52]. Calcium ions of the enamel crystallites dissolve in saliva more readily than phosphate ions. This leaves a greater proportion of phosphate ions on the enamel surface, giving it a negative charge. This negatively charged surface is coated in a layer of positively charged calcium counter-ions [47]. Charged groups of the proteins in the saliva are then electrostatically attracted to this ionic double layer. The thickness of this initial layer is in the region of 10 - 20 nm [47]. Those proteins that exhibit a high affinity for hydroxyapatite are known as '*pellicle precursor proteins*'; examples of such are statherin, histatin and proline-rich proteins (PRPs). Van der Waals and hydrophobic interactions also contribute to the adsorption of proteins to the enamel surface; these are driven thermodynamically due to a gain in entropy when a protein replaces an ordered water molecule at the surface [47, 53].

The second stage is a slower process comprising the adsorption of biopolymers from the saliva to the protein layer on the tooth surface. Protein aggregates are hypothesised to adsorb rather than single biopolymers, due to the pellicle's rapid increase in thickness between 30 – 90 min [47]. These have a globular structure with a hydrophobic core that resembles very closely that of micelles in milk, thus they are called '*micelle-like structures*' [47, 54, 55]. These give rise to a, globular, surface morphology characteristic of the salivary pellicle [56].

The pellicle protects the enamel surface against erosive demineralization by organic acids [57] and has been shown to have a protective effect against citric acid after only 3 minutes of maturation *in situ* [58]. The thickness of the acquired salivary pellicle is related to the severity of dental erosion [59]. After 60 min mean pellicle thickness, as measured by confocal laser scanning microscopy, ranges from 0.3 μm (on bovine enamel blocks mounted on appliances at palatal sites of upper incisors) to 1.06 μm (lingual sites of lower molars) [47, 59] and is known to reach a maximum in thickness after approximately 2 hours [47, 56].

1.3. Anti-Cariogenic and Anti-Erosion Effects of Milk Proteins and Peptides

1.3.1. Investigations into the Protective Effects of Diet and Milk

Milk and milk products exhibit both anticariogenic and anti-erosion properties. As early as 1945 (components of) milk were shown to have anticariogenic properties in the rat [60], and this was followed with work by Shaw in 1949 [61], Bavetta and McClure in 1957 [62] and Weiss and Bibby in 1966 showing that milk reduced the acid solubility of enamel [63]. Krobricka et al. [64] reported a protective effect of cheese snacks against coronal and root-surface caries in the rat and Gedalia et al. also reported a rehardening of the enamel surface by hard cheese after softening by a coca-cola type drink [24].

The great advantage of any protective agent found in milk is that this agent is not only perceived as a 'natural' product, but that it is also recommended and encouraged in the diet due to its nutritional properties [65].

1.3.2. Casein

Studies by Schweigert and co-workers [60] and Bavetta and McClure [62] demonstrated that altering the concentration of casein affected the incidence of caries, and furthermore that the effect was not due to an altered plaque level on the enamel surface. Weiss and Bibby noted that the protective effect of milk was due to casein or gastric mucin in the decalcifying buffer, which was far superior to what would be obtained from any common ion effect due to the associated calcium and phosphate [66]. It was suggested by both teams that the effect could be ascribed to a sort of '*physical barrier'* coating the tooth surface.

Much work has been done by Reynolds et al. investigating the anticariogenic effects of caseinates (in particular sodium caseinate) on the rat, either via addition to drinking water [67], or in confectionery [68]. Addition of sodium caseinate only successful reduced the cariogenicity of chocolate when added at 16.6 % w/w, which was considered to be unpalatable [69]. When a lower, palatable level was used (2% w/w), the addition of caseinate did not significantly alter the cariogenicity of the confectionery [69, 70]. For this reason, it was concluded that casein's undesired organoleptic properties and large quantities required for any significant efficacy to occur made it unsuitable for its use as an anticariogenic agent and casein largely fell out of favour [69].

The general conclusion of the mode of action of many studies investigating the protective effects of the casein proteins and peptides was due to some sort of protein layer being formed, acting as a physical and chemical protective layer.

1.3.3 CPP-ACP

There has been extensive research regarding the protective effects of CPP-ACP, much of which is by Reynolds and colleagues at the University of Melbourne and has utilized their commercial product 'Recaldent™'.

Most of the research has focused on the remineralizing properties of CPP-ACP with regard to carious lesions [71]. Studies have shown inhibition of caries in animal models [72], *in* situ [73], and promotion of enamel subsurface lesion remineralization *in vitro* [74, 75] and *in situ* [74, 76-79].

The main proposed protective mechanism of CPP-ACP is: '*the inhibition of enamel demineralization and enhancement of remineralization through the localization of ACP at the tooth surface*' [69]. Another protective mechanism could be ascribed to a physical and chemical protective layer being created due to a protein micellar layer forming on the tooth surface [80].

This lends itself to other work examining CPP-ACP looking at its ability to inhibit demineralization. Studies have shown that CPP-ACP has the ability to inhibit enamel or dentine demineralization *in vitro* [75, 81-85]. CPP-ACP has been shown to lead to an increase in the mechanical hardness of bovine enamel as determined by an ultrasonic device [82] and an increase in surface hardness of enamel previously softened by a cola drink [85].

1.3.4. CPP

The protective effects of casein phosphopeptide (CPP) are thought to stem from its ability to sequester amorphous calcium phosphate forming CPP-ACP. A major benefit of CPP is that it is not associated with any unpalatability, and so much larger amounts of CPP can be used than casein. The potential for anticariogenicity is therefore at least 10 times greater on a weight basis for CPP than casein [69, 86]. As far as the authors are aware, there is limited evidence showing the protective effects of CPP independently. Roberts [87] has conducted *in* vitro studies into the effects of different salts of 0.5% w/w in water of CPP (calcium, sodium and calcium phosphate) on demineralization from an acidic challenge, and found that all three were approximately one third as effective as 1000 μg/g fluoride. Roberts also carried out a human intra-oral caries model using a crude solution of 5% w/w CPP in a non-fluoridated dentifrice, and it was found that (in general agreement with his *in vitro* data and Reynolds' work) that again the CPP was about one third as effective as a 1000 μg/g fluoride solution at inhibiting the surface softening of enamel *in situ* [87].

1.3.5. Interaction with Plaque and Saliva on the Tooth Surface

Experiments by Rose [88] showed that CPP-ACP forms a calcium 'reservoir' by binding with plaque, resulting in a reduced calcium diffusion coefficient (slowing the diffusion of free calcium away so that it remains in the plaque for longer) which may assist in remineralization and inhibit demineralization [86]. Rose has shown, using equilibrium dialysis, that CPP-ACP and calcium compete for the same binding sites on plaque [88], and that roughly 3 calcium ions bind to the bacterial cells per unit CPP-ACP, corresponding to a calcium binding affinity of 0.3-0.5 mmol/L [89]. The phosphoseryl cluster sequence -Ser(P)-Ser(P)-Ser(P)-Glu-Glu- has a negative charge, and upon addition of acidic solutions can be neutralized by the positive protons, reducing the amount of bound ACP. This releases the ACP at the tooth surface where it is needed. At typical plaque pH values characteristic of a cariogenic or erosive attack, the reduced pH results in the loss of calcium binding sites and an increase in the calcium diffusion coefficient [88, 89]. The diffusion coefficient, however, is still markedly reduced due to the presence of CPP-ACP than without, decreasing the potential for demineralization to occur [89].

It has also been shown that casein and CPP can reduce the adherence of plaque bacteria on tooth surfaces. Vacca-Smith et al. [90] and Reynolds and Wong [91] have shown that *Streptococcus mutans* adherence to HA discs decreases on exposure to casein, and Vacca-Smith identified κ-casein as the fraction responsible. Studies by Guggenheim et al. [92] on *Streptococcus sobrinus* in rats and Schüpbach et al. [93] on *Streptococcus mutans in vitro* confirm these findings with respect to micellar casein and CPP respectively.

In another study, CPP and ACP were both found to be incorporated into plaque, and this resulted in a 144% increase in calcium and 160% increase in phosphate in the plaque (consistent with the ratio found in ACP) [69]. Whether the inhibition of anticariogenic streptococci or inhibition of demineralization is more significant is yet to be determined [69].

In terms of long-term adherence, work by Reynolds et al. [94, 95] has shown that CPP has a half-life of one to two hours when incorporated into plaque, which is consistent with previous studies showing the breakdown of CPP by bacterial phosphatase and peptidase [96].

1.3.6. Interaction of Casein and Related Proteins with Tooth Tissues

Pearce and Bibby hypothesized that adsorption of casein to enamel was indicative of monolayer adsorption due to a plateau in the adsorption isotherm. Reynolds and Wong attempted to identify the mechanism of adherence of proteins to HA surfaces [91]. They found that acidic proteins (such as casein) increase surface net negative charge. This increases electrical double layer repulsion (increasing the potential energy barrier) which hinders the approach of bacteria to the HA surface and preventing attachment (basic proteins, on the other hand, would increase surface net positive charge which could enhance bacterial approach) [91]. Although the zeta potential was measured, this was found not to correlate with adherence and so it was concluded that short-range hydrophobic and ionic interactions are overriding. It was also postulated that specific sequences and conformational specificity (β-turns) for ACP binding, and the location of hydrophobic residues, are both used as attachment '*anchors*' in plaque and oral mucosa at hydrophobic sites [69].

In the case of CPP, its soluble, flexible and mobile nature is also believed to facilitate its binding to a range of surfaces due to its ability to adapt its conformation [69]. The three contiguous –Ser(P)- residues are believed to be the main factor for high-affinity CPP binding to HA surfaces, and work by Reynolds et al. has shown that the loss of any one of these residues, by substitution with Glu or Asp, results in a lower affinity for a HA surface [69]. Studies on peptides α_{s1}-casein (f59-79) and β (1-25) binding to HA showed that the surface area occupied by the peptides on maximal binding was proportionate to the cross-section of their three-dimensional solution structures [69], suggesting a closely packed binding structure under saturation conditions [69]. Indeed, we [80] hypothesized that whole casein may also bind in this manner to HA surfaces (although notably not all the casein fractions contains the cluster sequence –Ser(P)-Ser(P)-Ser(P)-Glu-Glu-).

Reynolds also repeated the HA binding studies with saliva-coated HA (sHA). It was found that the peptide affinities for the sHA were slightly reduced by the presence of the salivary proteins but that the Ser(P) cluster motif was still the major factor in adsorption. Reynolds therefore concluded that the principal CPP interaction with both pellicle and plaque is most likely electrostatic and mediated by the Ser(P) cluster motif of the phosphopeptide [69].

Huq et al. [97] investigated computer simulations of the Ser(P) cluster motif binding onto three crystallographic planes of HA – [100], [010] and [001]. It was found that the Ser(P) cluster motif is most likely to bind to the [100] surface, followed by the [010] then [001] surface. This means that the Ser(P) cluster motif can bind to [100] and [010] surfaces, allowing for the deposition of calcium, phosphate and hydroxyl ions on the [001] surface and hence HA crystal growth along the c-axis only, confirming results of HA growth found *in vivo* [69]. By examining the computer simulation data, it was discovered that the carboxyl groups of the glutamyl residues and the phosphoryl groups of the phosphoseryl residues (of the –Ser(P)-Ser(P)-Ser(P)-Glu-Glu- motif) provide the greatest relative binding energy when they are positioned on the HA surface with maximal contact with the surface atoms [3, 97].

1.3.7. Casein and Fluoride

The combined effects of casein and fluoride were shown to be additive as early as 1966 by Weiss and Bibby [66]. Work by Reynolds has shown that CPP-ACP and fluoride exhibits a synergistic effect [72, 96]. A recent study by Reynolds et al. investigated these synergistic effects *in situ* [96]. Addition of 2 % CPP-ACP to a 450-µg/g-F mouthrinse was shown to increase the level of incorporated fluoride in plaque significantly, a toothpaste containing 2 % CPP-ACP showed remineralization levels similar to that of a 2800 µg/g F toothpaste, and a toothpaste containing 2 % CPP-ACP plus 1100 µg/g F was superior at exhibiting remineralization.

Cross [98] hypothesized that CPP-stabilized amorphous calcium fluoride phosphate (CPP-ACFP) may be formed when CPP-ACP is in the presence of fluoride ions. CPP-ACP/F nanocomplexes have been observed localized on the tooth surface and are thought to act in much the same way as CPP-ACP [96]. The levels of incorporated fluoride in enamel were found to be higher in the presence of CPP-ACP [96]. It has been observed that fluoride-alone treatments stimulate remineralization predominantly in the surface layer, whereas the CPP-ACP and CPP-ACFP treatments exhibit a more homogeneous remineralization throughout the entire body of a carious lesion [96].

2. Our Research

At Bristol, we have engaged in a series of studies to investigate the efficacy of whole casein, caseinphosphopeptide (CPP) and glycomacropeptide (GMP) as agents to inhibit dental erosion. There are two main applications of our research: the addition of an erosion-inhibitor to an acidic solution, rendering it less erosive, or the application of an erosion-inhibitor to the teeth prior to a later erosive challenge, thus providing a longer-lasting protection against erosion by any acidic solution. There is no calcium in any of the protein solutions that we studied and so the protective effects are ascribed to the proteins themselves, rather than any localization of calcium and phosphate at the tooth surface as proposed with the use of CPP-ACP.

2.1. Inhibition of Hydroxyapatite Dissolution by Whole Casein: The Effects of Ph, Protein Concentration, Calcium and Ionic Strength

In 2008 we reported a study whereby addition of casein to citric acid solutions reduced HA dissolution [99]. In this study, 0.2, 0.02 or 0.002 % w/v casein (Nutritional grade, BDH, Poole, UK) was added to citric acid solutions over a range of pH values designed to be representative of soft drinks [89]. Compressed HA discs were used as the test substrate, as they are known to have similar dissolution characteristics to human enamel [100] and are readily and cheaply available.

The HA dissolution rate before and after the addition of casein was assessed to measure its efficacy. The effect of the addition of calcium ions and the application of a salivary pellicle were also investigated. The dissolution rate was measured using a pH stat, which is an automatic titration system [100]. It measures the pH of the reaction solution continuously, and when the pH deviates from a pre-selected value, this triggers the addition of small aliquots of 50 mM HCl to maintain a constant pH. During HA dissolution, the rate of addition of H^+ ions is proportional to the dissolution rate.

Data were analysed using Kruskal–Wallis tests. Median dissolution rates are given. In the case of the pellicle part of the study, data were analysed using a Mann–Whitney two-tailed test. Differences were deemed to be significant when $p < 0.050$.

It was found that casein, when added to the citric acid solutions, reduced the dissolution rate of the HA discs by 51 ± 4 %. Casein concentration had no significant effect on dissolution rate, which is perhaps surprising as 0.002 % w/v is a relatively low concentration and corresponds to 2 μg of protein mm^{-2} of HA surface area. The density of the casein used is ~ 1.1 g/ml and so 2 μg of protein mm^{-2} corresponds to a thickness of 1.82 μm if all protein were deposited on the HA only (however it is likely there will be protein deposition on other surfaces too). In comparison, the pellicle layer formed per labial surface of a tooth over 2 h *in vivo* contains approximately 1 μg of protein and the thickness of the pellicle ranges from 0.1 – 1.0 μm after 60 min [101]. The addition of calcium to the casein-containing citric acid solution resulted in an additive reduction in dissolution rate for low calcium concentrations, but not for higher calcium concentrations.

Casein added to the citric acid solutions also significantly reduced the dissolution rate of the HA discs when they had been coated with a salivary pellicle. This is interesting as it tells us that the protein pellicle layer does not block or hinder casein and its protective mechanism, which is hypothesized to be the formation of a thin protein film on the mineral surface (due to the fact that the effect appears saturated even at 0.002 % w/v casein and so there must be very little protein).

This study therefore reveals the potential that casein has to inhibit dental erosion. Due to the low concentrations of casein required to elicit a protective effect, it could be used as an additive to soft drinks, reducing their inherent erosive potential. Casein could also be added to acidic medications, such as vitamin C, to reduce their potential to cause erosion. Care must be taken, however, as casein may reduce the shelf-life of any products that incorporate it, as the milk protein is likely to denature cause bacteria to grow. Further investigations in order to understand the protective mechanism for casein and how it interacts in details with the enamel and pellicle proteins should be taken.

2.2. Enamel Erosion in Dietary Acids: Inhibition by Food Proteins In Vitro

In 2010 [102], we reported another investigation into the effects of adding casein to acids. In this study we extended the investigation to encompass three common dietary acids and five commercially-available soft drinks. We looked at a range of pH values, typical of popular drinks. We employed a clinically more realistic substrate: human tooth enamel sectioned and polished to create a smooth, flat surface. We applied a simple *in vitro*-formed salivary pellicle so as to mimic the conditions *in situ* as closely as possible.

Treatment solutions of citric, malic or lactic acid with and without the addition of 0.2 % w/v casein (as above) were investigated. Five erosive soft drinks popular in the UK market with and without the addition of 0.2 % w/v casein were also investigated: Robinsons Apple 'Fruitshoot' Drink (Robinsons Soft Drinks, Chelmsford, UK); Capri Sun Orange Drink (Coca-Cola, Uxbridge, UK); Tropicana Pure Premium Orange Juice (Tropicana UK Ltd., Leicester, UK), Powerade Fruit Flavour Isotonic Arctic Shatter (Coca-Cola, Uxbridge, UK); Powerade Lemon and Grapefruit Isotonic Ice Storm (Coca-Cola, Uxbridge, UK).

Erosion was measured using a white-light, non-contact optical profilometer. The surface is probed using a beam of white light, which is split by means of a prism into its constituent colours. By recording the colour of the reflected light, one can measure the height of the surface to at least ± 0.4 μm. Enamel specimens were sectioned from whole, healthy human teeth using an annular diamond-tipped saw, mounted in resin, lapped and polished to obtain a flat surface. Two reference areas were created either side of an erosion 'window' by placing adhesive tape on the enamel surface. Height differences between the 'window' and the reference points indicate how much erosion has taken place (Figure 2).

One-way Analysis of Variance (ANOVA) followed by post-hoc Tukey's HSD statistical tests were conducted using XLstat® (Addinsoft®) and differences deemed to be significant when $p < 0.050$.

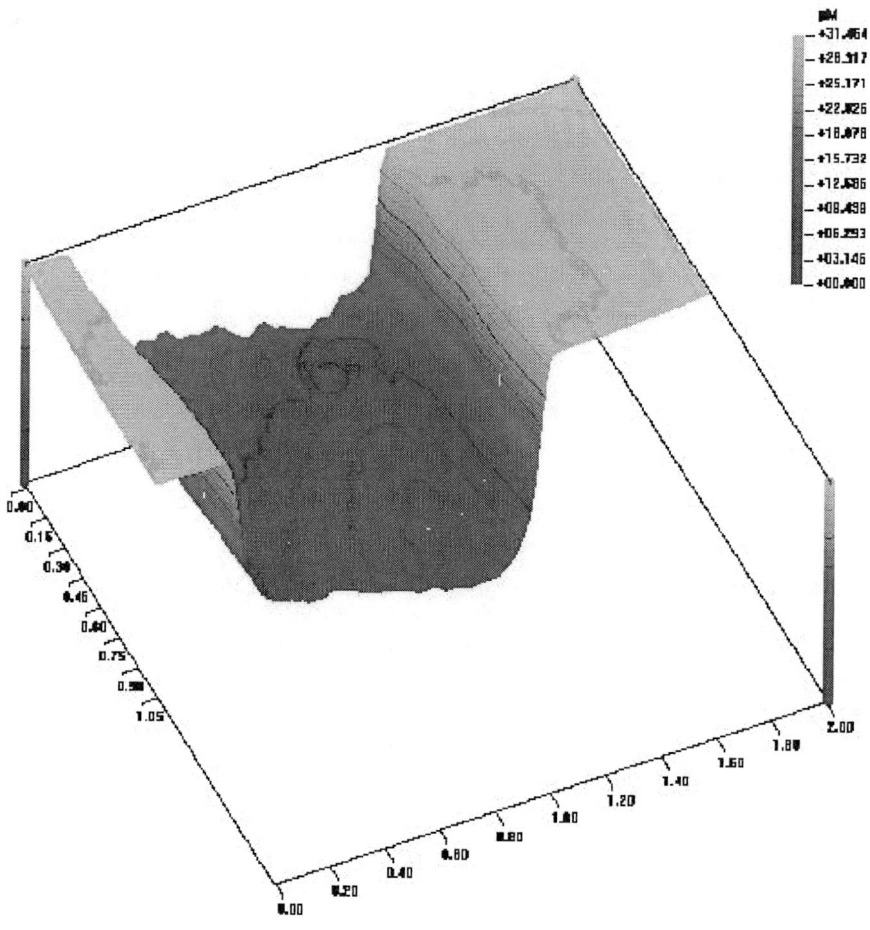

Figure 2. An example profilometer trace.

Under these experimental conditions, 10 min of exposure to acids and drinks caused between 1 and 11 μm of erosive enamel loss. It was found that addition of 0.2 % w/v casein to the citric acid reduced erosion by between 59-78 % over the pH range 2.8-3.8. Addition of 0.2 % w/v casein to malic acid reduced erosion by between 29-45 % over the pH range 2.8-3.2, and addition of 0.2 % w/v casein to lactic acid reduced erosion by 33-70 % over the pH range 2.8-3.8. Casein significantly reduced erosion in all drinks tested except Tropicana Orange (Figure 3); the average reduction in erosion by addition of casein was 36 %.

Addition of 0.2 % w/v of the individual sub-fractions of casein to the citric acid at pH 3.20 also resulted in a statistically significant inhibition of erosion, with alpha and kappa casein being the most effective (reducing erosion by 72 and 78 % respectively), but with all of the subfractions inhibiting erosion to a statistically similar degree to that of whole casein (76 % reduction).

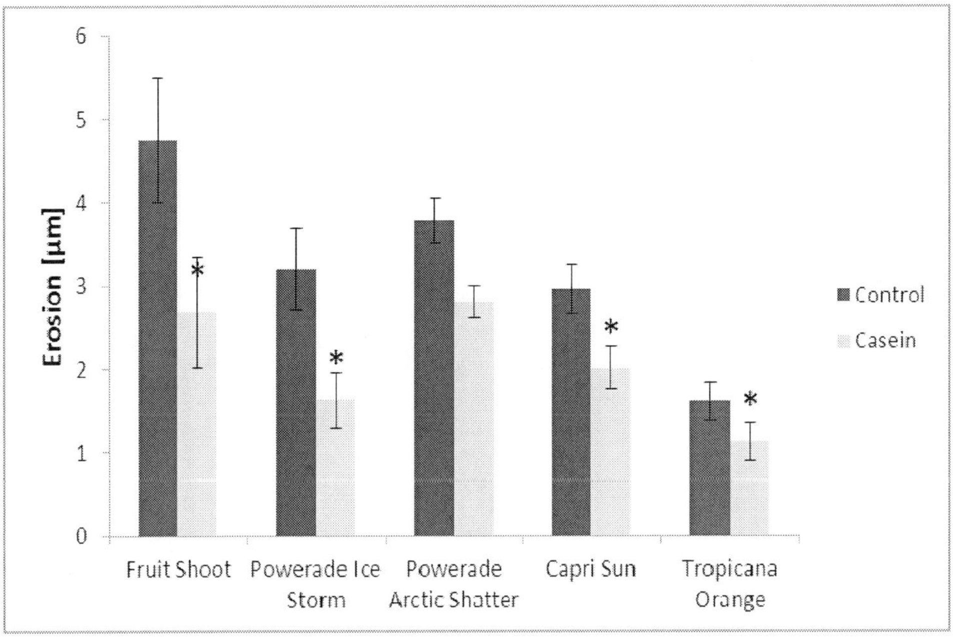

Figure 3. Inhibition of enamel erosion by casein when added to commercially available soft drinks in the presence of a salivary pellicle. Asterisks indicate statistically significant differences between the modified drink and the control.

Based on this and the previous study, we hypothesized that casein reduces erosion by interacting with the enamel surface, possibly by adsorption to the enamel surface and/or the pellicle proteins forming a casein-rich film or layer. This is in agreement with studies where evidence has been presented for formation of 'milk protein pellicles' on human enamel [103], and that pre-incubation of the enamel in saliva (forming a pellicle) did not inhibit or affect the 'milk protein pellicle' formation [103]. It has also been shown that there are micelle-like globules in the salivary pellicle similar to the milk protein micelles [54]. Furthermore, there is evidence that casein-derived proteins glycomacropeptide (GMP) and caseinphosphopeptide (CPP) incorporate into the salivary pellicle [104, 105], or displace salivary pellicle proteins [106] as well. This suggests that these proteins could form a layer with similarities to the salivary pellicle, displacing salivary proteins or incorporating into a hybrid pellicle.

The fact that there was minimal difference between the efficacies of the casein sub-fractions provides an insight into the interactions of casein with the enamel surface. The sub-fractions differ in size and molecular properties, and not all the casein subfractions contain the cluster sequence -Ser(P)-Ser(P)-Ser(P)-Glu-Glu-, which has been suggested to be important in the adsorption process, discussed above. The different sub-fractions, however, were all still effective at reducing erosion, suggesting that there may be multiple adsorption mechanisms taking place [107] and that the -Ser(P)-Ser(P)-Ser(P)-Glu-Glu- may not be critical for a favourable interaction.

The three different acids investigated, malic, lactic and citric, differed in baseline erosion figures even though the same concentration and pH values were used. This could be due to the differing buffer capacities of the acids, but also due to the differing motilities of the anions. The proportional reduction in erosion afforded by the casein also differed for each acid, however, with baseline erosion figures being similar for lactic and citric acid but casein reducing erosion significantly more in citric acid than in lactic acid (78 % compared to 33 % at pH 3.20 respectively), with casein having the least effect in malic acid (29 % at pH 3.20). This could be due to the buffering capacities of could be due to the malate, lactate or citrate ions interacting to differing degrees with the casein; citrate is a known component of casein micelles in milk and perhaps offers some stabilizing function.

In summary, within the limitations of an *in vitro* model, casein and its subfractions reduced HA dissolution (as a primary model) and enamel erosion (as a secondary model) in common dietary acids and in soft drinks, in the absence and presence of a salivary pellicle. Casein thus shows promise as an agent, which could be added to erosive products such as drinks to reduce their erosive potential.

2.3. Inhibition of Dental Erosion by Casein and Casein-Derived Proteins

These encouraging results on the reduction in erosive potential of acidic solutions when supplemented with casein lead to an attempt to discover the effect of applying a pre-treatment of casein, and to determine whether this provided protection against subsequent erosive attacks. In 2011 [108] we reported an investigation of the application of casein, CPP and GMP as agents which could be added to oral care products to provide erosion protection. 0.5 % w/v aqueous solutions of casein, CPP or GMP were investigated with regard to inhibition of dental erosion, measured by surface softening and bulk tissue loss, with and without the addition of 300 µg/g fluoride (F, as NaF) in comparison to a DIW negative control and a 300 µg/g fluoride positive control. Fluoride was included because it is used in the majority of oral care products.

Atomic force microscopy (AFM) nanoindentation was used to measure surface hardness, to assess the earliest stages of erosion, and non-contact optical profilometry to measure bulk tissue loss for the later stages of the erosion process. The latter is discussed in detail in section 2.2 above. Nanoindentation is in widespread use in the materials and engineering communities but is a relative newcomer to biomedical research. A diamond-tipped pyramidal indenter is incident on the surface of interest with a known load for a given time. Forces are typically in the range of 10-10000 nN. Displacement of the indenter tip is monitored continuously yielding a force-displacement curve, from which one can calculate a range of parameters. Hardness, in GPa, is the most useful to assessing dental erosion as the earliest detectable stage in enamel erosion is enamel dissolution, which is detectable by a softening of the enamel surface (due to thinning of the enamel crystals).

Treatment solutions were 0.5 % w/v casein (as above), CPP (Arla Foods®, Leeds, UK) and GMP (Davisco Foods®, Minnesota, USA), with and without 300 µg/g fluoride [F] (as NaF - Sigma Aldrich, Gillingham UK). A 0 µg/g F (DIW) solution was used as a negative control. The pH of all solutions was measured and found to be in the pH range 6-7.

For the nanoindentation part of the study, each treatment group was attached to an overhead rotating stirrer moving such that the specimens travelled at 0.25 m/s and then exposed to 150 mL treatment solution at room temperature for 10 minutes, rinsed in 150 mL DIW at 0.25 m/s for 10 s followed by exposure to 150 mL 0.3 % CA pH 3.20 for 1 minute at 0.25 m/s and rinsed again in 150 mL DIW at 0.25 m/s for 10 s. Specimens were allowed to dry in air for at least 24 h before the nanoindentation process was repeated. Percentage change in surface hardness was calculated for each specimen and a mean taken for each treatment group.

For the profilometry part of the study, specimens were exposed to the following regime: 10 min treatment – 10 s rinse – 10 min erosion – 10 s rinse, a total of 9 times. For "treatment" and "erosion", specimens were, as above, immersed in 150 mL of the relevant solution at room temperature at 0.25 m/s. For "rinse", specimens were immersed in DIW, to halt the reaction and rinse the treatment/erosion solution from the specimen. The treatment and erosion solutions were the same for both parts of the study.

One-way Analysis of Variance (ANOVA) followed by a post-hoc Newman-Keuls statistical tests were conducted using XLstat® (Addinsoft®) and differences deemed to be significant when $p < 0.050$.

After short acid exposures, three of the treatment solutions STATISTICALLY significantly inhibited enamel softening: 0.5 % w/v casein, 300 μg/g F and 0.5 % w/v casein + 300 μg/g F (Figure 4).

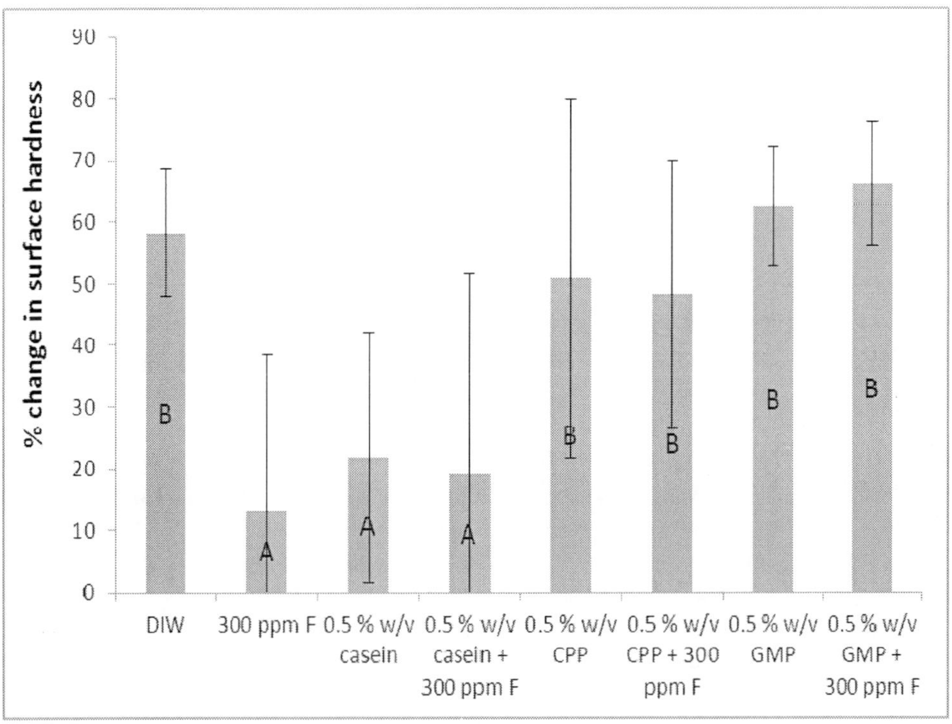

Figure 4. Enamel hardness as a percentage of baseline as measured by nanoindentation. Error bars are standard deviations and lettering indicates statistical homogeneity.

After longer acid exposures all treatment solutions resulted in a statistically significant reduction in enamel erosion compared to DIW (Figure 5). The 0.5 % w/v casein + 300 μg/g F solution resulted in a significantly greater reduction in erosion compared to all other treatment solutions.

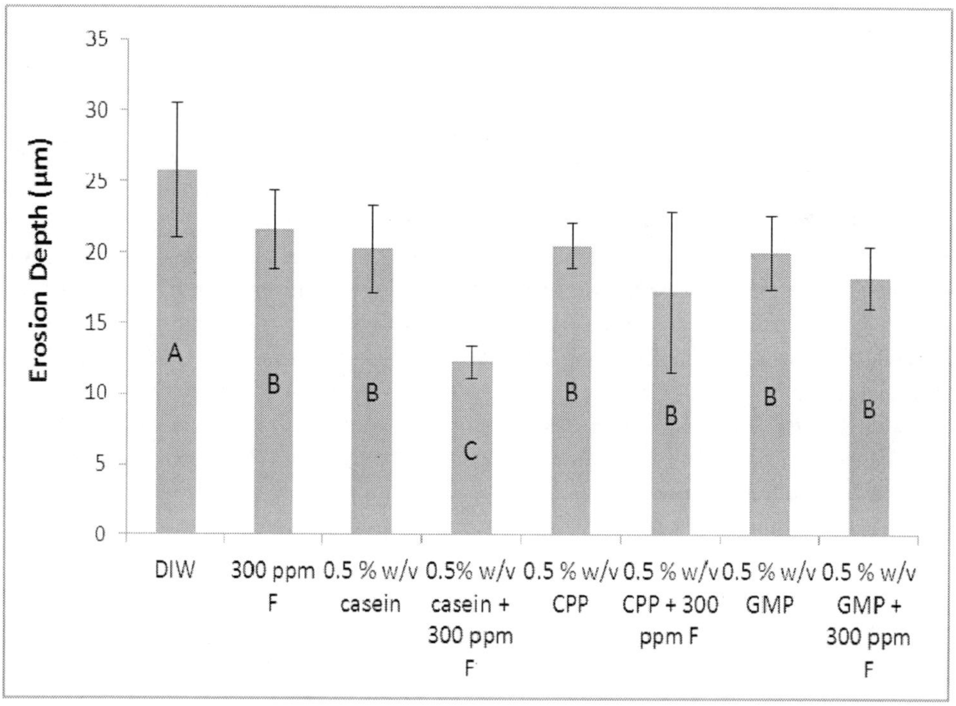

Figure 5. Enamel erosion as measured by profilometry. Error bars are standard deviations and lettering indicates statistical homogeneity.

Again, we hypothesize that the proteins adhered to the enamel surface to form a protective layer, which inhibits dissolution. Previously we knew that casein only exhibits a protective effect when applied in acidic solutions. We now know that casein exhibits a protective effect when applied at neutral pH. This is interesting because at acidic pH, casein has a positive charge (isoelectric point 4.6 [80]) whereas at neutral pH casein is negatively charged. It was also thought that acidic pH that the acidic anion may play a role in the protective mechanism of casein; it was known that citrate is a component of casein micelles. At neutral pH there were no such anions and yet casein still exerts a protective effect.

In previous studies we also knew that casein was effective when added to the erosion medium; the casein was effective at time of deposition. We now know that a pre-treatment of casein deposits casein in such a way that it remains on the enamel surface, even in the presence of an erosion attack, and that it remains on the enamel surface for at least 10 minutes (from the profilometry part of the study). This is even longer than the salivary pellicle, which is known to exert a protective effect to citric acid for 5 min but not 10 min [109].

The difference between the efficacy of the casein and the smaller molecular weight proteins CPP and GMP could be ascribed to the casein forming a more coherent layer (due to its size and possibly conformation).

The combination of casein and fluoride reduced erosion more than either agent alone, but no such effect was observed for CPP or GMP with fluoride. It is possible that fluoride could be incorporated in the protein layer, being released at the lower pH of the erosive challenge, in such a way as it was possible with the larger casein but not the smaller peptides. This could explain why casein and fluoride together were so effective in the 'cycling' profilometry experiment but not in the nanoindentation experiment which utilized only one acid exposure.

We have shown here that casein is at least as effective at preventing enamel surface softening and bulk tissue loss (erosion) as the current major accepted technology, fluoride. This has huge implications. Dental fluorosis can occur when teeth are exposed to too much fluoride. Casein offers an alternative to patients suffering from this condition. When casein is used in combination with fluoride, the effect is additive, resulting in an unprecedented level of protection to the teeth. Casein can therefore see applications in a wide range of oral healthcare products.

2.4. Mechanism of Action

A number of hypotheses may be put forward for the erosion protection observed. It could be simply a physical layer which blocks the access of al lions, including H^+ ions, to the mineral surface and restricts the detachment of Ca^{2+} and PO_4^{3-} ions.

The protection could also be ascribed to a buffering effect; phosphoserine, histidine, glutamate and aspartate are all known to have H^+ accepting groups and are found in the proteins studies here [110], which would remove H^+ ions from the acidic solution and hence raising the local pH.

Finally, the adsorbed protein layer could act as an ion-retarding membrane, restricting the access of certain ions e.g. H^+ and detached Ca^{2+} ions and allowing others to pass through, much like "permselectivity" of the salivary pellicle [111].

As discussed above, the three contiguous –Ser(P)- residues in CPP have been investigated and are believed to be the main factor for high-affinity CPP binding to HA surfaces. This could be the case for casein as well [69], but not for GMP, which lacks this motif but has also been visualized on enamel surfaces in a 'globular-micellar' form, with similarities to the salivary pellicle [93]). The casein subfractions do not all contain the –Ser(P)- motif either, yet exhibit a protective effect. It seems that perhaps an alternative adherence mechanism for the proteins studied here exists. They are amphiphiles and perhaps hydrophobic parts of the proteins localize at the mineral surface, with the hydrophilic parts exposed.

We suggest that both electrostatic/Van der Waals and hydrophobic adherence mechanisms are both taking place for the casein and casein-derived proteins studied here.

2.5. Film analysis

We have repeatedly ascribed the protective effect of casein and related proteins to the formation of a film or layer on the surface of the mineral. All of the experiments reported above provide indirect evidence of this layer – the outcome of its proposed presence, not a direct visualization of its veracity. It is, therefore, interesting to probe the protein layer directly.

AFM Analysis

We used AFM to image casein films formed on enamel, glass and freshly cleaved mica. Glass and mica were selected due to their smoothness; it would be near impossible to obtain enamel or HA that is comparably smooth and flat.

Mica was exposed to a solution of 0.5 % w/v casein for 2 h and then rinsed in DIW for 10 s. It was desirable to achieve maximal film thickness on the mica; exposure times could always be reduced if a film was found. 2 h was chosen as this is the time that the pellicle takes to reach maturity in the oral cavity, and, although not being under the same conditions in terms of proteins concentrations or environment, was felt sufficiently long for a film to form (indeed it has been shown earlier that protection is afforded from casein after as little as 30 s).

When imaging casein on mica both tapping and contact modes were trialed. Tapping modes, most commonly used when imaging softer surfaces was found to reveal the best images and also allowed the simultaneous acquisition of height and phase images of the same region.

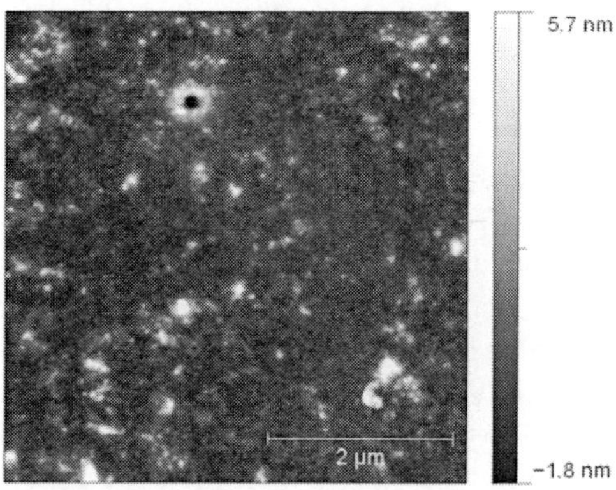

Figure 6. Tapping mode AFM height image of freshly cleaved mica exposed to a 0.5 % w/v casein solution for 2 h followed by a 10 s rinse in DIW.

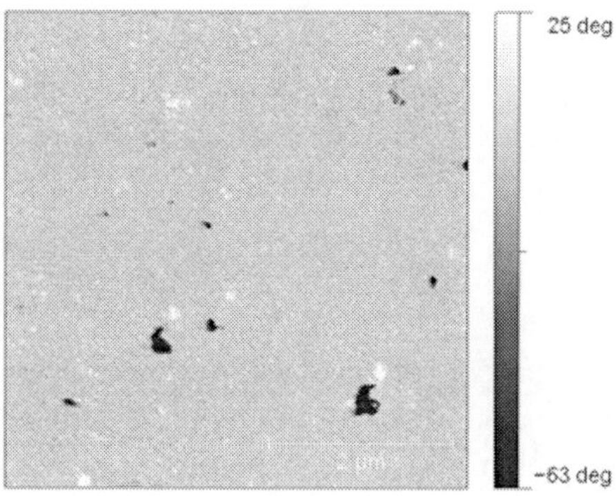

Figure 7. Tapping mode AFM phase image (of previous figure) of freshly cleaved mica exposed to a 0.5 % w/v casein solution for 2 h followed by a 10 s rinse in DIW.

Figure 6 shows a series of structures on the mica surface that vary in size and shape. It is clear that a uniform layer has not formed, with what appears to be aggregations of protein in some regions and less in others. This could be due to the charging of the mica surface. The phase image of the same region (Figure 7) does not really reveal anything more than the height image does. In the top half of the height image, there appears to be a hole, or 'volcano' – a raised rim surrounding a hollow core. This region is examined more closely in Figure 8.

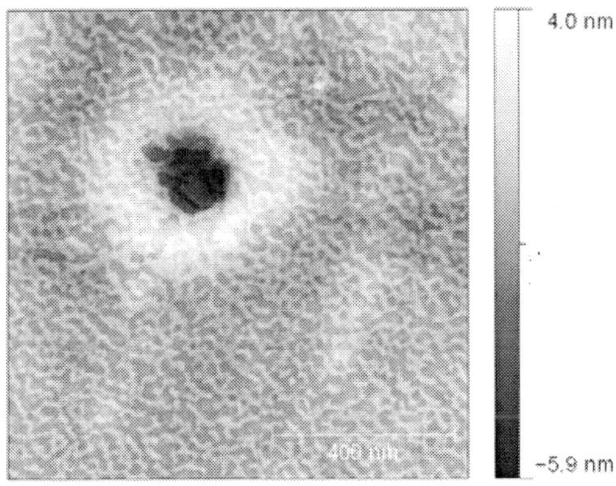

Figure 8. Tapping mode AFM height image of selected region shown in Figure 6.

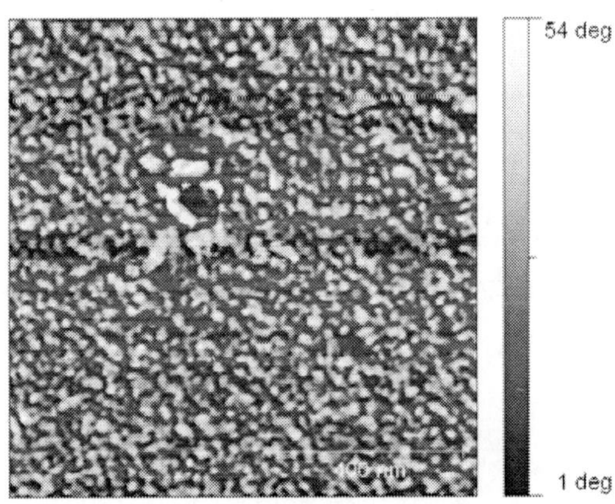

Figure 9. Tapping mode AFM phase image (of previous figure) of selected region shown in Figure 6.

Figure 8 and Figure 9 show the 'volcano' mentioned earlier. A raised rim surrounding a hollow core is revealed. Surrounding the volcano is a 'sea' of what looks like dried salt – this could be dried sodium caseinate. The height difference between the mean 'sea' height and the bottom of the 'volcano' is approximately 6-7 nm (Figure 10), suggesting that a 6-7 nm dried casein film has formed on the mica. The 'volcano' itself is approximately 200 nm in diameter (rim to rim). Casein micelles are roughly spherical structures and are polydisperse in size, with diameters as large as 600 nm, although approximately half of casein micelles are found to have a diameter of 140-250 nm, with a weight-averaged radius of 140 nm [112-114]. This 'volcano' could then be the remnant of a casein micelle, which has dried and collapsed (possibly due to the charge on the mica surface) leaving a hollow micelle core.

What is interesting is that casein adsorbs to the mica surface at all; freshly cleaved mica has a negative charge [115], and casein at neutral pH (isoelectric point 4.6 [80]) is also negatively charged.

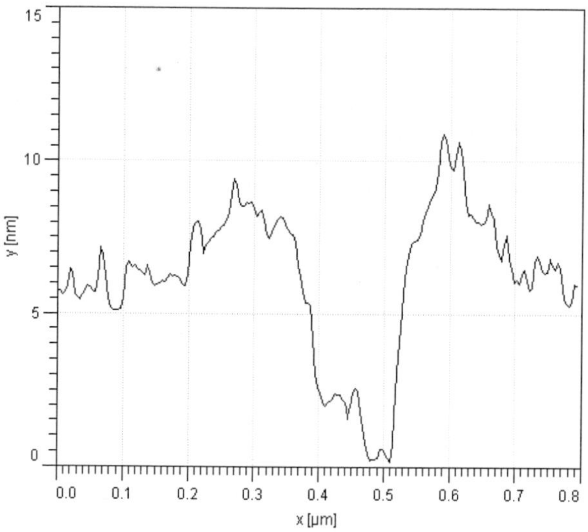

Figure 10. Cross-section through 'volcano' shown in figure 8.

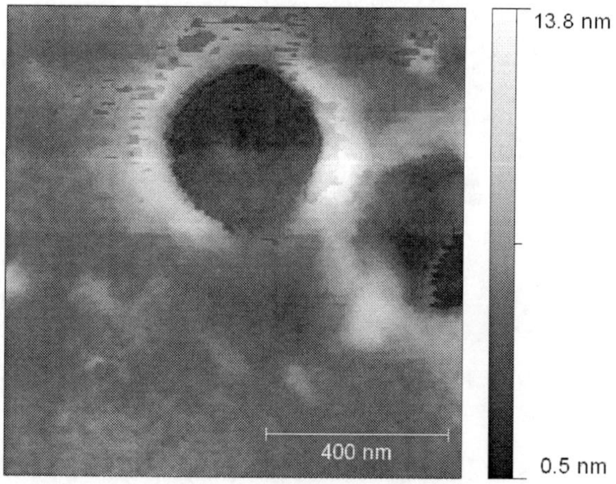

Figure 11. Tapping mode AFM height image of freshly cleaved mica exposed to a 0.5 % w/v casein solution for 2 h followed by a 10 s rinse in DIW. This image shows another 'volcano'.

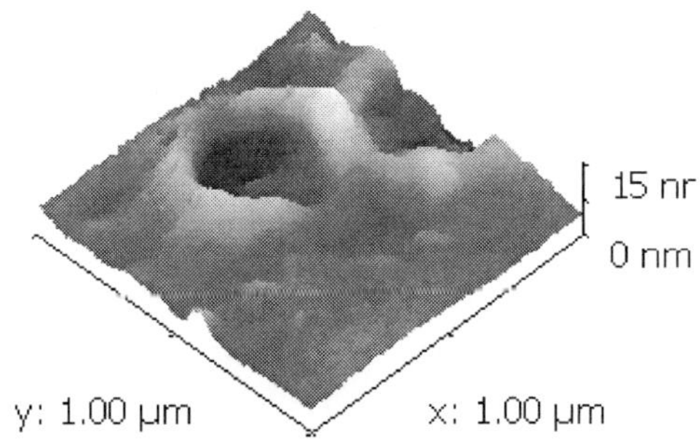

Figure 12. Tapping mode AFM phase image (of previous figure) of freshly cleaved mica exposed to a 0.5 % w/v casein solution for 2 h followed by a 10 s rinse in DIW.

Figure 11 shows another mica sample exposed to casein, and again reveals another 'volcano'. A three-dimensional representation of this image is shown in Figure 12.

Figure 13 and Figure 14 show a glass slide exposed to a 0.5 % w/v casein solution for 2 h followed by a 10 s rinse in DIW (the same exposure conditions as for the mica specimens). It is evident from the height image that a more uniform film is formed on the glass surface, with the occasional 'volcano'. The general distribution on the glass surface appears far more ordered. The phase image shown in Figure 14 reveals far more information when used in combination with the height image shown in Figure 13. Spherical aggregates, of a completely different phase to the rest of the surrounding material are seen, but these aggregates are the same height as the surrounding material. These aggregates are approximately 200 - 400 nm in diameter and could be the casein micelles.

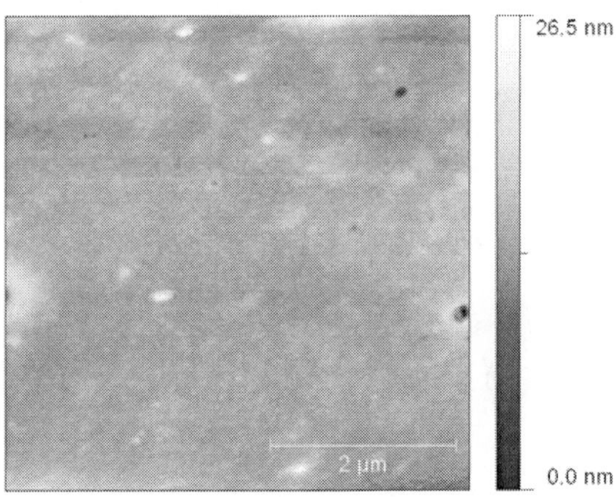

Figure 13. Tapping mode AFM height image of a glass slide exposed to 0.5 % w/v casein for 2 h followed by 10 s rinse in DIW.

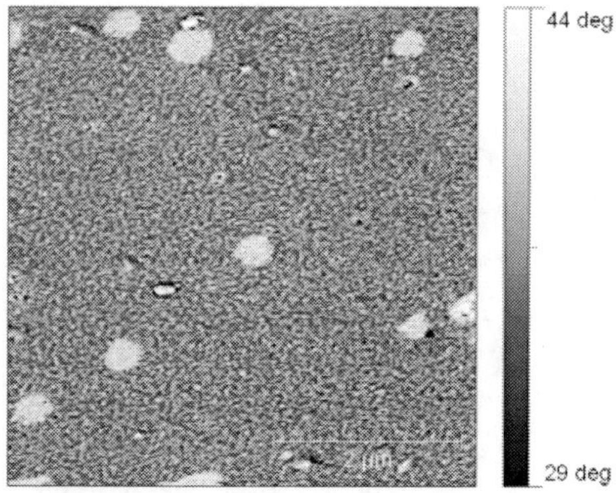

Figure 14. Tapping mode AFM phase image of a glass slide exposed to 0.5 % w/v casein for 2 h followed by 10 s rinse in DIW.

These casein micelles may be slightly larger than those seen on mica and do not reveal themselves in the height image due to the lack of surface charging; the charged mica may be attracting the casein to the surface quite strongly which acts to compress the casein film and collapse the micelles. The casein adsorbing to the glass could be due to hydrophobic interactions, interactions that are less strong than the electrostatic interactions experienced on mica, and so compress the protein film and micelles less.

Further imaging on glass revealed similar uniform surfaces on the glass, but with some interesting features – patches of casein with little channels in between (Figure 15 - Figure 17). These channels were analyzed by means of a two-dimensional cross-section (Figure 18 and Figure 19) at multiple locations, and revealed consistent height differences on the order of 14 - 18 nm that 'bottomed out' to the same point which is, most likely, the underlying substrate (glass).

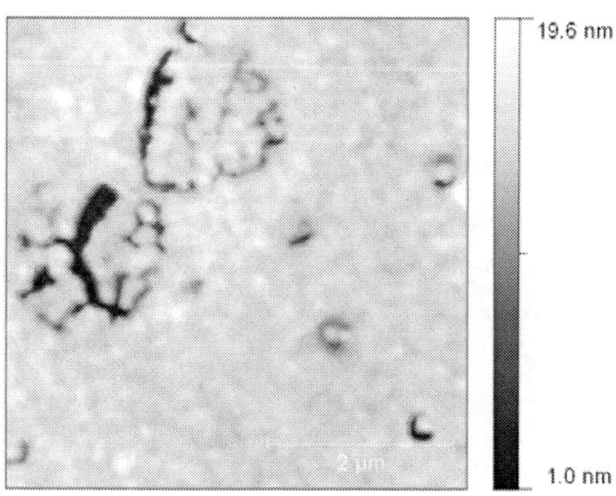

Figure 15. Tapping mode AFM height image of a glass slide exposed to 0.5 % w/v casein for 2 h followed by 10 s rinse in DIW.

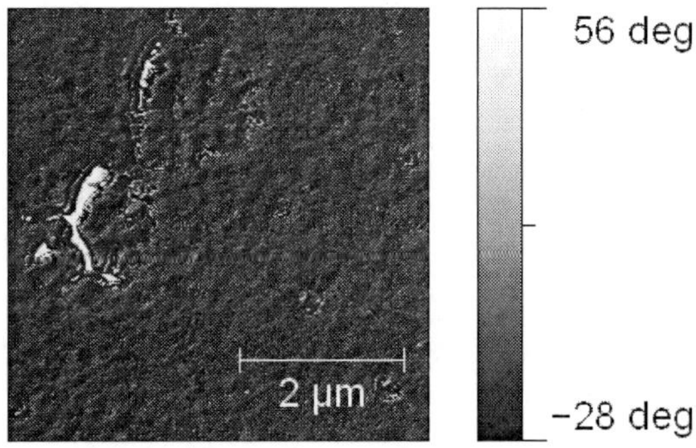

Figure 16. Tapping mode AFM phase image (of previous image) of a glass slide exposed to 0.5 % w/v casein for 2 h followed by 10 s rinse in DIW.

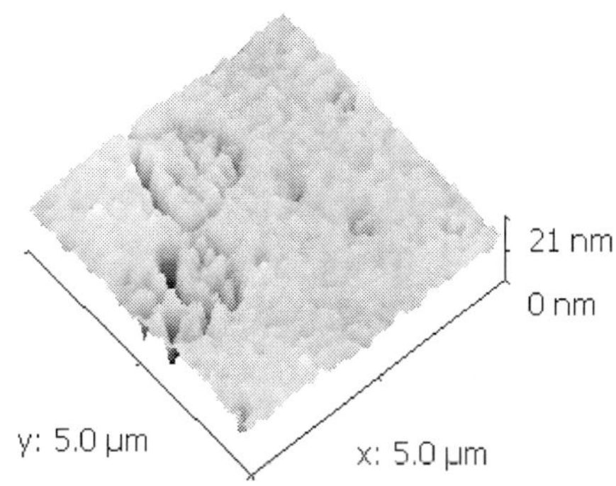

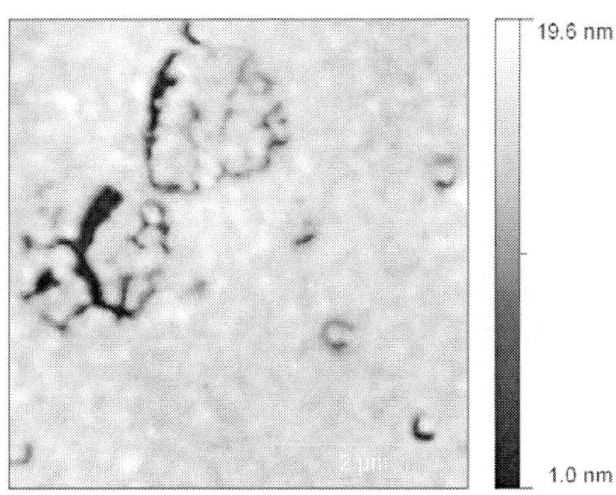

Figure 17. 3d representation of

Figure .

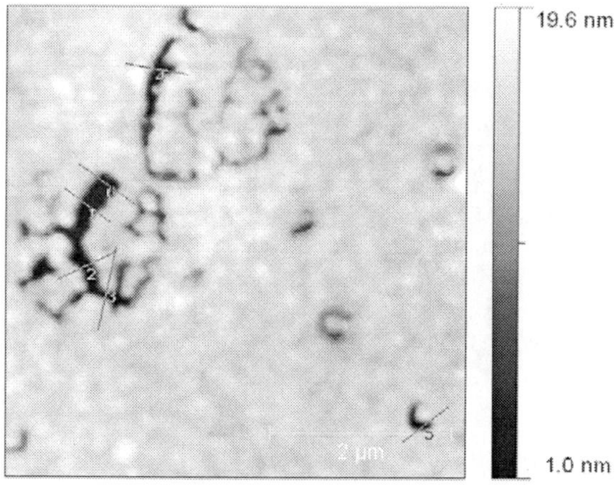

Figure 18. Diagram showing where the 2D cross sections were taken on AFM image.

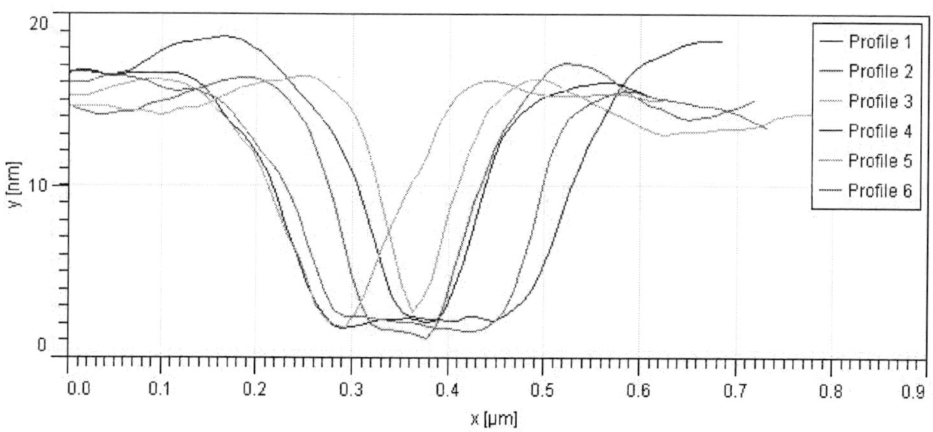

Figure 19. 2D cross-section profiles.

X-ray Reflectometry

X-ray reflectometry (XRR) was also used to investigate these thin protein films on mica. Grazing angle incidence XRR has, over the past decades, been applied to obtain information about nanofilm structures, and following a recent publication by Briscoe [115], has been applied to mica using a simple 'bending mica' technique.

For these XRR experiments, casein films were formed in the same way as for the AFM experiments above. Briefly, freshly cleaved mica sheets of ~ 1 cm × 3 cm (width × length) were incubated in the 0.5w/v casein solution for three different incubation times: 30 s, 10 min and 2 h, before rinsed with ~ 25 ml DIW and then dried naturally overnight in a dust free laminar flow hood (LFH) fitted with hyperfilter. These casein coated mica sheets were then stored in capped glass vials for later measurements, where they would be mounted in a custom designed chamber implementing the "bending mica" method (Briscoe et al., 2011, to be submitted).

The XRR measurements were performed at synchrotron X-ray beamlines I07 (UK Diamond) and BM28 (European Synchrotron Radiation Facility, ESRF, Grenoble). Figure 20 shows an example XRR reflectivity curve for the sample with 30 s incubation time, collected at Diamond I07. For this measurement, the energy for the X-ray beam E = 20 keV, corresponding to an X-ray wavelength $\lambda = hc/E = 0.620$ Å where $h = 6.626 \times 10^{-34}$ J s is Planck's constant and c the speed of light. Specular reflection was collected with an avalanche photo diode (APD) for the incident and reflection angle θ in the range of ~ 0 – 2.5 degrees. In the figure, the reflectivity (empty circles O) is plotted, on a log-linear scale, vs. the momentum transfer $Q = 4\pi\sin(\theta)/\lambda$. The intensity oscillations in the reflectivity curve, called Keissig fringes, could be analysed to yield the structural information about the casein film, such as the thickness, roughness and surface coverage. Such an analysis is very challenging due to mica's layered crystal structure which gives rise to surface diffraction peaks (Briscoe et al., 2011, to be submitted). Here in order to obtain the casein film thickness, we analysed the XRR curve assuming a simple one-layer model for the film and fitted the curve using the Parratt algorithm and the Motofit software package (motofit.sourceforge.net). The solid curve in Figure 20 was obtained this way, giving a thickness of 6.5 ± 0.1 nm for these 30 s incubation films, in close agreement with the AFM measurements above. Furthermore, our XRR measurements on the casein films of 10 min and 2 h incubation at ESRF BM28 showed that the film thickness was very similar, in the range of 6.4 – 6.6 nm. This indicates very fast casein adsorption kinetics at the mica surface, with the film reaching a saturation thickness after just 30 s.

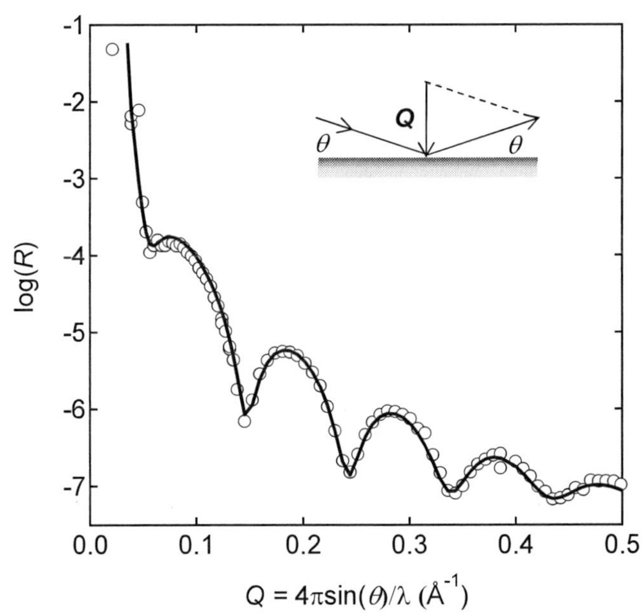

Figure 20. Reflectivity vs momentum transfer Q on a log-linear plot for 2 h casein exposure specimen. The substrate is muscovite mica, and the measurement was made at synchrotron beamline I07 (UK Diamond) using a simple and unique method that enables XRR to be applied to the mica substrate.

Discussion

The XRR and AFM data have provided direct and corroborating evidence of a protein film of approximately 6 - 18 nm thickness, deposited on a range of substrates including mica, glass and tooth enamel. This supports our earlier hypothesis of the formation of a protein layer on the tooth surface, which is responsible for the erosion reduction effects we observed in a range of models. It is somewhat peculiar that casein, slightly negatively charged at neutral pH, would adsorb onto highly negatively charged mica (~ 50 $Å^2$ per charge when fully charged). The glass surface again represents different surface chemistry, prominently coated with hydroxyl groups. There are different from the HA or enamel surfaces. Casein may employ different mechanisms to facilitate adsorption to these different surfaces, including hydrophobic interactions, electrostatic forces, hydrogen bonding and specific ion binding. We suggest that this might be related to casein's lack of secondary structures, which may afford the flexibility and versatility to employ these different binding mechanisms. Thus it may be considered as a biocompatible surface coating material, and, in our case, it readily forms a protective layer on the tooth surface to prevent decay, reducing dental erosion.

3. Conclusion

Casein and related proteins and sub-fractions can reduce dental erosion in in vitro models. The efficacy is compatible with existing fluoride and calcium technologies for reducing erosion, and is not compromised by the presence of a salivary pellicle. This effect can be ascribed to the formation of a protein film, approximately 6-10 nm thick forming on the mineral surface.

Casein and casein-derived proteins may therefore find applications as topical agents to inhibit dental erosion.

4. Acknowledgements

The authors would like to thank O. Bikondoa at ESRF BM28, T. Arnold at I07 UK Diamond, P. Cresswell and G. Pilkington at Bristol University School of Chemistry for their help with XRR measurements and F. Speranza at Bristol University School of Chemistry for her help with sample preparation.

References

[1] Featherstone, J.D.B., The science and practice of caries prevention. *JADA*, 2000. 131: p. 887.

[2] Kidd, E.A.M. and O. Fejerskov, What constitutes dental caries? Histopathology of carious enamel and dentin related to the action of cariogenic biofilms. *J Dent Res*, 2004. 83(Spec Iss C C35-C38).

[3] Cross, K.J., N.L. Huq, and E.C. Reynolds, Anticariogenic Peptides, in Nutraceutical Proteins and Peptides in Health and Disease, Y. Mine and F. Shahidi, Editors. 2006, CRC Press: Boca Ranton.

[4] Smith, B.G.N., D.W. Bartlett, and N.D. Robb, The prevalence, etiology and management of tooth wear in the United Kingdom. *Journal of Prostheitc Dentistry*, 1997. 78(4): p. 367-372.

[5] Pickles, M.J., Tooth Wear. *Monographs of Oral Science*, 2006. 19: p. 86-104.

[6] Imfeld, T., Dental erosion. Definition, classification and links. *Eur J Oral Sci*, 1996. 104: p. 151-155.

[7] Cate, J.M.t. and T. Imfeld, Dental erosion, summary. *Eur J Oral Sci*, 1996. 104: p. 241-244.

[8] Moss, S.J., Dental Erosion. *International Dental Journal*, 1998. 48: p. 529-539.

[9] Attin, T., et al., Correlation of microhardness and wear in differently eroded bovine dental enamel. *Archives of Oral Biology*, 1997. 42(3): p. 243-250.

[10] Attin, T., Methods for Assessment of Dental Erosion. *Monographs Oral Science*, 2006. 20: p. 152-172.

[11] Barbour, M.E. and J.S. Rees, The laboratory assessment of enamel erosion: a review. *Journal of Dentistry*, 2004. 32: p. 591-602.

[12] Davis, W.B. and P.J. Winter, The Effect of Abrasion on Enamel and Dentine after Exposure to Dietary Acid. *British Dental Journal*, 1980. 148: p. 253-256.

[13] Nunn, J.H., et al., Dental erosion - changing prevalence? A review of British national childrens' surveys. International *Journal of Paediatric Dentistry*, 2003. 13: p. 98-105.

[14] Steele, J. and D. Lader, Social factors and oral health in children. Office for National Statistics, 2004.

[15] Shaw, L. and A.J. Smith, Dental erosion - the problem and some practical solutions. *British Dental Journal*, 1998. 186(3): p. 115.

[16] Al-Dlaigan, Y.H., L. Shaw, and A. Smith, Dental erosion in a group of British 14-year-old, school children. Part I: Prevalence and influence of differing socioeconomic backgrounds. *British Dental Journal*, 2001. 190(3): p. 145-149.

[17] Al-Dlaigan, Y.H., L. Shaw, and A.J. Smith, Dental erosion in a group of British 14-year-old, school children. Part III: Influence of oral hygiene practises. *British Dental Journal*, 2002. 192(9): p. 526-530.

[18] Al-Dlaigan, Y.H., L. Shaw, and A. Smith, Dental erosion in a group of British 14-year-old school children. Part II: Influence of dietary intake. *British Dental Journal*, 2001. 190(5): p. 258-261.

[19] Zero, D.T., Etiology of dental erosion - extrinsic factors. *Eur J Oral Sci*, 1996. 104: p. 162-177.

[20] Mandel, L., Dental erosion due to wine consumption. JADA, 2005. 136: p. 71-75.

[21] Bartlett, D., Intrinsic Causes of Erosion. *Monographs of Oral Science*, 2006. 20: p. 119-139.

[22] Meurman, J.H. and J.M.t. Cate, Pathogenesis and modifying factors of dental erosion. *Eur J Oral Sci*, 1996. 104: p. 199-206.

[23] Kelleher, M. and K. Bishop, Tooth surface loss: an overview.

[24] Gedalia, I., et al., Tooth enamel softening with a cola type drink and rehardening with hard cheese or stimulated saliva in situ. *Journal of Oral Rehabilitation*, 1991. 18: p. 501-506.

[25] Amaechi, B.T. and S.M. Higham, Dental erosion: possible approaches to prevention and control. *Journal of Dentistry*, 2005. 33: p. 243-252.
[26] Grenby, T.H., Lessening dental erosive potential by product modification. *Eur J Oral Sci*, 1996. 104: p. 221-228.
[27] Waltimo, A. and M. Könönen, A novel bite force recorder and maximal isometric bite force values for healthy young adults. Scandinavian *Journal of Dental Research*, 1993. 101: p. 171-5.
[28] He, L.H. and M.V. Swain, Understanding the mechanical behaviour of human enamel from its structural and compositional characteristics. *Journal of the Mechanical Behaviour of Biomedical Materials*, 2007. 2008: p. 18-29.
[29] Cate, A.R.t., Oral Histology: Development, Structure and Function. 1998.
[30] Barbour, M.E. and G.D. Rees, The Role of Erosion, Abrasion and Attrition in Tooth Wear. *J Clin Dent*, 2006. 17(Special Issue): p. 88-93.
[31] Hara, A.T., A. Lussi, and D.T. Zero, Biological Factors. *Monographs of Oral Science*, 2006. 20: p. 88-99.
[32] Zero, D.T. and A. Lussi, Behavioural Factors. Monographs of Oral Science, 2006. 20: p. 100-105.
[33] Meurman, J.H., et al., Experimental sports drinks with minimal dental erosion effect. *Scand J Dent Res*, 1990. 98: p. 120-128.
[34] Attin, T., et al., Effect of mineral supplements to citric acid on enamel erosion. *Archives of Oral Biology*, 2003. 48: p. 753-759.
[35] Reussner, G.H., G.C. Jr, and R.T. Jr, Effects of Phosphates in Acid-Containing Beverages on Tooth Erosion. J Dent Res, 1975. 54: p. 365-370.
[36] Beiraghi, S., et al., Effect of calcium lactate in erosion and S. mutans in rats when added to Coca-Cola®. *Pediatric Dentistry*, 1989. 11(4): p. 312-315.
[37] Hughes, J.A., et al., Development and evaluation of a low erosive blackcurrant juice drink in vitro and in situ 1. Comparison with orange juice. *Journal of Dentistry*, 1999. 27: p. 285-289.
[38] West, N.X., et al., Development and evaluation of a low erosive blackcurrant juice drink 2. Comparison with a conventional blackcurrant juice drink and orange juice. *Journal of Dentistry*, 1999. 27: p. 341-344.
[39] Hughes, J.A., et al., Development and evaluation of a low erosive blackcurrant juice drink 3. Final drink and concentrate, formulae comparisons in situ and overview of the concept. *Journal of Dentistry*, 1999. 27: p. 345-350.
[40] West, N.X., et al., Development of low erosive carbonated fruit drinks 2. Evaluation of an experminetal carbonated blackcurrant drink compared to a conventional carbonated drink. *Journal of Dentistry*, 2003. 31: p. 361-365.
[41] Hunter, M.L., et al., Development of low erosive carbonated fruit drinks 1. Evaluation of two experminetal orange drinks in vitro and in situ. *Journal of Dentistry*, 2003. 31: p. 253-260.
[42] Jensdottir, T., A. Bardow, and P. Holbrook, Properties and modification of soft drinks in relation to their erosive potential in vitro. *Journal of Dentistry*, 2005. 33: p. 569-575.
[43] Barbour, M.E., et al., An investigation of some food-approved polymers as agents to inhibit hydroxyapatite dissolution. *European Journal of Oral Science*, 2005. 113: p. 457-461.

[44] Hall, A.F., et al., The effect of saliva on enamel and dentine erosion. *Journal of Dentistry*, 1999. 27: p. 333-339.
[45] DePaola, D.P., F.A. Curro, and D.T. Zero, Saliva: The precious body fluid. *J Am Dent Assoc*, 2008. 139: p. 5s-10s.
[46] Amaechi, B.T. and S.M. Higham, Eroded enamel lesion remineralization by saliva as a possible factor in the site-specifity of human dental erosion. *Archives of Oral Biology*, 2001. 46: p. 697-703.
[47] Hannig, M. and A. Joiner, The Strucure, Function and Properties of the Acquired Pellicle. *Monographs of Oral Science*, 2006. 19: p. 29-64.
[48] Dawes, C., Salivary flow patterns and the health of hard and soft oral tissues. *J Am Dent Assoc*, 2008. 139(S): p. 18S-24S.
[49] Lendenmann, U., J. Grogan, and F.G. Oppenheim, Saliva and Dental Pellicle - A Review. *Adv Dent Res*, 2000. 14: p. 22-28.
[50] Baier, R.E. and P.O. Glantz, Characterization of oral in vivo films formed on different types of solid surfaces. *Acta Odontologica Scandinavica*, 1978. 36: p. 289-301.
[51] Dawes, C., Salivary flow patterns and the health of hard and soft oral tissues. *J Am Dent Assoc*, 2008. 139: p. 18S-24S.
[52] Hannig, M., Ultrastructural investigation of pellicle morphogenesis at two different intraoral sites during a 24-h period. *Clinical Oral Investigations*, 1999. 3(2): p. 88-95.
[53] Vassilakos, N., et al., Adsorption from salivary fractions at solid/liquid and air/liquid interfaces. *Archives of Oral Biology*, 1992. 37(7): p. 549-557.
[54] Rykke, M., et al., Micelle-like structures in human saliva. *Colloids and Surfaces B: Biointerfaces*, 1995. 4: p. 33-44.
[55] Rølla, G. and M. Rykke, Evidence for presence of micelle-like protein globules in human saliva. *Colloids and Surfaces B: Biointerfaces*, 1994. 3: p. 177-182.
[56] Mann, A.B. and M.E. Dickinson, Nanomechanics, Chemistry and Structure at the Enamel Surface. *Monographs of Oral Science*, 2006. 19: p. 105-131.
[57] Nekrashevych, Y. and L. Stösser, Protecive Influence of Experimentally Formed Salivary Pellicle on Enamel Erosion. *Caries Research*, 2003. 37: p. 225-231.
[58] Hannig, M., et al., Pretective effect of the in situ formed short-term salivary pellicle. *Archives of Oral Biology*, 2004. 49: p. 903-910.
[59] Amaechi, B.T., et al., Thickness of Acquired Salivary Pellicle as a Determinant of the Sites of Dental Erosion. *J Dent Res*, 1999. 78(12): p. 1821-1828.
[60] Schweigert, B.S., et al., Dental Caries in the Cotton Rat VI. The Effect of the Amount of Protein, Fat and Carbohydrate in the Diet on the Incidence and Extent of Carious Lesions. *Journal of Nutrition*, 1945.
[61] Shaw, J.H., Effects of Dietary Composition on Tooth Decay in the Albino Rat. *Journal of Nutrition*, 1949.
[62] Bavetta, L.A. and F.J. McClure, Protein Factors and Experimental Rat Caries. *Journal of Nutrition*, 1957.
[63] Weiss, M.E. and B.G. Bibby, Effects of Milk on Enamel Solubility. *Archives of Oral Biology*, 1966. 11: p. 49-57.
[64] Krobicka, A., et al., The Effects of Cheese Snacks on Caries in Desalivated Rats. *J Dent Res*, 1987. 66(6): p. 1116-1119.
[65] Grenby, T.H., et al., Dental caries-protective agents in milk and milk products: investigations in vitro. *Journal of Dentistry*, 2001. 29: p. 83-92.

[66] Weiss, M.E. and B.G. Bibby, Some Protein Effects on Enamel Solubility. *archives of oral Biology*, 1966. 11: p. 59-63.

[67] Reynolds, E.C. and A.d. Rio, Effect of Casein and Whey-Protein Solutions on Caries Experience and Feeding Patterns of the Rat. *Archives of Oral Biology*, 1984. 29(11): p. 927-933.

[68] Reynolds, E.C. and C.L. Black, Reduction of Chocolate's Cariogenicity by Supplementation with Sodium Caseinate. *Caries Research*, 1987. 21: p. 445-451.

[69] Reynolds, E.C., Anticariogenic complexes of amorphous calcium phosphate stabilized by casein phosphopeptide: A review. *Special Care in Dentistry*, 1998. 18(1): p. 8-16.

[70] Reynolds, E.C. and C.L. Black, Cariogenicity of a Confection Supplemented with Sodium Caseinate at a Palatable Level. *Caries Research*, 1989. 23: p. 368-370.

[71] Reynolds, E.C., Casein Phosphopeptide-Amorphous Calcium Phosphate: The Scientific Evidence. *Adv Dent Rs*, 2009. 21: p. 25-29.

[72] Reynolds, E.C., et al., Anticariogenicity of Calcium Phosphate Complexes of Tryptic Casein Phosphopeptides in the Rat. *J Dent Res*, 1995. 74(6): p. 1272-1279.

[73] Walker, G.D., et al., Casein Phosphopeptide-Amorphous Calcium Phosphate Incorporated into Sugar Confections Inhibits the Progression of Enamel Subsurface Lesions in situ. *Caries Research*, 2010. 44: p. 33-40.

[74] Cochrane, N.J., et al., Enamel subsurface lesion remineralisation with casein phosphopeptide stabilised solutions of calcium, phosphate and fluoride. *Caries Research*, 2008. 42(2): p. 88-97.

[75] Oshiro, M., et al., Effect of CPP-ACP paste on tooth mineralization: an FE-SEM study. *Journal of Oral Science*, 2007. 49(2): p. 115-120.

[76] Shen, P., et al., Remineralization of Enamel Subsurface Lesions by Sugar-free Chewing Gum Containing Casein Phosphopeptide-Amorphous Calcium Phosphate. *J Dent Res*, 2001. 80(12): p. 2066-2070.

[77] Cai, F., et al., Remineralization of enamel subsurface lesions in situ by sugar-free lozenges containing casein phosphopeptide-amorphous calcium phosphate. *Australian Dental Journal*, 2003. 48(4): p. 240-243.

[78] Shen, P., et al., Enamel remineralization by a mouthrinse containing casein phosphopeptide-amorphous calcium phosphate and fluoride in an in situ model. *Australian Dental Journal ADRF* Special research Supplement, 2004. 49(4): p. S19.

[79] Manton, D.J., et al., Remineralization of enamel subsurface lesions in situ by the use of three commercially available sugar-free gums. International *Journal of Paediatric Dentistry*, 2008. 18(4): p. 284-290.

[80] Barbour, M.E., et al., Inhibition of hydroxyapatite dissolution by whole casein: the effects of pH, protein concentration, calcium, and ionic strength. *European Journal of Oral Science*, 2008. 116: p. 1-6.

[81] Mazzaoui, S.A., et al., Incorporation of Casein Phosphopeptide-Amorphouse Calcium Phosphate into a Glass-ionomer Cement. *J Dent Res*, 2003. 82(11): p. 914-918.

[82] Yamaguchi, K., et al., Effect of CPP-ACP paste on mechanical properties of bovine enamel as determined by an ultrasonic device. *Journal of Dentistry*, 2006. 34: p. 230-236.

[83] Yamaguchi, K., et al., Ultrasonic determination of the effect of casein phosphopeptide-amorphous calcium phosphate paste on the demineralization of bovine dentin. *Caries Research*, 2007. 41(3): p. 204-207.

[84] Rees, J., T. Loyn, and B. Chadwick, Pronamel and tooth mousse: An initial assessment of erosion prevention in vitro. *Journal of Dentistry*, 2007. 35: p. 355-357.

[85] Tantbirojn, D., et al., Change in surface hardness of enamel by a cola drink and a CPP-ACP paste. Journal of Dentistry, 2008. 36: p. 74-79.

[86] Azarpazhooh, A. and H. Limeback, Clinical efficacy of casein derivatives. *JADA*, 2008. 139: p. 915-924.

[87] Roberts, A.J., Role of Models in Assessing New Agents for Caries Prevention - Non-Fluoride Systems. *Adv Dent Rs*, 1995. 9(3): p. 304-311.

[88] Rose, R.K., Binding Characteristics of Strptococcus mutans for Calcium and Casein Phosphopeptide. *Caries Research*, 2000. 34: p. 427-431.

[89] Rose, R.K., Effects of an anticariogenic casein phosphopeptide on calcium diffusion in streptococcal model dental plaques. *Archives of Oral Biology*, 2000. 45: p. 569-575.

[90] Vacca-Smith, A.M., et al., The Effect of Milk and Casein Proteins on the Adherence of Streptococcus Mutans to Saliva-Coated Hydroxyapatite. *Archives of Oral Biology*, 1994. 39(12): p. 1063-1069.

[91] Reynolds, E.C. and A. Wong, Effect of Adsorbed Protein on Hydroxyapatite Zeta Potential and Streptococcus mutans Adherence. *Infection and Immunity*, 1983. 39(3): p. 1285-1290.

[92] Guggenheim, B., et al., Powdered Milk Micellar Casein Prevents Oral Colonization by Streptococcus sobrinus and Dental Caries in Rats: A Basis for the Caries-Protective Effect of Dairy Products. *Caries Research*, 1999. 33: p. 446-454.

[93] Schüpbach, P., et al., Incorporation of Caseinoglycomacropeptide and Caseinophosphopeptide into the Salivary Pellicle Inhibits Adherence of Mutans Streptococci. *J Dent Res*, 1996. 75(10): p. 1779-1788.

[94] Reynolds, E.C., et al., Retention in Plaque and Remineralization of Enamel Lesions by Various Forms of Calcium in a Mouthrinse or Sugar-free Chewing Gum. *J Dent Res*, 2003. 82(3): p. 206-211.

[95] Reynolds, E.C. and P.F. Riley, Protein Dissimilation by Human Salivary-sediment Bacteria. *J Dent Res*, 1989. 68(2): p. 124-129.

[96] Reynolds, E.C., et al., Fluoride and Casein Phosphopeptide-Amorphous Calcium Phosphate. *J Dent Res*, 2008. 87(4): p. 344-348.

[97] Huq, N.L., K.J. Cross, and E.C. Reynolds, Molecular Modelling of a Multiphosphorylated Sequence Motif Bound to Hydroxyapatite Surfaces. *J. Mol. Model.*, 2000. 6: p. 35-47.

[98] Cross, K.J., et al., NMR studies of a novel calcium, phosphate and fluoride delivery vehicle-αS1-casein(59-79) by stabilized amorphous calcium fluoride phosphate nanocomplexes *Biomaterials*, 2004. 25: p. 5061-5069.

[99] Barbour, M.E., et al., Inhibition of hydroxyapatite dissolution by whole casein: the effects of pH, protein concentration, calcium, and ionic strength. *European Journal of Oral Sciences*, 2008. 116(5): p. 473-478.

[100] Shellis, R.P., et al., Effects of pH and acid concentration on erosive dissolution of enamel, dentine, and compressed hydroxyapatite. *European Journal of Oral Science*, 2010. 118: p. 475-482.

[101] Yao, Y., et al., Compositional analysis of human acquired enamel pellicle by mass spectroscopy. *Archives of Oral Biology*, 2001. 46: p. 293-303.

[102] Hemingway, C.A., et al., Enamel Erosion in Dietary Acids: Inhibition by Food Proteins in vitro. *Caries Research*, 2010. 44: p. 525-530.
[103] Devold, T.G., et al., In vitro studies of adsorption of milk proteins onto tooth enamel. *International Dairy Journal*, 2006. 16: p. 1013-1017.
[104] Vacca-Smith, A.M., et al., The effect of milk and casein proteins on the adherence of Streptococcus mutans to saliva-coated hydroxyapatite. *Archives of Oral Biology*, 1994. 39(12): p. 1063-1069.
[105] Schüpbach, P., et al., Incorporation of caseinoglycomacropeptide and caseinophosphopeptide into the salivary pellicle inhibits adherence of mutans streptococci. *Journal of Dental Research*, 1996. 75: p. 1779-1788.
106] Devold, T.G., et al., In vitro studies of adsorption of milk proteins onto tooth enamel. *International Dairy Journal*, 2006. 16: p. 1013-1017.
[107] Coultate, T.P., Proteins, in Food: The Chemistry of its Components2002, The Royal Society of Chemistry: Cambridge, UK. p. 126-174.
[108] White, A.J., L.H. Gracia, and M.E. Barbour, Inhibition of Dental Erosion by Casein and Casein-Derived Proteins. *Caries Research*, 2010. 45: p. 13-20.
[109] Nekrashevych, Y., M. Hannig, and L. Stösser, Assessment of Enamel Erosion and Protective Effect of Salivary Pellicle by Surface Roughness Analysis and Scanning Electron Microscopy. *Oral Health Prev dent*, 2004. 2(1): p. 5-11.
[110] Reynolds, E.C. and E. Storey, A Review of the Effect of Milk on Dental Caries. *The Australian Journal of Dairy Technology,* 1979: p. 175-180.
[111] Zahradnik, R.T., E.C. Moreno, and E.J. Burke, Effect of Salivary Pellicle on Enamel Subsurface Demineralization In Vitro. J Dent Res, 1976. 55: p. 664-670.
[112] Coultate, T.P., Proteins, in Food: The Chemistry of its Components, The Royal Society of Chemistry, Cambridge, UK. 2002: p. 126-174.
[113] Tuinier, R. and G.d. Kruif, Stability of casein micelles in milk. *Journal of Chemical Physics*, 2002. 117(3): p. 1290-1295.
[114] Holt, C., Structure and Stability of Bovine Casein Micelles. *Advances in Protein Chemistry*, 1992. 43: p. 63-151.
[115] Bezanilla, M., et al., Adsorption of DNA to Mica, Silyated Mica, and Minerals: Characterization by Atomic Force Microscopy. *Langmuir*, 1995. 11: p. 655-659.
[116] Briscoe, W.H., et al., Applying grazing incidence X-ray reflectometry (XRR) to characterising nanofilms on mica. *Journal of Colloid and Interface Science*, 2007. 306: p. 459-463.

In: Casein: Production, Uses and Health Effects
Editors: A. M. Ventimiglia, et. al, pp. 59-81
ISBN: 978-1-62100-129-4

Chapter 3

GENETICS OF CASEINS IN DOMESTIC RUMINANTS

M. Amills[1,4]***, J. M. Serradilla***[2]***, A. Clop***[3]***,***
A. Zidi[1] ***and J. Jordana***[4]

[1] Departament de Genètica Animal, Centre de Recerca en Agrigenòmica (CRAG), Universitat Autònoma de Barcelona, 08193 Bellaterra, Spain.
[2] Departamento de Producción Animal, Campus de Rabanales, Universidad de Córdoba, 14071 Córdoba, Spain.
[3] Department of Medical and Molecular Genetics, King's College London, Great Maze Pond, SE1 9RT. London, United Kingdom.
[4] Departament de Ciència Animal i dels Aliments, Universitat Autònoma de Barcelona, 08193 Bellaterra, Spain.

ABSTRACT

Milk caseins are the major protein source for the newborn, providing amino acids that are essential to sustain growth as well as peptides with immunomodulatory activities. Caseins have also a remarkable impact on the dairy industry, being the main determinant of the efficiency with which milk is transformed into cheese. Casein genes have been thoroughly studied in cattle, sheep and goats, allowing for the stablishment of to causal relationships between their polymorphism and milk composition. In cattle, polymorphism in the β-casein (*CSN2*) and κ-casein (*CSN3*) gene has been reported to have effects on the composition and rheological properties of milk. Similarly, the variability of the goat α_{S1}-casein (*CSN1S1*) gene has been shown to determine the synthesis rate of this protein as well as to have effects, at least in certain breeds, on milk composition and cheese yield and organoleptic properties. Even more, the caprine *CSN3* gene has been associated with milk protein and casein levels. These findings confirm domestic ruminants as a unique model to elucidate the genetic factors that modulate the amount of synthesized caseins and milk composition.

1. An Introduction to the Biology of Caseins

Milk is a biological fluid secreted by the mammary gland and composed primarily of water, carbohydrates, lipids, proteins and minerals. It is an important component of the human diet, being a rich source of animal proteins and calcium. About 83% of the milk consumed in the world is produced by cattle, while buffalo, sheep, goat and camel have a more secondary role. Lactoproteins differ in their physico-chemical properties as well as in their biological functions. Three main classes are usually distinguished: caseins (representing approximately 80% of the total protein content), whey proteins (around 19%) and milk fat globule membrane proteins (about 1%).

Caseins are synthesized exclusively in the mammary gland and can be found, for the most part, forming multi-molecular aggregates in colloidal suspension that receive the name of "micelles" (Grosclaude 1988). At a biochemical level, caseins are defined as insoluble phosphoproteins that, at room temperature (~ 20 °C) and pH = 4.6 (isoelectric point), precipitate from skim milk (Thompson et al. 1965). In ruminants, four casein types have been defined so far (reviewed in Martin et al. 2002): i.e. α_{S1}-casein (CSN1S1), α_{S2}-casein (CSN1S2), β-casein (CSN2) and κ-casein (CSN3). The relative contents of these caseins vary among ruminant species, as can be seen in Table 1. Whilst CSN1S1, CSN1S2 and CSN2 have a high phosphate content and precipitate in the presence of calcium (Grosclaude 1991), i.e. they are calcium sensitive caseins, CSN3 has a completely unrelated evolutionary origin, as we will discuss later, and happens to be a calcium insensitive casein (Grosclaude 1988). Importantly, CSN3 has a stabilizing role in the micelle complexes by establishing a hydrophilic coat that prevents their aggregation and precipitation. The enzymatic hydrolysis of CSN3 by chymosin involves the destabilization of micelles and the subsequent coagulation of milk, which is the initial step in the manufacturing of cheese (Jollès 1975, Remeuf et al. 1991).

Table 1. Milk protein composition in human and ruminant species

Milk component (g/l)	Human	Cow	Sheep	Goat
Total protein	9 - 15	32 - 34	50 - 54	28 - 32
Casein	2.0 - 2.5	26 - 37	40 - 44	22 - 28
CSN1S1	-	11 - 15	14 - 16	6 -10
CSN1S2	-	3 - 4	5 - 6	3 - 4
CSN2	1.5	9 - 11	17 - 18	11 - 21
CSN3	0.5	2 - 4	4 - 5	2 - 4

As soon as 1961 it was discovered that the *CSN2* gene is polymorphic in cows (Grosclaude 1988), initiating a scientific venture that, throughout the decades, has resulted in the identification of many causal mutations affecting the rate of casein synthesis as well as milk and cheese properties. The goal of the current review is to give an updated perspective about the genetic factors that regulate the total amount and composition of the casein fraction in domestic ruminants, as well as to define their impact on the quality of dairy products.

2. Caseins and Human Health

Milk industrial processing and digestion in the gastrointestinal tract promote the degradation of proteins into multiple short bioactive peptides which have many beneficial effects on human health (Korhonen and Pihlanto-Leppälä 2001). These peptides might have hypotensive, antithrombotic, antioxidative, antimicrobial, and immunomodulatory activities (Fitzgerald and Meisel 2004, Korhonen and Pihlanto 2001), they may enhance mineral absorption (Meisel 1998), opioid agonism / antagonism (Smacchi and Gobbetii 1998), and they might even display antitumoral properties (Matar et al. 2003). Bioactive peptides are generated during food processing to produce cheese, yogurt, or sour milk, for example, by enzymatic hydrolysis or by physico-chemical triggers such as heat and pH. Also, the industrial addition of micro-organisms (such as different strains of *Lactobacillus* or streptococcal bacteria) or certain plant extracts promote proteolytic activity. Finally, digestion generates bioactive peptides via proteolysis, an enzymatic reaction mostly led by pepsin, trypsin, and other aminopeptidases.

Despite these beneficial effects, milk is the most frequent cause of food allergy in children. This allergenic reaction is mostly provoked by α_S-caseins and β-lactoglobulin (Park 1994). Goat milk was suggested as a good alternative to cow milk in children with allergy to the latter (Haenlein 2004), but this tenet is highly controversial since intolerance to goat and sheep milk proteins have been reported both in children and adults (Restani et al. 2004). Even more, individuals tolerant to bovine milk and cheese but sensitized to goat or sheep dairy products have been reported (Martins et al. 2005). Genetic polymorphisms in the caprine *CSN1S1* gene have been pointed as one of the causes of the variability observed in the tolerance/sensitivity response to goat milk (Ballabio et al. 2011). In a study feeding goat milk to guinea pigs, it was found that milk lacking CSN1S1 generated less allergic reactions than its normal counterpart (Bevilacqua et al. 2001). Moreover, allergology studies identified IgE and IgG binding epitopes against CSN2 and CSN3 domains in patients with intolerance to bovine milk (Chatchatee et al. 2001[a], Chatchatee et al. 2001[b]). Genetic variation affecting IgE-binding epitopes might be used as a source of information to select alleles that are hypoallergenic. Although likely to have a significant impact on human health, the precise role of genetic variation in bioactive and allergenic activity of milk proteins has not been fully elucidated yet.

3. Genomic Organization and Evolution of the Casein Gene Cluster

Casein genes are tightly clustered in a 250-350 kb region located at chromosome 6 of cattle (Threadgill and Womack 1990, Ferretti et al. 1990), sheep (De Gortari et al. 1998) and goats (Hayes et al. 1993, Martin et al. 2002). Structural features of the casein genes have been extensively reviewed by Rijnkels et al. (2002) and the general organization of the casein cluster is displayed at Figure 1. One important difference between artiodactyla and the human/rodent clades, that diverged around 80 MYR, is that in the former there is a single-copy *CSN1S2* gene while in the latter this locus is duplicated. As previously mentioned, two main casein types have been identified so far i.e. those that are calcium-sensitive (CSN1S1,

CSN1S2 and CSN2), that descend from a common ancestral gene, and CSN3 that is calcium insensitive and it has an independent origin. Given that CSN3 plays an essential role in micelle stabilization and milk clotting, it was proposed that it might be evolutionarily related with the γ-fibrinogen (*FGG*) gene, one of the components of the blood coagulation pathway (Jollès et al. 1978). This relationship was established on the basis of the physiological role and putative amino acid sequence identities (31.4-34.1%) between the human *FGG* gene and cow and sheep *CSN3* loci. Recent data have challenged this view by showing that caseins are members of the secretory calcium-binding phosphoprotein (*SCPP*) gene family (Kawasaki and Weiss 2006). Other molecules comprised in this family are enamel matrix proteins, that organize hydroxyapatite crystallization in the enamel organ, as well as diverse salivary proteins regulating the spontaneous precipitation of Ca-phosphate salts on enamel surface (Kawasaki and Weiss 2003). In a very recent work, the ancestry of the four casein genes was traced back to the odontogenic ameloblast associated (*ODAM*) gene, that encodes a molecule deeply involved in the process of teeth formation (Kawasaki et al. 2011). In this way, the SCPP-Pro-Gln-rich 1 (*SCPPPQ1*) and the follicular dendritic cell secreted peptide (*FDCSP*) genes arose from *ODAM* before synapsid-sauropsid divergence i.e. about 315 MYR (Kawasaki et al. 2011). Subsequently, *SCPP1Q1* gave origin to *CSN1/2*, the putative common ancestor of calcium-sensitive caseins (*CSN1S1, CSN1S2* and *CSN2*), whilst *CSN3* likely descends from *FDCSP*. The formation of the primitive casein micelles might have been prompted by the association of CSN3 and CSN1/2 or, conversely, they might have been exclusively constituted by CSN3, being CSN1/2 co-oopted later to perform this function (Kawasaki et al. 2011).

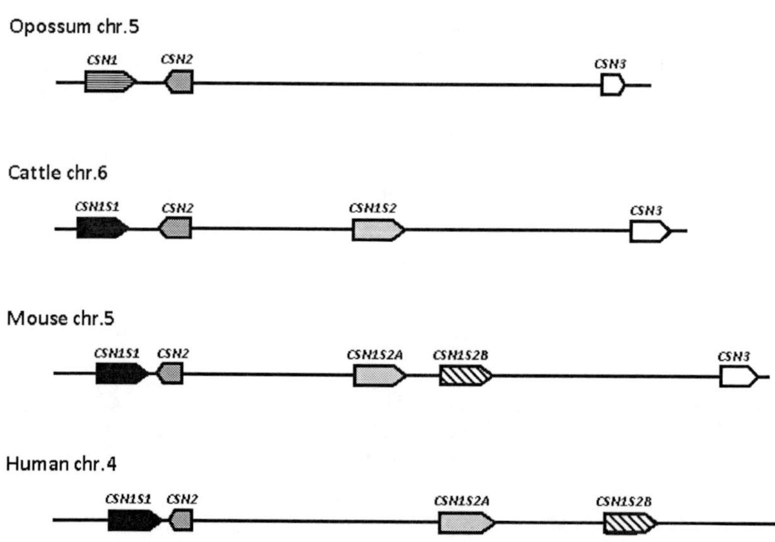

Based on Lefèvre et al. 2010.

Figure 1. Genomic organization of the casein cluster in diverse mammalian species.

Casein gene expression is coordinately regulated by multiple lactogenic hormones such as prolactin, glucocorticoids, insulin and growth hormone (Martin et al. 2002). Genes encoding calcium-sensitive caseins share common regulatory motifs in their 5'ends while the organization of the 5' flanking region of *CSN3* is remarkbly different confirming its independent origin. There is growing evidence that a core set of transcription factors is required for the expression of milk protein genes (Martin et al. 2002). For instance, STAT5-binding sites are always present in the promoters of milk protein genes because this molecule is an essential mediator of prolactin signaling, thus playing a major role not only in mammary gland protein synthesis but also in the growth and alveolar differentiation of epithelial cells (Rosen et al. 1999). Interestingly, there is evidence that lactogenic hormones not only induce the transcriptional activation of casein genes but their re-localization from the periphery to the inner part of the nucleus, thus facilitating their expression (Ballester et al. 2008). Post-transcriptional changes are also known to have a strong impact on casein expression (Martin et al. 2002). In this regard, it is worth to mention that alternative splicing and exon skipping are quite frequent in the casein genes of ruminants (Martin et al. 2002), leading to the synthesis of multiple isoforms that might differ in their functional properties. Moreover, differences in the stability (Martin et al. 2002) and translatability (Bevilacqua et al. 2006) of casein mRNA transcripts have been also reported. Another mechanism finely tuning casein levels involves epigenetic modifications such as dynamic DNA methylation and changes in histone acetylation (Singh et al. 2010).

4. Genetic Architecture of Milk Casein Content

Heritability estimates provide a meaningful insight about the proportion of phenotypic variability of milk casein content explained by genetic additive factors. Estimates of total casein heritability have been obtained in Dutch ($h^2 = 0.41$), Canadian ($h^2 = 0.26$) and Italian ($h^2 = 0.35$) Holstein Friesian cattle (Hayes et al. 1984, Cassandro et al. 2008, Schopen et al. 2009[a]), Churra sheep ($h^2 = 0.21$, Othmane et al. 2002) and Murciano-Granadina goats ($h^2 = 0.13$, Benradi 2007). Heritabilities of specific casein fractions have been also estimated in Holstein cattle, providing values that go from 0.25 (*CSN2*) to 0.73 (*CSN1S2*), and in goats (Benradi 2007), with values that range between 0.09 (*CSN1S2*) and 0.25 (*CSN1S1*). These results allow us to infer that a substantial amount of genetic variability of casein content is explained by alleles that act in an additive manner, implying that the milk content of the four casein types can be modified through selection. Interestingly, Schopen et al. (2009[a]) demonstrated that casein genotypes explain an important part of the polygenic additive variance of casein levels. For instance, polygenic additive variance of this trait was reduced from 0.19 to 0.12 when *CSN3* genotypes were taken into account (Schopen et al. 2009[a]). In consequence, a substantial fraction of alleles with effects on casein content might map to casein genes, a feature that should facilitate their identification and subsequent use in gene assisted selection schemes. Noteworthy, genetic correlations between casein fractions are mostly negative or close to zero (Schopen et al. 2009[a]) e.g. $r_a = -0.49$ between CSN1S1 and CSN1S2, $r_a = -0.56$ between CSN1S1 and CSN3, $r_a = -0.30$ between CSN1S2 and CSN2 etc., with the only exception of CSN1S2 and CSN3 contents that are moderately and positively correlated ($r_a = 0.11$). Besides, milk content of the four casein types is positively correlated with the total milk casein content (particularly for CSN1S1 and CSN3, where genetic correlations are quite strong). The existence of negative correlations between casein levels might be explained by the existence of compensatory mechanisms that act in response to increases in a specific casein fraction by downregulating the remaining ones, thus keeping total casein levels stable. This conclusion stresses the need of identifying casein alleles or haplotypes that have a global impact on all casein fractions rather than limited effects on a specific casein type.

Recent studies have depicted the genomic landscape of casein variation (Schopen et al. 2009[b]). A Dutch Holstein cattle population with registers for casein fractions and other milk traits was typed for 1,341 single nucleotide polymorphisms (SNP) including several genetic markers located within casein genes. As expected, significant quantitative trait loci (QTL) for the four casein types mapped to the bovine chromosome 6 (BTA6) region that contains the casein gene cluster. However, clear evidences that levels of specific casein fractions are regulated *in trans* by additional genomic regions were also obtained. For instance, significant *trans*-QTL were detected for CSN1S1 (BTA9), CSN1S2 (BTA1, BTA10 and BTA17), CSN2 (BTA3) and total caseins (BTA11) concentrations. In general, the percentage of the phenotypic variance explained by individual QTL was relatively small, ranging from 2.7% to 10.4%. Correction of the BTA6 data for *CSN2* and *CSN3* genotypes decreased the percentage of phenotypic variance explained by the QTL in a proportion of 50% and 59%, respectively. This result indicates that part of the observed QTL effects are produced by well-characterized casein alleles genotyped in the course of the aforementioned experiment, whilst others correspond to unidentified casein polymorphisms in linkage disequilibrium with their known counterparts. Quantitative trait loci with effects on protein percentage and mapping to a genomic region close to the casein complex have been also reported in the Churra sheep (Barillet et al. 2005). This finding supports the notion that many polymorphisms modulating casein levels have *cis*-regulatory effects.

5. Casein Polymorphisms and their Association with Milk Quality and Rheological Traits

Many studies have been performed to characterize the variability of casein genes in domestic ruminants (reviewed at Martin et al. 2002, Moioli et al. 2007, Ibeagha-Awemu et al. 2008 and Caroli et al. 2009). Milk protein polymorphism in cattle and associated genotyping techniques have been recently reviewed by Caroli et al. (2009), so these issues will not be covered in this chapter. Interested readers are encouraged to consult the aforementioned review, plus the one authored by Ibeagha-Awemu et al. (2008). Moioli et al. (2007) have also published a very comprehensive review about candidate genes with effects on milk quality of goats and sheep that gives an overview of casein gene variability in these two species. Herewith, rather than describing an extensive catalog of casein alleles in domestic ruminants, we will exclusively focus on the ones that have quantitative effects on casein content and milk composition (Table 2).

Table 2. A representative list of association studies relating casein polymorphisms with milk quality traits

Gene	Phenotype	Goat	Ovine	Bovine
CSN1S1	Milk yield	Avondo et al. (2009); Barbieri et al. (1995); Pagano et al. (2010) ; Hayes et al. (2006)	Ramos et al. (2009)	Kučerová et al.(2006); Ng-Kwai-Hang et al. (1984)
	Fat,%	Albenzio et al. (2009); Avondo et al. (2009); Barbieri et al. (1995); de la Torre et al. (2009); Pagano et al. (2010); Pirisi et al. (1994); Hayes et al. (2006)		Zullo et al.(2005); Ng-Kwai-Hang et al. (1984)
	Protein, %	Albenzio et al. (2009); Avondo et al. (2009); Barbieri et al. (1995); Yue et al. (2011); de la Torre et al. (2009); Pagano et al. (2010); Pirisi et al. (1994); Mahé et al. (1994); Hayes et al. (2006)	Pirisi et al. (1999); Wessels et al. (2004)	Zullo et al. (2005); Ng-Kwai-Hang et al. (1984)
	Casein, %	Albenzio et al. (2009); Avondo et al. (2009); de la Torre et al. (2009); Pagano et al. (2010); Pirisi et al. (1994); Greppi and Roncada (2005)	Pirisi et al. (1999)	Zullo et al. (2005)
	Lactose, %	Avondo et al. (2009); Pagano et al. (2010)		
	Fatty acid profile, %	Valenti et al. (2010); Chilliard et al. (2006)		
CSN2	Milk yield		Corral et al. (2007, 2010)	Kučerová et al. (2006); Heck et al. (2009), Morris et al. (2005); Ng-Kwai-Hang et al. (1984); Comin et al. (2008); Heck et al. (2009); Nilsen et al. (2009); Ikonen et al. (1999, 2001); Bech et al. (1990) ; Bovenhuis and Weller (1994)
	Fat,%		Wessels et al. (2004)	Kučerová et al. (2006); Ng-Kwai-Hang et al. (1984); Haenlein et al. (1987) ; Bovenhuis and Weller (1994)
	Protein, %		Wessels et al. (2004)	Ng-Kwai-Hang et al. (1984); Gonyon et al. (1987); Bovenhuis and Weller (1994); Ikonen et al. (1999, 2001)
	Casein, %			Morris et al. (2005)
	Lactose, %			
	Fatty acid profile, %			Morris et al. (2005)

Table 2. (Continued).

Gene	Phenotype	Goat	Ovine	Bovine
CSN3	Milk yield		Corral et al. (2007, 2010)	Kučerová et al. (2006); Heck et al. (2009), Morris et al. (2005); Ng-Kwai-Hang et al. (1984); Comin et al. (2008); Heck et al. (2009); Nilsen et al. (2009); Ikonen et al. (1999, 2001); Bech et al. (1990) ; Bovenhuis and Weller (1994)
	Fat,%		Wessels et al. (2004)	Kučerová et al. (2006); Ng-Kwai-Hang et al. (1984); Haenlein et al. (1987) ; Bovenhuis and Weller (1994)
	Protein, %		Wessels et al. (2004)	Ng-Kwai-Hang et al. (1984); Gonyon et al. (1987); Bovenhuis and Weller (1994); Ikonen et al. (1999, 2001)
	Casein, %			Morris et al. (2005)
	Lactose, %			
	Fatty acid profile, %			Morris et al. (2005)

5.1. Cattle

About three decades ago, close relationships between casein genotypes and milk composition traits were observed in several independent studies (Ng-Kwai-Hang et al. 1984, Lin et al. 1987, Gonyon et al. 1987, Haenlein et al. 1987). These results prompted the performance of many other experiments aimed to identify casein alleles associated with milk quality. Throughout the time, two of these associations have been consistently confirmed in different bovine breeds, clearly suggesting that the involved polymorphisms have causal effects. In this way, the A_2-allele of the *CSN2* locus has been associated with higher milk and protein yields (when compared to its A_1 counterpart) in Italian (Comin et al. 2008) and Dutch (Heck et al. 2009) Holstein, Norwegian Red (Nilsen et al. 2009), Finnish Ayrshire (Ikonen et al. 1999, 2001), Danish Jersey (Bech et al. 1990), Red Danish Dairy Cattle (Bech et al. 1990) and Black and White Dairy Cattle (Bech et al. 1990). So far, the main difference between both alleles appears to be a His (A_1) to Pro (A_2) substitution at position 67, but a functional link between this polymorphism and milk and protein yield remains to be established. Interestingly, Bonfatti et al. (2010) showed that the B-allele of the *CSN2* locus is associated with a shorter rennet coagulation time and an increased curd firmness. While this latter effect might be explained by differences in CSN2 content, the former one might be attributed to the two amino acid substitutions that characterize this allele i.e. His_{67} and Arg_{122}. Bonfatti et al. (2010) have hypothesized that these two positively charged amino acids might decrease repulsion forces between micelles thus favouring their aggregation.

Another association that has emerged frequently, when examining the polymorphism of bovine casein genes, involves the A and B *CSN3* alleles. These two variants just differ at two amino acid positions 136 (A: Thr, B: Ile) and 148 (A: Asp, B: Ala) plus a synonymous substitution at codon 168 (Caroli et al. 2009). The B-allele has been associated with a higher milk protein content in Dutch Holstein (Bovenhuis and Weller 1994, Heck et al. 2009) and Finnish Ayrshire (Ikonen et al. 1999 and 2001) cattle. Moreover, a consistent relationship between the B-allele of the *CSN3* gene and a shorter rennet coagulation time and a stronger curd firmness have been highlighted by Bonfatti et al. (2010), whilst Walsh et al. (1998), Di Stasio and Mariani (2000) and Bonfatti et al. (2011) showed that BB (*CSN3*) milk has better rennet coagulation properties, lower micelle size and higher cheese yield than the AA one. Even more, rare *CSN3* alleles, such as E and G, have been linked to unfavourable milk coagulation properties (Caroli et al. 2009, Mariani et al. 2001). The close association between *CSN3* genotype and milk rheological properties is not surprising given the key role of this protein in micelle stabilization. Interestingly, Bonfatti et al. (2010) demonstrated that when statistical models account for variation in protein fractions content or protein composition, the association between the B (*CSN3*) allele and milk coagulation properties becomes non-significant. This result suggests that the B-allele achieves an effect on curd firmeness and rennet coagulation time by modifying protein composition i.e. *CSN3*-mediated effects are produced by differences in the synthesis rate of the A and B alleles rather than by a change in protein function. This conclusion is supported by the observation that in AB heterozygous cattle the levels of B-allele mRNA are 13.4% higher than the ones corresponding to the A-allele (Debeljak et al. 2000). Intriguingly, comparison of promoter sequences corresponding to the A and B *CSN3* alleles did not evidence the existence of any mutation candidate to explain such effect (Debeljak et al. 2000, Keating et al. 2007). Conceivably, the causal mutation might lie in a distal regulatory element yet to be characterized.

Evidence about the implication of bovine *CSN1S1* genetic variability on milk quality is scarcer than that of *CSN2* and *CSN3*, but promoter *CSN1S1* haplotypes with effects on protein percentage (Prinzenberg et al. 2003) have been identified. Besides, a polymorphic retrotransposon insertion at exon 19 (typical of allele G) has been associated with lower milk CSN1S1 levels (Rando et al. 1998). It is hypothesized that this insertion might exert such an effect by reducing the stability of the *CSN1S1* transcript (Rando et al. 1998).

5.2. Buffalo and Yak

Casein genetic variation has been poorly studied in Bovinae species other than cattle. Polymorphisms have been reported in the buffalo *CSN1S2* gene, i.e. a G to C transversion that inactivates the intron 7 splice site thus leading to the skipping of exon 7 (Cosenza et al. 2009), and in the *CSN3* gene (Sulimova et al. 1996, Mitra et al. 1998). Polymorphisms have been also detected in the *CSN1S1* (Bai et al. 20101) and *CSN3* (Bai et al. 2008, Prinzenberg et al. 2008) genes of the Chinese yak. However, to the best of our knowledge no association studies between specific casein variants and milk yield and production traits have been carried to date in buffalos or yaks.

5.3. Goats

Variability of the goat *CSN1S1* gene has been convincingly shown to alter CSN1S1 expression in the mammary gland (Martin et al. 2002). According to the levels of synthesis they are associated with, goat *CSN1S1* alleles can be classified as follows (reviewed in Moioli et al. 2007): strong (3.5 g CSN1S1/L/allele, A, B1, B2, B3, B4, C, H, L, M), medium(1.1 g CSN1S1/L/allele, E and I), weak (0.45 g CSN1S1/L/allele, D, F and G) and null (0 g CSN1S1/L/allele, 01, 02 and N) variants. Regulatory polymorphisms at the caprine *CSN1S1* gene exert their effects by using diverse transcriptional and post-transcriptional mechanisms. For instance, the E-allele has been shown to contain a long interspersed nucleotide element insertion at exon 19 that probably reduces the stability of the corresponding transcript and causes a 3-fold decrease in *CSN1S1* mRNA levels (Pérez et al. 1994). As mentioned in the previous paragraph, the G-allele of the bovine *CSN1S1* locus contains the same type of mutation and is also associated with a CSN1S1 reduction, suggesting a common mechanism of synthesis reduction in both species. In contrast, the D, F and G alleles of the goat *CSN1S1* locus diminish CSN1S1 synthesis through a defective mechanism of alternative splicing that leads to the skipping of exon 9 (allele D), exons 9 to 11 (alelle F) or exon 4 (allele G) and, consequently, to the synthesis of internally deleted CSN1S1 proteins (Leroux et al. 1992). According to Leroux et al. (1992), a cytosine deletion at exon 9 and two 11-bp and 3-bp insertions at intron 9 might be involved in the out-splicing of the three consecutive 9, 10 and 11 exons that are skipped from the allele-F mRNA. Loss of these exons yields a protein 37 amino acids shorter and lacking a multiple phosphorylation site. Finally, the 01 and N alleles are characterized by a 8.5 kb deletion that goes from intron 12 onwards and a premature stop codon at exon 12, respectively (Cosenza et al. 2003, Ramunno et al. 2005). As expected, both alleles are associated with undetectable levels of CSN1S1 in milk.

The aforementioned polymorphisms have been widely reported to have major effects on the synthesis rate of CSN1S1 both in French (Martin et al. 2002) and Spanish (Caravaca et al. 2008) goat breeds. Moreover, Mahé et al. (1994) have described consistent differences in the milk protein yield between strong and weak alleles in the Alpine breed, whilst differences in protein and fat content were less obvious. Similarly, Barbieri et al. (1995) showed that milk protein, fat and CSN1S1 contents decrease in the order AA > AE > AF > EE > EF > FF in Alpine goats. However, analysis of French Saanen and Spanish Murciano-Granadina goats did not show such a clear trend (Manfredi et al. 1995, Caravaca et al. 2009). In the case of the Murciano-Granadina breed, convincing evidence that *CSN1S1* genotype affects the concentration of CSN1S1 in milk was obtained, but association with protein and casein contents did not happen to be significant. Probably, the effect of the *CSN1S1* genotype on protein synthesis is affected by polygenic and environmental factors that vary from breed to breed, thus yielding inconsistent results when different populations are compared. In this sense, it is worth to mention that significant interactions between *CSN1S1* genotype and diet have been detected in Italian (Pagano et al. 2010) and Spanish (De la Torre et al. 2008) goat breeds.

In summary, all studies performed so far coincide in the observation that *CSN1S1* genotype affects CSN1S1 levels in milk. On the contrary, associations with total protein or casein contents have been detected in certain breeds but not in others. A global effect of the *CSN1S1* genotype on total casein content might be explained by the fact that CSN1S1 seems to be a key player in the intracellular transport of caseins from the endoplasmic reticulum to the Golgi compartment (Chanat et al. 1999). In this way, low levels of CSN1S1 cause a retention of immature caseins in the cisternae of the endoplasmic reticulum because they cannot be efficiently exported to the Golgi complex. It remains to be established if this perturbation also has consequences on the intracellular transport of lipids.

Effects of *CSN1S1* genotype on parameters related with cheese manufacturing have been widely reported in the scientific literature. This issue is particularly relevant, from an economic perspective, because, in small ruminants, milk is fundamentally devoted to the elaboration of cheese. In French breeds, strong *CSN1S1* alleles have been associated with an increased cheese yield (Vassal et al. 1994), with differences that go from 8.2% (AA vs EE) to 17.7% (AA vs FF). This higher cheese yield might be explained by an augment in the retention of dry matter (Delacroix-Buchet et al. 1996, Lamberet et al. 1996). Differences in the processes of proteolysis and lipolysis as well as in fat composition have been also observed when comparing AA and FF cheeses (Delacroix-Buchet et al. 1996). Hardness and plasticity happened to be 44% and 71% higher in fresh cheeses from AA vs FF goats, but this relationship was reversed (FF > AA) after 1-2 weeks of ripening (Pirisi et al. 1994). In Murciano-Granadina goats, cheese yield does not seem to be affected by *CSN1S1* genotype whilst the curdling rate seems to be higher in EE than in BB goats (Caravaca et al. 2011). With regard to cheese organoleptic features, cheese from FF individuals has a more intense goat flavour than that from AA individuals, possibly because differences in the lipolysis rate (Heil and Dumont 1993).

Although polymorphism of the *CSN1S1* gene seems to be the major factor determining casein synthesis in goats, variants with quantitative effects have been also identified in the *CSN1S2* and *CSN2* genes. With regard to *CSN1S2*, seven alleles associated with three levels of casein synthesis have been described so far (Bouniol et al. 1994, Ramunno et al. 2001[a], 2001[b]): high (2.5 g/l/allele, A, B, C, E and F), medium (D variant) and null (0) variants. The main molecular feature of the D allele is a 106 bp deletion encompassing 11 bp exon 11 and 95 bp intron 11. In contrast, the null allele contains a premature stop codon at exon 11 that hinders the synthesis of CSN1S2. Null alleles have been also reported for the *CSN2* gene: namely alleles 0 and 01 that contain premature stop codons at exon 7 (Persuy et al. 1999, Cosenza et al. 2005). These two variants induce the synthesis of truncated CSN2 at levels that are 100 and 10 fold lower than the A variant (Martin et al. 2002). A third null allele at the promoter region of the *CSN2* gene has been reported by Cosenza et al. (2007).

Last but not least, the *CSN3* gene has been shown to be extremely polymorphic in goats (Prinzenberg et al. 2005), with at least 13 amino acid variants (from A to M). Performance of an association study between CSN3 isoelectrofocusing variants A and B in Orobica goats revealed that the B variant correlates with higher milk protein and casein contents (Chiatti et al. 2007). Consistently, the B-allele (at the DNA level) has been associated with higher protein and casein percentages as well as with a shorter rennet coagulation time in Murciano-Granadina goats (Caravaca et al. 2009, 2011). Consistent with findings reported above, genotyping of 39 SNP, distributed in the four casein genes, in Norwegian goats revealed significant associations between genetic variation at the *CSN1S1* and *CSN3* loci and fat and protein percentages (Hayes et al. 2006). These results suggest that both *CSN1S1* and *CSN3* genotypes might be used as potential markers in gene-assisted selection schemes aimed to improve milk quality. Of course, the existence of significant associations with milk phenotypes of economic interest should be validated in each specific breed before undertaking such strategy, since casein allele effects vary from population to population on the basis of multiple genetic and environmental factors.

5.4. Sheep

Polymorphisms at the ovine casein genes were first described by Arave et al. (1973) and Di Gregorio et al. (1991). In strong contrast with cattle and goats, the amount of data relating casein genotypes with milk quality is very scarce in sheep. Wessels et al. (2004) reported that *CSN1S1* and *CSN2* genotypes affect protein content while fat content displayed an association with *CSN2* polymorphism. In contrast, Corral et al. (2010) have reported associations between *CSN2-CSN3* composite genotypes and milk yield and protein, fat and lactose contents. Hitherto, molecular mechanisms explaining these associations have not been proposed, so it remains to be established if they reflect a causal relationship or they have been just produced by the existence of unknown causal mutations in linkage disequilibrium with the genotyped ones.

6. Selection of Casein Genotypes as a Tool to Improve Milk Quality

Increasing milk protein yield and/or content is an important economic goal of the dairy industry. A high casein content is specially beneficial to optimize the transformation of milk into cheese. On the contrary, manufacturing of hypoallergenic milk would imply to decrease its protein content, specially of those proteins that are not present in human milk (Caroli et al.

2009). Despite the broad variability present in ruminant casein genes and its association with milk composition and renneting traits, casein genotype information has been scarcely used for selection purposes. Using favourable caseins genotypes as markers to assist selection of dairy bulls has been often proposed, since fast genotyping techniques are available (Martin et al., 2002; Comin et al., 2008; Caroli et al. 2009). In a recent review, Caroli et al. (2009) justified the use of variation found in genes, proteins and peptides as a tool to produce milk with different technological and nutraceutical properties i.e. in the context of "designing milk with different protein structures appropiate for its specific use". Several authors (Ikonen, 2000) recognize the intrinsic difficulties of using casein genotypes as selection criteria in official breeding programs and they propose, e.g. for Finish dairy cattle, that this information should be available just for farmers willing to improve milk coagulation properties. Currently, the genotyping of progeny tested bulls for *CSN2* and *CSN3* as well as *LGB* has been a frequent practice in testing stations. Holstein Association USA, for example offers to members a *CSN3* and *LGB* genotyping service (Holstein USA, 2011). However, casein genotypes, as well as *LGB* genotypes, are not generally used as main selection criteria, and they are just provided, on the basis of farmer demands, when cheese production is a major economic goal. Improvement of cheese yields is also expected to be reached by using information about the genetic merit of bulls for fat and protein yields and contents, combined in an index of cheese merit, which is included in bulls catalogs (Cole et al., 2010).

Research on casein polymorphism in sheep has been mainly focused on CSN1S1. The D (or Welsh) variant of this casein is associated to a lower casein content and worst renneting properties of milk (Pirisi et al., 1999). However, its low frequency in most of the European breeds studied (Barillet et al., 2005) does not justify genotyping animals for selection (Barillet, 2007; Carta et al., 2009). With regard to goats, CSN1S1 is the casein showing the higher polymorphism and the strongest incidence on milk rheology and cheese yield. These facts, together with the relatively low frequency of "strong" alleles explains why information on *CSN1S1* genotypes is taken into account in the selection programs of French Alpine and Saanen breeds since 1991 (Pailhière et al., 2003). Increasing the frequency of these alleles was expected to augment the levels of CSN1S1 and protein in French goats (Grosclaude et al., 1994; Barbieri et al., 1995), as thus has been. Frequencies of Alpine bucks, entering the INRA artificial insemination center from 1997 to 2003, that are homozygous for the strong alleles of *CSN1S1* raised from 0.51 to 0.70 (and from 0.05 to 0.12 in the case of Saanen bucks). In the Saanen breed, frequency of bucks homozygous for weak alleles decreased from 0.53 to 0.30 in the same period of time. (Manfredi, 2003). However, these changes are really slow, particularly if we take into account that, before using any type of molecular information, the selection of young artificial insemination bucks already yielded a positive evolution of frequencies of favourable *CSN1S1* alleles (Manfredi et al., 1995). Limited *CSN1S1* genotyping might be the major cause of this paradoxical situation. In this sense, optimal strategies for the use of gene-assisted selection were investigated by Manfredi et al. (1998), but the current strategy applied to breeding schemes, that was originally described by Piacère et al. (1997), is based on planned matings and within full sib selection of favourable *CSN1S1* genotypes. All 1 month-old male kids candidates to progeny testing, and born from assortative mating of best goats with artificial insemination males (both with known *CSN1S1*genotypes), are blood sampled in farms and genotyped for *CSN1S1* in order to select, among full-brothers, individuals carrying the most favourable genotypes (Palhière et al., 2003). *CSN1S1* genotypes are not integrated in the model to genetically evaluate protein contents and yields. On the contrary, the genetic evaluation of these traits is still exclusively based on phenotypic quantitative information. One of the reasons of the limited use of *CSN1S1* genotypes in breeding schemes is partly a consequence of the insufficient knowledge about the effects of *CSN1S2*, *CSN2* and *CSN3* polymorphisms on milk quality. These polymorphisms might be co-selected with the *CSN1S1* ones due to close linkage, maybe leading to undesirable consequences. In fact, gene-assisted selection should be based on casein haplotypes rather than on specific alleles. Another reason that limited the use of *CSN1S1*-based selection was that strong *CSN1S1* alleles seem to have slight unfavourable effects on cheese sensorial properties (Manfredi, 2003). Sánchez-Palma et al. (2005) studied three alternative ways of using *CSN1S1* genotypes for selection in a simulated population of dairy goats. The first of these alternatives (S1) coincided more or less with the way *CSN1S1* genotypes are employed in French breeding schemes i.e. within full-sib selection of male kids issued from mating sires and of dams with the best breeding values (that are predicted with a conventional animal model). In the second alternative (S2), genotype information was taken into account only as a fixed effect in the model used to predict breeding values. The third option (S3) was a combination of the two former alternatives. Results clearly showed that, despite of starting with a frequency of strong alleles higher than that of French breeds, all three methods produced higher genetic gains for protein yield and contents, both in the short and long terms, than the option of not selecting favourable *CSN1S1* genotypes. Genotype selection (S1 and S3) generated a faster rate of fixation of strong alleles than S2, particularly when the effect of the *CSN1S1* gene on the trait was not large (less than one phenotypic SD), as is the case for protein yield. Genetic gain reached in the long-term when using *CSN1S1* genotype information only for genotypic selection of male kids (S1) was lower than that obtained when this information was also used in the model for indexing sires and dams (S3); although differences were not significant. Practical implications of this latter result are important, since the first of these two methods requires the genotyping of only a few males offspring, issued from the best indexed sires and dams, while the other ones require the genotyping of a large number of goats. We can conclude that using *CSN1S1* genotypes for selection purposes seems to be a good strategy, even when the starting frequencies of favourable alleles is not very low. However, recent results on the effect of other casein genes, particularly *CSN3*, suggests the convenience of further studying the effect of haplotypes of the whole cluster of caseins in order to use them as a complementary source of information to conventional selection.

7. Final Remarks and Prospects

Considerable advances have been made in the identification of gene variation at casein loci of cattle and goats, whilst the state of knowledge is much more limited in other dairy ruminant species such as sheep, buffalo, yak and camel. The recent implementation of next generation sequencing technologies combined with capture array hybridization tools might allow the massive sequencing of the whole casein cluster in comprehensive pools of breeds/individuals at an affordable price. This strategy might be particularly suitable for species where reference genomes are available (e.g. cattle) or underway (e.g sheep and goats), and it should have two important advantages. First of all, large collections of SNP could be generated in a single sequencing experiment without the need of employing low throughput strategies that are very time consuming. Second, it might allow to explore the variation of not only the casein genes but also of the intergenic regions that surround them, and that might contain important regulatory elements such as enhancers and repressors. These SNP might be used subsequently to genotype reference populations with records for milk traits in order to identify those variants that have a global positive effect on milk quality.

Of course, polymorphisms affecting casein content are not only located at the casein cluster but also in other genomic regions (Schopen et al. 2009[b]). An example that illustrates this concept is the osteopontin gene, that seems to have a major role in mammary involution as well as in the regulation of *CSN2* and *CSN3* gene expression (Sheehy et al. 2009). To identify genetic variation that regulates casein expression *in trans* might be a difficult endeavour, although the combination of massive whole-genome association analysis and transcriptomic techniques might yield fruitful results in the form of expression QTL for casein phenotypes. The molecular basis of these eQTL might be further explored to identify the corresponding causal mutations. Since the effects of SNP are shaped by both genetic and environmental factors, associations with traits of economic interest might vary from one population to another. This caveat needs to be taken into account when implementing gene-assisted selection schemes based on molecular information obtained from breeds different than the one that needs to be improved.

Genomic selection is now a reality in cattle (Hayes and Goddard 2010), where genomic breeding values are predicted according to genotypic information provided by thousands of markers distributed along the bovine genome. In cattle, an accuracy of 0.71 was obtained when predicting genomic breeding values on the basis of a reference population of 3,576 bulls genotyped for 38,416 SNP (Van Raden et al. 2009). Given the effective size of the Holstein breed and its level of linkage disequilibrium, around 30,000 markers should be used to implement genomic selection schemes (Hayes and Goddard 2010). Although high density SNP chips are also available in sheep (Illumina Ovine SNP50 beadchip) and will be soon available in goats, their high cost hinders the immediate application of genomic selection schemes in small ruminants. An affordable and more realistic alternative, at least in the near future, might consist in the implementation of reduced SNP chips just containing markers that have shown a consistent association with dairy traits. These SNP might be generated by using massive sequencing techniques, on a breed-specific basis, and further used in association analysis to explore their effects. Subsequently, those that present the most relevant effects would be included in the design of a reduced SNP chip. A potential pitfall of this approach is that the genetic architecture of milk traits is distributed amongst many SNP with small effects. In consequence, the use of marker-assisted selection, rather than genomic selection, might limit to a certain extent the response to selection (Hayes and Goddard 2010).

Besides its potential application in animal breeding and selection, the dissection of the genetic variation at ruminant casein genes, as well as in other unrelated loci with effects on milk casein content, will provide a wealth of biological information about the mechanisms and gene networks that finely tune casein synthesis in the mammary gland. The functional characterization of casein variants that lack specific exons or domains might also help to understand how protein structure correlates with biological function and micelle formation. With no doubt, in the forthcoming years domestic ruminants will constitute an irreplaceable biological model to understand how the genetic architecture of casein and milk traits is built.

REFERENCES

Albenzio M, Santillo A, d'Angelo F, Sevi A. 2009 .Focusing on casein gene cluster and protein profile in Garganica goat milk. *J Dairy Res.* 76:83-9.

Arave CW, Gillett TA, Price DA, Matthews DH. 1973. Polymorphisms in caseins of sheep milk. *J Anim Sci.* 36:241-4.

Avondo M, Pagano RI, Guastella AM, Criscione A, Di Gloria M, Valenti B, Piccione G, Pennisi P. 2009. Diet selection and milk production and composition in Girgentana goats with different α_{S1}-casein genotype. *J Dairy Res.* 76:202-9.

Bai WL, Yin RH, Zhao SJ, Zheng YC, Zhong JC, Zhao ZH. 2008. Characterization of a κ-casein genetic variant in the Chinese yak, *Bos grunniens*. *J Dairy Sci.* 91:1204-8.

Bai WL, Yin RH, Dou QL, Yang JC, Zhao SJ, Ma ZJ, Yin RL, Luo GB, Zhao ZH. 2010. A single nucleotide polymorphism and sequence analysis of *CSN1S1* gene promoter region in Chinese *Bos grunniens* (yak). *Anim Biotechnol.* 21:36-41.

Ballabio C, Chessa S, Rignanese D, Gigliotti C, Pagnacco G, Terracciano L, Fiocchi A, Restani P, Caroli AM. 2011. Goat milk allergenicity as a function of αs_{S1}-casein genetic polymorphism. *J Dairy Sci.* 94:998-1004.

Ballester M, Kress C, Hue-Beauvais C, Kiêu K, Lehmann G, Adenot P, Devinoy E. 2008. The nuclear localization of WAP and CSN genes is modified by lactogenic hormones in HC11 cells. *J Cell Biochem.* 105:262-70.

Barbieri ME, Manfredi M, Elsen JM, Ricordeau G, Bouillon J, Grosclaude F, Mahé MF, Bibé B. 1995. Influence du locus de la caséine α_{S1} sur les performances laitières et les paramètres genétiques des chèvres de race Alpine. *Genet Sel Evol.* 27:437-50.

Barillet F, Arranz JJ, Carta A. 2005. Mapping quantitative trait loci for milk production and genetic polymorphisms of milk proteins in dairy sheep. *Genet Sel Evol.* 37 (Suppl.1):S109-23.

Barillet F. 2007. Genetic improvement for dairy production in sheep and goats. *Small Rumin Res.* 70:60-75.

Bech AM, Kristiansen KR. 1990. Milk protein polymorphism in Danish dairy cattle and the influence of genetic variants on milk yield. *J Dairy Res.* 57:53-62.

Benradi Z. 2007. Análisis genético de los caracteres de rendimiento, composición, reología y rendimiento quesero de la leche en las razas caprinas Malagueña y Murciano-Granadina. Master Thesis. International Center for Advanced Mediterranean Studies.

Bevilacqua C, Martin P, Candalh C, Fauquant J, Piot M, Roucayrol AM, Pilla F, Heyman M. 2001. Goats' milk of defective α_{S1}-casein genotype decreases intestinal and systemic sensitization to β-lactoglobulin in guinea pigs. *J Dairy Res.* 68:217-27.

Bevilacqua C, Helbling JC, Miranda G, Martin P. 2006. Translational efficiency of casein transcripts in the mammary tissue of lactating ruminants. *Reprod Nutr Dev.* 5:567-78.

Bonfatti V, Di Martino G, Cecchinato A, Degano L, Carnier P. 2010. Effects of β-κ-casein (*CSN2-CSN3*) haplotypes, β-lactoglobulin (*BLG*) genotypes, and detailed protein composition on coagulation properties of individual milk of Simmental cows. *J Dairy Sci.* 93:3809-17.

Bonfatti V, Cecchinato A, Di Martino G, De Marchi M, Gallo L, Carnier P. 2011. Effect of κ-casein B relative content in bulk milk κ-casein on Montasio, Asiago, and Caciotta cheese yield using milk of similar protein composition. *J Dairy Sci.* 94:602-13.

Bouniol C, Brignon G, Mahé MF, Printz C. 1994. Biochemical and genetic analysis of variant C of caprine α_{S2}-casein (*Capra hircus*). *Anim Genet.* 25:173-7.

Bovenhuis H, Weller JI. 1994. Mapping and analysis of dairy cattle quantitative trait loci by maximum likelihood methodology using milk protein genes as genetic markers. *Genetics* 137:267-80.

Caravaca F, Amills M, Jordana J, Angiolillo A, Agüera P, Aranda C, Menéndez-Buxadera A, Sánchez A, Carrizosa J, Urrutia B, Sànchez A, Serradilla JM. 2008. Effect of α_{S1}-casein (*CSN1S1*) genotype on milk CSN1S1 content in Malagueña and Murciano-Granadina goats. *J Dairy Res.* 75:481-4.

Caravaca F, Carrizosa J, Urrutia B, Baena F, Jordana J, Amills M, Badaoui B, Sànchez A, Angiolillo A, Serradilla JM. 2009. Short communication: Effect of α_{S1}-casein (*CSN1S1*) and κ-casein (*CSN3*) genotypes on milk composition in Murciano-Granadina goats. *J Dairy Sci.* 92:2960-4.

Caravaca F, Ares JL, Carrizosa J, Urrutia B, Baena F, Jordana J, Badaoui B, Sànchez A, Angiolillo A, Amills M, Serradilla JM. 2011. Effects of α_{S1}-casein (*CSN1S1*) and κ-casein (*CSN3*) genotypes on milk coagulation properties in Murciano-Granadina goats. *J Dairy Res.* 78:32-7.

Caroli AM, Chessa S, Erhardt GJ. 2009. Milk protein polymorphisms in cattle: effect on animal breeding and human nutrition. *J Dairy Sci.* 92:5335-52.

Carta A, Casu S, Salaris S. 2009. Current state of genetic improvement in dairy sheep. *J Dairy Sci.* 92:5814-33.

Cassandro M, Comin A, Ojala M, Dal Zotto R, De Marchi M, Gallo L, Carnier P, Bittante G. 2008. Genetic parameters of milk coagulation properties and their relationships with milk yield and quality traits in Italian Holstein cows. *J Dairy Sci.* 91:371-6.

Chanat E, Martin P, Ollivier-Bousquet M. 1999. α_{S1}-casein is required for the efficient transport of β- and κ-casein from the endoplasmic reticulum to the Golgi apparatus of mammary epithelial cells. *J Cell Sci.* 112:3399-412.

Chatchatee P, Jarvinen KM, Bardina L, Vila L, Beyer K, Sampson HA. 2001[a]. Identification of IgE and IgG binding epitopes on beta- and kappa-casein on cow's milk allergic patients. *Clin Exp Allergy* 31:1256-65.

Chatchatee P, Jarvinen KM, Bardina L, Vila L, Beyer K, Sampson HA. 2001[b]. Identification of IgE and IgG binding epitopes on α_{S1}-casein: differences in patients with persistent and transient cow's milk allergy. *J Allergy Clin Immunol.* 107:379-83.

Chiatti F, Caroli A, Chessa S, Bolla P, Pagnacco G. 2005. Relationships between goat k-casein (*CSN3*) polymorphism and milk composition. FAO The role of biotechnology, Villa Gualino, pp.163-164.

Chiatti F, Chessa S, Bolla P, Cigalino G, Caroli A, Pagnacco G. 2007. Effect of κ-casein polymorphism on milk composition in the Orobica goat. *J Dairy Sci.* 90:1962-6.

Chilliard Y, Rouel J, Leroux C. 2006. Goat's alpha-s1 casein genotype influences its milk fatty acid composition and delta-9 desaturation ratios. *Anim Feed Sci Tech.* 131:474-87

Cole JB, VanRaden PM. 2010. Net merit as a measure of lifetime profit: 2010 revision. United States Department of Agriculture. Agricultural Research Service. Animal Improvement Programs (http://aipl.arsusda.gov/reference/nmcalc.htm)

Comin A, Cassandro M, Chessa S, Ojala M, Dal Zotto R, De Marchi M, Carnier P, Gallo L, Pagnacco G, Bittante G. 2008. Effects of composite β- and κ -casein genotypes on milk coagulation, quality, and yield traits in Italian Holstein cows. *J Dairy Sci.* 91:4022-7

Corral JM, Izquierdo M, Mateos S, Parejo JC, Salazar J, Rabasco A, Martínez-Trancón, M, Sansinforiano ME, Hernández FI, Portilla FJ, Padilla JA. 2007. Asociación de los polimorfismos del exón 7 del gen de la β-caseína (*CSN2*) con la producción y composición de la leche en ovino Merino. Book of Abstracts of the XII Jornadas sobre Producción Animal, Zaragoza, Spain, 16-17 May, pp.477-479

Corral JM, Padilla JA, Izquierdo M. 2010. Associations between milk protein genetic polymorphisms and milk production traits in Merino sheep breed. *Livest Sci.* 129:73-9.

Cosenza G, Illario R, Rando A, Di Gregorio P, Masina P, Ramunno L. 2003. Molecular characterization of the goat *CSN1S1* (01) allele. *J Dairy Res.* 70:237-40.

Cosenza G, Pauciullo A, Gallo D, Berardino DD, Ramunno L. 2005. A *Ssp* I PCR-RFLP detecting a silent allele at the goat *CSN2* locus. *J Dairy Res.* 72:456-9.

Cosenza G, Pauciullo A, Colimoro L, Mancusi A, Rando A, Di Berardino D, Ramunno. 2007. An SNP in the goat *CSN2* promoter region is associated with the absence of β-casein in milk. *Anim Genet.* 38:655-8.

Cosenza G, Pauciullo A, Feligini M, Coletta A, Colimoro L, Di Berardino D, Ramunno L. 2009. A point mutation in the splice donor site of intron 7 in the α_{S2}-casein encoding gene of the Mediterranean River buffalo results in an allele-specific exon skipping. *Anim Genet.* 40:791.

De Gortari MJ, Freking BA, Cuthbertson RP, Kappes SM, Keele JW, Stone RT, Leymaster KA, Dodds KG, Crawford AM, Beattie CW. 1998. A second-generation linkage map of the sheep genome. *Mamm Genome* 9:204-9.

De la Torre G, Serradilla JM, Gil Extremera F, Sanz Sampelayo MR. 2008. Nutritional utilization in Malagueña dairy goats differing in genotypes for the content of α_{S1}-casein in milk. *J Dairy Sci.* 91:2443-8.

De la Torre G, Morales ER, Manrique JM, Extremera FG, Sanz Sampelayo MR. 2009. Milk production and composition in Malagueña dairy goats. 2009. Effect of genotype for synthesis of α_{S1}-casein on milk production and its interaction with dietary protein content. *J Dairy Res.* 2009 76:137-43.

Debeljak M, Susnik S, Marinsek-Logar R, Medrano JF, Dovc P. 2000. Allelic differences in bovine *κ-CN* gene which may regulate gene expression. *Pflugers Arch.* 439(3 Suppl):R4-6.

Delacroix-Buchet A, Degas C, Lamberet G, Vassal L. 1996. Influence des variants AA et FF de la caséine α_{S1} caprine sur le rendement fromager et les caractéristiques sensorielles des fromages. *Le Lait* 76:217-41

Di Gregorio P, Rando A, Pieragostini E, Masina. 1991. DNA polymorphism at the casein loci in sheep. *Anim Genet.* 22:21-30.

Di Stasio L, Mariani P. 2000. The role of protein polymorphism in the genetic improvement of milk production. *Zoot Nutr Anim.* 26:69-90.

Ferretti L, Leone P, Sgaramella V. 1990. Long range restriction analysis of the bovine casein genes. *Nucleic Acids Res.* 18:6829-33.

Fitzgerald RJ, Murray BA, Walsh DJ. 2004. Hypotensive peptides from milk proteins. *J Nutr.* 134:980S-88S.

Gonyon DS, Mather RE, Hines HC, Haenlein GF, Arave CW, Gaunt SN. 1987. Associations of bovine blood and milk polymorphisms with lactation traits: Holsteins. *J Dairy Sci.* 70:2585-98.

Greppi GF, Roncada P. 2005. La componente proteica del latte caprino. In: L'alimentazione della capra da latte (Ed. Pulina G.). Avenue Media, Bologna, pp.71-99.

Grosclaude F. 1988. Le polymorphisme génétique des principales lactoprotéines bovines. *INRA Prod Anim.* 1:5-17.

Grosclaude F. 1991. Structure, déterminisme génétique et polymorphisme des 6 lactoprotéines principales des bovins, des caprins et des ovins. Journeés sur la qualité des laits à la production et aptitude fromagère. Colloque INRA-ENSA (Rennes, France).

Grosclaude F, Ricordeau G, Martin P, Remeuf F, Vassal L, Bouillon J. 1994. Du gène au fromage : le polymorphisme de la caséine α_{S1} caprine, ses effets, son évolution. *INRA Prod Anim.* 7:3-19.

Haenlein GFW, Gonyon DS, Mather RE, Hines HC. 1987. Associations of bovine blood and milk polymorphisms with lactation traits: Guernseys. *J Dairy Sci.* 70:2599-609.

Haenlein GFW. 2004. Goat milk in human nutrition. *Small Rumin Res* 51:155-63.

Hayes JF, Ng-Kwai-Hang KF, Moxley JE. 1984. Heritability of milk casein and genetic and phenotypic correlations with production traits. *J Dairy Sci.* 67:841-6.

Hayes H, Petit E, Bouniol C, Popescu P. 1993. Localization of the α_{S2}-casein gene (CASAS2) to the homoeologous cattle, sheep, and goat chromosomes 4 by in situ hybridization. *Cytogenet Cell Genet.* 64:281-5.

Hayes B, Hagesaether N, Adnøy T, Pellerud G, Berg PR, Lien S. 2006. Effects on production traits of haplotypes among casein genes in Norwegian goats and evidence for a site of preferential recombination. *Genetics* 174:455-64.

Hayes B, Goddard M. 2010. Genome-wide association and genomic selection in animal breeding. *Genome* 53:876-83.

Heck JM, Schennink A, van Valenberg HJ, Bovenhuis H, Visker MH, van Arendonk JA, van Hooijdonk AC. 2009. Effects of milk protein variants on the protein composition of bovine milk. *J Dairy Sci.* 92:1192-202.

Heil F, Dumont JP. 1993. Caractéristiques organolpetiques des fromages de chèvre fabriqués à partir de laits contenant des variants génétiques différents de la caséine α_{S1}. *Le Lait* 73:559-65.

Holstein USA. 2011. Holstein Association USA. Programs and Services. Milk Proteins. http://www.holsteinusa.com/programs_services/genetic_services.html.

Ibeagha-Awemu EM, Kgwatalala P, Zhao X. 2008. A critical analysis of production-associated DNA polymorphisms in the genes of cattle, goat, sheep, and pig. *Mamm Genome* 19:591-617.

Ikonen T, Ojala M, Ruottinen O. 1999. Associations between milk protein polymorphism and first lactation milk production traits in Finnish Ayrshire cows. *J Dairy Sci.* 82:1026-33.

Ikonen T. 2000. Possibilities of genetic improvement of milk coagulation properties of dairy cows. Thesis. Faculty of Agriculture and Forestry of the University of Helsinki. pp.34.

Ikonen T, Bovenhuis H, Ojala M, Ruottinen O, Georges M. 2001. Associations between casein haplotypes and first lactation milk production traits in Finnish Ayrshire cows. *J Dairy Sci.* 84:507-14.

Jollès P. 1975. Structural aspects of the milk clotting process. Comparative features with the blood clotting process. *Mol Cell Biochem.* 7:73-85.

Jollès P, Loucheux-Lefebvre MH, Henschen A. 1978. Structural relatedness of κ-casein and fibrinogen γ-chain. *J Mol Evol.* 11:271-7.

Kawasaki K. and Weiss K.M. 2003. Mineralized tissue and vertebrate evolution: The secretory calcium-binding phosphoprotein gene cluster. *Proc Natl Acad Sci. USA* 100:4060-65.

Kawasaki K, Weiss KM. 2006. Evolutionary genetics of vertebrate tissue mineralization: the origin and evolution of the secretory calcium-binding phosphoprotein family. *J Exp Zool B Mol Dev Evol.* 306:295-316.

Kawasaki K, Lafont AG, Sire JY. 2011. The evolution of milk casein genes from tooth genes before the origin of mammals. *Mol Biol Evol.* (in press)

Keating AF, Davoren P, Smith TJ, Ross RP, Cairns MT. 2007. Bovine κ-casein gene promoter haplotypes with potential implications for milk protein expression. *J Dairy Sci.* 90:4092-9.

Korhonen H, Pihlanto A. 2001. Food-derived bioactive peptides-opportunities for designing future foods. *Current Pharmaceutical Design* 9:1297-1308.

Korhonen H, Pihlanto-Leppälä A. 2001. Milk protein-derived bioactive peptides-novel opportunities for health promotion. *IDF Bulletin* 363:17-26.

Kučerová J, Matějíček A, Jandurová OM, Sorensen P, Němcová E, Štípková M, Kott T, Bouška J, Frelich J. 2006. Milk protein genes *CSN1S1*, *CSN2*, *CSN3*, *LGB* and their relation to genetic values of milk production parameters in Czech Fleckvieh. *Czech J Anim Sci.* 51:241-47.

Lamberet G, Degas C, Delacroix-Buchet A, Vassal L. 1996. Influence des caractères liés aux allèles A et F de la caséine α_{S1} caprine sur la flaveur chèvre : fabrications fromagères avec échange de protéines et de matières grasses. *Le Lait* 76:349-61.

Lefèvre MC, Menzies K, Sharp JA, Nicholas KR. 2010. Comparative genomics and transcriptomics of lactation. In: Evolutionary Biology: concepts, molecular and morphological evolution (Pontarotti P., Ed.). Springer, Berlin, pp.120

Leroux C, Mazure N, Martin P. 1992. Mutations away from splice site recognition sequences might cis-modulate alternative splicing of goat α_{S1}-casein transcripts. Structural organization of the relevant gene. *J Biol Chem.* 267:6147-57.

Lin CY, McAllister AJ, Ng-Kwai-Hang KF, Hayes JF, Batra TR, Lee AJ, Roy GL, Vesely JA, Wauthy JM, Winter KA. 1987. Association of milk protein types with growth and reproductive performance of dairy heifers. *J Dairy Sci.* 70:29-39.

Mahé MF, Manfredi E, Ricordeau G, Piacère A, Grosclaude F. 1994. Effets du polymorphisme de la caséine α_{S1} caprine sur les performances laitières: analyse intradescendance de boucs de race Alpine. *Genet Sel Evol.* 26:151-7.

Manfredi E, Ricordeau G, Barbieri ME, Amigues Y, Bibé B. 1995. Genotype caséine α_{s1} et sélection des boucs sur descendance dans les races Alpine et Saanen. *Genet Sel Evol.* 27:451-58.

Manfredi E, Barbieri ME, Fournet F, Elsen JM.1998. A dynamic deterministic model to evaluate breeding strategies under mixed inheritance. *Genet Sel Evol.* 30:127-148.

Manfredi E, Serradilla JM, Leroux C, Martin P, Sánchez A. 2000. Genetics for milk production. Book of Abstracts of the 7th International Conference on Goats, Paris (France), pp.191-6.

Manfredi E. 2003. The tale of goat α_{s1}-casein. Proceedings of the International Workshop on Major Genes and QTL in Sheep and Goat, Toulouse (France).

Mariani BP, Summer A, Di Gregorio PD, Rando A, Fossa E, Pecorari M. 2001. Effects of the *CSN1A(G)* allele on the clotting time of cow milk and on the rheological properties of rennet-curd. *J Dairy Res.* 68:63-70.

Martin P, Szymanowska M, Zwierzchowski L, Leroux C. 2002. The impact of genetic polymorphisms on the protein composition of ruminant milks. *Reprod Nutr Dev.* 42:433-59.

Martins P, Borrego LM, Pires G, Pinto PL, Afonso AR, Rosado-Pinto J. 2005. Sheep and goat's milk allergy - a case study. *Allergy* 60:129-30.

Matar C, LeBlanc JG, Martin L, Perdigon G. 2003. Biologically active peptides released in fermented milk: Role and functions. In: Handbook of fermented functional foods. Functional foods and nutraceuticals series. Ed. by ER Farnworth. CRC press, Florida (USA), pp.177-201.

Meisel H. 1998. Overview on milk protein-derived peptides. *Int Dairy J.* 8:363-73.

Mitra A, Schlee P, Krause I, Blusch J, Werner T, Balakrishnan CR, Pirchner F. 1998. κ-casein polymorphisms in Indian dairy cattle and buffalo: a new genetic variant in buffalo. *Anim Biotechnol.* 9:81-7.

Moioli B, D'Andrea M, Pilla F. 2007. Candidate genes affecting sheep and goat milk quality. *Small Rumin Res.* 68:179-192.

Morris CA, Hickey SM, Cullen NG, Prosser CG, Anderson RM, Tate ML. 2005. Associations between β-casein genotype and milk yield and composition in grazing dairy cows. *New Zealand J Agri Res.* 48:441- 50.

Ng-Kwai-Hang KF, Hayes JF, Moxley JE, Monardes HG. 1984. Association of genetic variants of casein and milk serum proteins with milk, fat, and protein production by dairy cattle. *J Dairy Sci.* 67:835-40.

Nilsen H, Olsen HG, Hayes B, Sehested E, Svendsen M, Nome T, Meuwissen T, Lien S. 2009. Casein haplotypes and their association with milk production traits in Norwegian Red cattle. *Genet Sel Evol.* 41:24.

Othmane MH, De La Fuente LF, Carriedo JA, San Primitivo F. 2002. Heritability and genetic correlations of test day milk yield and composition, individual laboratory cheese yield, and somatic cell count for dairy ewes. *J Dairy Sci.* 85:2692-8.

Pagano RI, Pennisi P, Valenti B, Lanza M, Di Trana A, Di Gregorio P, De Angelis A, Avondo M. 2010. Effect of *CSN1S1* genotype and its interaction with diet energy level on milk production and quality in Girgentana goats fed ad libitum. *J Dairy Res.* 77:245-51.

Pailhière I, Astruc JM, Piacère A. 2003. Use of major genes in small ruminants selection: the French examples of α_{s1}-casein in goats and *PrP* gene in sheep. 54th Annual Meeting of the European Association for Animal Production, Rome.

Park YW. 1994. Hypo-allergenic and therapeutic significance of goat milk. *Small Rumin Res.* 14:151-59.

Pérez MJ, Leroux C, Bonastre AS, Martin P. 1994. Occurrence of a LINE sequence in the 3' UTR of the goat α_{S1}-casein E-encoding allele associated with reduced protein synthesis level. Gene 147:179-87.

Persuy MA, Printz C, Medrano JF, Mercier JC. 1999. A single nucleotide deletion resulting in a premature stop codon is associated with marked reduction of transcripts from a goat β-casein null allele. *Anim Genet.* 30:444-51.

Piacère A, Bouloc-Duval N, Sigwald JP, Larzul C, Manfredi E. 1997. Utilisation de l'index combiné caprin et du polymorphisme de la caséine alpha s1 dans le schéma de sélection caprin. *Renc Rech Rumin* 4:187-190.

Pirisi A, Colin O, Laurent F, Scher J, Parmentier M. 1994. Comparison of milk composition, cheesemaking properties and textural characteristics of the cheese from two groups of goats with a high or low rate α_{S1}-casein synthesis. *Int Dairy J.* 4:329-45.

Pirisi A, Piredda G, Papoff CM, Di Salvo R, Pintus S, Garro G, Ferranti P, Chianese L. 1999. Effects of sheep α_{S1}-casein CC, CD and DD genotypes on milk composition and cheesemaking properties. *J Dairy Res.* 66:409-19.

Prinzenberg EM, Weimann C, Brandt H, Bennewitz J, Kalm E, Schwerin M, Erhardt G. 2003. Polymorphism of the bovine *CSN1S1* promoter: linkage mapping, intragenic haplotypes, and effects on milk production traits. *J Dairy Sci.* 86: 2696-705.

Prinzenberg EM, Gutscher K, Chessa S, Caroli A, Erhardt G. 2005. Caprine κ-casein (*CSN3*) polymorphism: new developments in molecular knowledge. *J Dairy Sci.* 88:1490-8.

Prinzenberg EM, Jianlin H, Erhardt G. 2008. Genetic variation in the κ-casein gene (*CSN3*) of Chinese yak (*Bos grunniens*) and phylogenetic analysis of *CSN3* sequences in the genus *Bos*. *J Dairy Sci.* 91:1198-203.

Ramos AM, Matos CAP, Russo-Almeida PA, Bettencourt CMV, Matos J, Martins A, Pinheiro C, Rangel-Figueiredo T. 2009. Candidate genes for milk production traits in Portuguese dairy sheep. *Small Rumin Res.* 82:117-121

Ramunno L, Cosenza G, Pappalardo M, Longobardi E, Gallo D, Pastore N, Di Gregorio P, Rando A. 2001[a]. Characterization of two new alleles at the goat *CSN1S2* locus. *Anim Genet.* 32:264-8.

Ramunno L, Longobardi E, Pappalardo M, Rando A, Di Gregorio P, Cosenza G, Mariani P, Pastore N, Masina P. 2001[b]. An allele associated with a non-detectable amount of α_{S2}-casein in goat milk. *Anim Genet.* 32:19-26.

Ramunno L, Cosenza G, Rando A, Pauciullo A, Illario R, Gallo D, Di Berardino D, Masina P. 2005. Comparative analysis of gene sequence of goat *CSN1S1* F and N alleles and characterization of *CSN1S1* transcript variants in mammary gland. *Gene* 345:289-99.

Rando A, Di Gregorio P, Ramunno L, Mariani P, Fiorella A, Senese C, Marletta D, Masina P. 1998. Characterization of the *CSN1A G* allele of the bovine α_{S1}-casein locus by the insertion of a relict of a long interspersed element. *J Dairy Sci.* 81:1735-42.

Remeuf F, Cossin V, Dervin C, Lenoir J, Tomassone R. 1991. Relations entre les caractères physico-chimiques des laits et leur aptitude fromagère. *Le Lait* 71:397-421.

Restani P. 2004. Goat milk allergenicity. *J Pediatr Gastroenterol Nutr.* 39:323-4.

Rijnkels M. 2002. Multispecies comparison of the casein gene loci and evolution of casein gene family. *J Mammary Gland Biol Neoplasia* 7:327-45.

Rosen JM, Wyszomierski SL, Hadsell D. 1999. Regulation of milk protein gene expression. *Annu Rev Nutr.* 19:407-36.

Sánchez Palma A, Hilahi H, Manfredi E. Serradilla J.M. 2005. Potential benefit from using the α_{S1}-casein genotype information in a selection scheme for dairy goats. *J Anim Breed Genet.* 122:21-29.

Schopen GC, Heck JM, Bovenhuis H, Visker MH, van Valenberg HJ, van Arendonk JA. 2009[a]. Genetic parameters for major milk proteins in Dutch Holstein-Friesians. *J Dairy Sci.* 92:1182-91.

Schopen GC, Koks PD, van Arendonk JA, Bovenhuis H, Visker MH. 2009[b]. Whole genome scan to detect quantitative trait loci for bovine milk protein composition. *Anim Genet.* 40:524-37.

Sheehy PA, Riley LG, Raadsma HW, Williamson P, Wynn PC. 2009. A functional genomics approach to evaluate candidate genes located in a QTL interval for milk production traits on *BTA6*. *Anim Genet.* 40:492-8.

Singh K, Erdman RA, Swanson KM, Molenaar AJ, Maqbool NJ, Wheeler TT, Arias JA, Quinn-Walsh EC, Stelwagen K. 2010. Epigenetic regulation of milk production in dairy cows. *J Mammary Gland Biol Neoplasia* 15:101-12.

Smacchi E, Gobbetti M. 1998. Peptides from several Italian cheeses inhibitory to proteolytic enzymes of lactic acid bacteria, *Pseudomonas fluorescens* ATCC 948 and to the angiotensin I-converting enzyme. *Enzyme Microb Technol.* 22:687-94.

Sulimova GE, Badagueva IuN, Udina IG. 1996. Polymorphism of the κ-casein gene in populations of the subfamily Bovinae. *Genetika* 32:1576-82.

Thompson MP, Tarassuk NP, Jenness R, Lillevik HA, Ashworth US, Rose D. 1965. Nomenclature of the proteins of cow's milk--second revision. report of the committee on milk protein nomenclature, classification, and methodology of the manufacturing section of ADSA for 1963-64. *J Dairy Sci.* 48:159-69.

Threadgill DW, Womack JE. 1990. Genomic analysis of the major bovine milk protein genes. *Nucleic Acids Res.* 18:6935-42.

Valenti B, Pagano RI, Pennisi P, Lanza M, Avondo M. 2010. Polymorphism at α_{s1}-casein locus. Effect of genotype × diet interaction on milk fatty acid composition in Girgentana goat. *Small Rumin Res.* 94:210-13.

VanRaden PM, Van Tassell CP, Wiggans GR, Sonstegard TS, Schnabel RD, Taylor JF, Schenkel FS. 2009. Invited review: reliability of genomic predictions for North American Holstein bulls. *J Dairy Sci.* 92:16-24.

Vassal L, Delacroix-Buchet A, Bouillon J. 1994. Influence des variants AA, EE et FF de la caséine αs1 sur le rendement fromager et les caractéristiques sensorielles des fromages traditionnels : prèmieres observations. *Le Lait* 74:89-103.

Walsh CD, Guinee TP, Reville WD, Harrington D, Murphy JJ, O'Kennedy BT, FitzGerald RJ. 1998. Influence of κ-casein genetic variant on rennet gel microstructure, cheddar cheesemaking properties and casein micelle size. *Int Dairy J.* 8:707-14

Wessels G, Hamann H, Erhardt G, Distl O. 2004. Genotype effects of milk protein polymorphisms on milk production in East Friesian dairy sheep. *Berl Munch Tierarztl Wochenschr.* 117:414-9.

Yue XP, Zhang XM, Wang W, Ma RN, Deng CJ, Lan XY, Chen H, Li F, Xu XR, Ma Y, Lei CZ. 2011. The *CSN1S1* N and F alleles identified by PCR-SSCP and their associations with milk yield and composition in Chinese dairy goats. *Mol Biol Rep.* 38:2821-5.

Zullo A, Barone CMA, Chianese L, Colatruglio P, Occidente M, Matassino D. 2005. Protein polymorphisms and coagulation properties of Cilentana goat milk. *Small Rumin Res.* 58:223-30.

In: Casein: Production, Uses and Health Effects
Editors: A. M. Ventimiglia, et. al, pp. 83-100
ISBN: 978-1-62100-129-4

Chapter 4

PHOSPHOPEPTIDES OF CASEIN: MINERAL CARRIERS AND POTENTIAL NUTRACEUTICALS

Amelia Fiorilli, Silvia Perego and Anita Ferraretto
Dipartimento di Chimica, Biochimica e Biotecnologie per la Medicina, Università degli Studi di Milano, Milano, Italy.

ABSTRACT

The gastrointestinal digestion of casein gives rise to a series of peptides displaying particular features and biological activity, such as the immunostimulant, ACE inhibitory, antihypertensive, opioid agonist/antagonist, antithrombotic and antimicrobial peptides. A last group of peptides, named casein phosphopeptides, are defined mineral carriers and originate from the proteolytic cleavage of αs_1-, αs_2- and β-casein. These casein phosphopeptides are characterized by the presence of phosphate linked to serine residues all over the amino acid sequence but, above all, localized in a particular Ser(P)-Ser(P)-Ser(P)-Glu-Glu sequence, known as the "acidic motif". Although casein-derived phosphopeptides (CPPs) correspond to different phosphorylated regions of αs_1-, αs_2- and β-caseins, all CPPs present the "acidic motif", which was highly conserved among species and centuries. The presence of phosphate confer them a high negative charge responsible for their further proteolysis and undiscussed ability to bind minerals such as calcium, iron and zinc. Numerous studies aimed to demonstrate the possibility to use the CPP mineral binding capacity for improving the absorption of calcium and iron at intestinal level, as well as for enhancing remineralization of tooth enamel. Besides the controversial results obtained with calcium transport studies performed using rat intestinal preparations, the use of in vitro human intestinal cell models do confirm that CPPs induce calcium uptake in differentiated intestinal cells and this action is correlated to their ability to form aggregates of defined size with calcium ions. Also in human primary osteoblast like cells CPPs increase the intracellular calcium concentration and the in vitro cell mineralization. In all these events the calcium bound to CPPs represents the fraction of soluble mineral involved in the process. Finally, new and interesting potentialities for CPPs come from recent data displaying a modulation of the intestinal immune system by triggering cytokine secretion and stimulating IgA production, the release of IL-6 cytokine in human epithelial intestinal cell lines, the modulation of cell viability, i.e. proliferation and apoptosis in different human cell cultures. Taken together these findings claimed for a real possibility to use CPPs as functional food, due to their presence in dairy products and their physiological formation in vivo, and for nutraceuticals, due to possibility to add to mouth rinse solutions, toothpaste and chewing gum, but also to confectionary products (breakfast foods, sweets, cakes, ice-cream, milk, powdered milk, yogurt, cheese, sport drinks, mayonnaise..) due to their good solubility in water, which will allow to use these peptides as an easy-to-formulate ingredient in food.

INTRODUCTION

Casein does not display any bioactivity, being a reservoir of amino acids for our health, but peptides coming from the in vitro and/or in vivo casein digestion can display particular bioactivities. Among these peptides, casein phosphopeptides (CPPs) represent a family characterized by a particular chemical-physical structure correlated to precise functioning in the body. The primary structure of CPPs is characterized by the presence of phosphorylation sites linked to serine amino acid, Ser(P), moreover in determined region of the amino acids sequence, a motif constituted of five amino acids, Ser(P)Ser(P)Ser(P)GluGlu, named "acidic motif", is present. This motif was found to be conserved among centuries and species, thus the studies initially addressed to understand its reason and utility. CPPs and their properties were discovered in the Fifties [1], and since then a potentiality in enhancing calcium intestinal absorption was speculated for them. This hypothesis was based on the fact that the presence of phosphorylated serines can be correlated to the possibility to bind and maintain calcium in a soluble state, a prerequisite for its absorption by the intestine [2] and was supported by the notion that, after a casein containing meal, a higher concentration of soluble calcium was found in the small intestinal tract than with the other proteins [3]. Indeed, first works tried to realize this possibility, employing intestinal preparations such as ligated loop of rat small intestine [4], ligated segments of rat ileum [5, 6] or through intestinal calcium transport studies in the rat ileum [7] obtaining positive results. Other studies were performed on animals (pigs, rats) fed with CPP-supplemented diets in which calcium absorption and bone mineralization were monitored, in some cases giving rise to contradictory results [8-12]. The contradiction found in these works reflects the heterogeneity in: i) the used CPPs (commercial preparation or single purified CPP); ii) variables in the animal experimentations; iii) the source of calcium (calcium lactate or calcium carbonate); iv) diet formulation. In the meantime a great interest arose in understanding the physical structure of the binding of CPPs with minerals, especially calcium, an issue well explored and now well comprised, see the section Casein phosphopeptides – calcium aggregates in this chapter. Due to the stabilization of calcium and phosphate in solution carried on by CPPs, the complexes made by CPPs and calcium phosphate, CPP-ACP, amorphous calcium-phosphate nanoclusters, were suggested as a new remineralization technology in dental caries [13].

In the last decades some studies were also conducted on human volunteers and even in this case the results found were not totally favorable to encourage CPPs as ingredient in food. In fact, in one study the absorption of calcium monitored in normal postmenopausal women was significantly better when calcium was co-ingested with CPPs [14], while more recent clinical studies failed to recognize a CPP effect on calcium absorption and balance [15-17].

On the contrary, in vitro cell studies demonstrated with no doubts that CPPs can affect diverse biological function, and the authors claimed for a CPP role as nutraceuticals. For this reason these in vitro works and their results will be reviewed in this chapter, waiting for a clear and definite CPP role in *in vivo* calcium absorption mechanism.

PRODUCTION AND PROPERTIES OF CASEIN PHOSPHOPEPTIDES

As a rule, biologically active peptides can be obtained from precursor milk proteins by the following means: a) enzymatic hydrolysis using digestive enzymes [18, 19]; b) proteolytic starter cultures for the fermentation of milk [20-23], and/or c) proteolysis by enzymes derived from microorganisms or plants [24].

Bioactive peptides are most commonly produced by the enzymatic hydrolysis of whole proteins, and are generated by means of gastrointestinal enzymes such as pepsin and trypsin, and enzymes found in bacteria and fungi.

Many dairy starter cultures are highly proteolytic, and lactic acid bacteria use milk proteins, mainly caseins, as their prime source of essential and growth-stimulating amino acids [25]. Starter and non-starter bacteria (Lactobacilli, Lactococci and Streptococci), which are used in the manufacture of fermented dairy products, can generate bioactive peptides [20, 26-28].

Both live microorganisms and proteolytic enzymes are able to release bioactive peptides from milk proteins. Indeed, Ueno and colleagues [29] not only purified and characterized an endopeptidase from Lactobacillus helveticus CM4, they also demonstrated that this peptidase can generate anti-hypertensive peptides using synthetic pro-peptides as substrates. Given the low cost and safety of commercially available microbially-derived food grade proteinases, their use to hydrolyze casein is very advantageous [30].

During human gastrointestinal digestion physiologically active peptides are produced from several food proteins [31-34]. Indeed, during the various stages of alimentation, like ingestion, digestion and absorption, the proteins and peptides in food undergo hydrolysis [35, 36]. Ingested proteins are first hydrolyzed by proteinases (such as pepsin, trypsin and chymotrypsin) in the gastrointestinal tract and peptides of various size are produced. These peptides are then further digested by brush border peptidases at the surface of intestinal epithelial cells, producing amino acids, while some oligopeptides remain intact. However, the quantity of peptides that can be released is unknown, because the broad specificity of the range of proteolytic enzymes precludes an accurate prediction of how great the liberation of bioactive fragments is [37, 38].

Several casein-derived phosphopeptides (CPPs), corresponding to different phosphorylated regions of αs_1-, αs_2- and β-caseins, are produced in vivo and/or in vitro and now are well characterized [39, 40].

The high negative charge of these phosphopeptides renders them partially resistant to further proteolysis, and confer them the ability to bind cations as Ca, Fe, Mn, Cu and Se, therefore they possibly act as carriers for a variety of minerals [4, 41]. The proteolysis resistance has led to a number of CPPs still being detectable in the faeces of experimental animals [42].

The *in vitro* production of CPPs generally involves the enzymatic hydrolysis of sodium caseinate [43], and such hydrolysis is carried out using a whole range of proteinase activities of animal, plant, fungal and bacterial origin [40, 41, 44-46]. Hydrolysates are usually prepared at neutral to alkaline pH, which is then adjusted to pH 4.6, resulting in the formation of insoluble precipitates that are removed by successive centrifugations.

Briefly, the three main routes used for the separation/enrichment of CPPs are:

a) CPP aggregation in the presence of mineral salts at neutral pH, followed by ultrafiltration through defined molecular mass cut-off membranes
b) mineral salt aggregation of CPPs followed by solvent precipitation
c) chromatographic separation on ion-exchange type matrices [43]

The isolated CPPs or CPP-enriched fractions are then subjected to either freeze-drying or spray-drying; according to the literature the CPP yields range from 6-20%. In the initial hydrolysis step the yield appears to be affected by many factors, including the specificity of the enzyme preparation [40, 44, 47-49].

The quality of the CPPs generated from the above protocols depends greatly on minimizing CPP exposure to high temperatures during manufacture, because CPPs are sensitive to high temperature-induced dephosphorylation [50].

Different strategies have been employed to identify bioactive peptides in milk proteins:

1) isolation and characterization of bioactive peptides from their *in vivo* formation in the gastrointestinal tract [39, 51]
2) a search for bioactive sequences, in the isolation and characterization of bioactive peptides from *in vitro* protein digestion [52, 53]
3) peptide synthesis to elucidate the bioactive structures of distinct amino acid sequences [52, 54, 55]

Many known bioactive peptides are biologically very potent, and even trace amounts of liberated peptides might be sufficient to cause physiological effects.

Casein Phosphopeptides – Calcium Aggregates

In bovine milk casein is present as micelles, stable structures composed of hundreds of smaller aggregates, named calcium phosphate nanoclusters, or nanocomplexes, of about 100 nm radius, which enclose about two thirds of calcium and one half of inorganic phosphate [56]. Calcium phosphate nanoclusters (or complexes) were also recognized to be formed by CPPs [57, 58], namely by β-CN(1-25)4P and β-CN(1-42)5P, corresponding to the first 25 or 42 amino acids of β-casein, respectively, and by α_{S1}-CN(59-79)5P, corresponding to the sequences 59-79 of α_{S1}-casein [59, 60]. A subsequent effort was made by investigating the correlation between CPP – calcium aggregates and their biological activity, when the experimental conditions used to detect the biological effect of CPPs were adopted [61]. Laser light scattering experiments clearly demonstrated the occurrence of CPP self-aggregation in solution, analogously to casein behavior in milk, with precise features: very rapid occurrence; 60 nm hydrodynamic radius; absolute need of concomitant presence of Ca^{2+} and CPPs for optimal aggregation. Moreover, the ability to aggregate, measured as dimension and concentration of aggregates, relies on the chemical structure of CPPs, analogously to their bioactivity, in particular the "acidic motif" pentapeptide does not aggregate and does not display any bioactivity, as well as the removal of phosphate from serines brings to a lack of function. The last notion points out to the important role of serine-linked phosphate in binding calcium, inducing a suitable aggregation of CPPs in order to elicit the biological effect. The aggregate formation is optimal for a precise CPP/calcium ratio (4mM Ca^{2+}/200 μM β-CN(1-25)4P or 4mM Ca^{2+}/1280 μM CPP mixture), but occurs with a different shape for β-CN(1-25)4P compared to α_{S1}-CN(59-79)5P, a result also confirmed by the induced calcium uptake in HT-29 cells, a feature probably residing in the different structural changes induced in the two CPPs by the binding with calcium and already demonstrated for $Fe^{2+/3+}$ [62]. An explanation of the calcium aggregation ability demonstrated by CPPs can be found in the coexistence of hydrophilic and hydrophobic regions in the single CPP monomers, and in the occurring of the hydrophobic interactions between monomers which give rise to the aggregation. In addition, calcium divalent counter-ions, carrying two positive charges, may facilitate the aggregation by coordinating two charges belonging to different molecules of CPPs. The final result is that aggregated forms are the active forms of a bioactive CPP like β-CN(1-25)4P.

Mineral Absorbing Properties

The mineral absorbing property of CPPs has been observed in several *in vitro* studies [61, 63, 64]. It was reported by Hansen et al. [11] that addition of 14 micromoles of CPPs to Caco2 cells has a positive effect on zinc binding and zinc uptake from phytate containing solutions, while higher levels of CPPs (36 and 73 micromoles) cause inhibition. Zinc bioavailability in human milk containing CPPs was investigated by Etcheverry et al. [65] in Caco2 cells using a variety of calcium salts like $Ca_3(PO_4)_2$, $CaCl_2$ and calcium glycerophosphate/calcium gluconate. Higher zinc uptakes were obtained with human milk added CPPs, calcium glycerophosphate and calcium gluconate than with other calcium forms. The studies indicate that important factors in zinc bioavailability are zinc levels, the type of calcium salt and the CPP dose [63].

Casein Phosphopeptide Effects on In Vitro Cell Cultures

All the effects displayed by CPPs in *in vitro* cell cultures together with the relative references are summarized in Table 1.

Effects of CPPs on In Vitro Intestinal Cells

Intestinal cells represent the first contact and space of interaction between nutrients, in this case CPPs, and our organism. For this reason the possible CPP effects as nutraceuticals start from the comprehension of their interaction with intestinal cells, i.e. are they internalized by the cells or not? And if not, what is their contribution to the intestinal cell functions? At present no data are available about the possibility that CPPs could entry into intestinal cells, anyway, CPPs strongly interact with intestinal cells insomuch to differently modulate cell proliferation and apoptosis in differentiated and tumor cells [66]. The first direct evidence of an interaction between intestinal cells and CPPs came from the observation that in presence of CPPs and extracellular calcium, HT-29 and Caco2 cells, two known human intestinal in vitro cell models, are induced to uptake calcium ions from the extracellular milieu [63]. This activity was subsequently recognized to be due to the presence of a structural CPP conformation, conferred by both the phosphorylated 'acidic motif' and the preceding N-terminal portion, necessary for the interaction with the cell plasma membrane [64]. Instrumental to this CPP aptitude is the formation of stable aggregates with calcium ions (as above described), in fact the cellular response, i.e, the increased intracellular calcium concentration, follows a dose-response relationship similar to that reported for the formation of aggregates, with a maximum at 4 mM extracellular calcium concentration and a CPP dose of 1280 μM, if a purified commercial mixture, or 200 μM, if a single purified CPP [61]. The use of commercial CPP mixtures instead of single purified CPPs is another important aspect to be considered, since differences were determined as concerned properties and bioactivity among: i) the diverse single purified CPPs, for instance β-CN(1-25)4P, deriving from the first 25 amino acids of bovine β casein, and α_{S1}-CN(59-79)5P, deriving from the α_{s1} casein; ii) the CPP mixtures from diverse commercial sources; iii) the single purified CPPs compared to the commercial mixtures [61, 67]. In brief, the concentration of the CPPs necessary to monitor a consistent intracellular calcium increase is about six times higher, in the case of CPP mixtures, than that of a single purified peptide, in the case of β-CN(1-25)4P. CPP α_{S1}-CN(59-79)5P displays a reduced aggregate formation and a parallel lower bioactivity, in agreement with its calcium binding constant of 0.85 mM^{-1} with respect to that of 0.63 mM^{-1} determined for β-CN(1-25)4P a pH 7 [68]. More recent studies have also provided results concerning differences between the intestinal cell lines used, HT-29 and Caco2, in their response to CPPs, evidencing that, regardless of the slight differences (mostly at the quantitative levels) between the two cell lines, the ability of these cells to take up extracellular calcium under CPP stimulation is a molecular attitude expressed upon differentiation [69]. Although the last information, together with the demonstration that the fraction of calcium bound to CPP is directly involved in the mineral uptake by the cells [67], is of fundamental importance to understand the nature of the interaction between CPPs and intestinal cells, at present the molecular mechanism underlying this event is not definite, even if in differentiated HT-29 cells CPPs seem to modulate the activity of the voltage operated L-type calcium channels, known to activate calcium entry into the cells under depolarization and to exert a mitogenic effect [66].

Changes in the cell division and cellular kinetics were also described in the IEC-6 cells derived from intestines of 18-24 d-old rats when incubated in presence of peptides produced from casein by yogurt bacteria [70, 71]. A result which speculate on a role for casein-derived peptides in the control of colon tumor.

Another demonstration of the interaction between CPPs and intestinal cells can be derived by the works of Kawahara and colleagues [72], Kitts and Nakamura [73] who presented experimental evidences that a particular CPP sequence, commercially available as CPP-III, (Meiji Seika, Tokyo, Japan), consisting of about 90% casein phosphopeptides such as bovine α_{S2}-casein (1-32) and β-casein (1-28) and enriched in calcium, modulates the intestinal immune system by: i) enhancing the mRNA expression of TNF-α as well as IL-6 in a dose dependent manner, while the expression of IL-1 was not affected. TNF-α has been reported to mediate the production of IL-6 in Caco2 cells [74]. Thus, CPP-III is presumed to stimulate the mRNA expression of IL-6 in a direct way or in a TNF- α mediated indirect way; ii) triggering IL-6 cytokine release from intestinal 407 (Int-407) cell line. IL-6 is a cytokine with functions of modulator of cellular differentiation, proliferation and apoptosis, all cellular functions involved in the progression and/or regression of tumor [61] together with an important role for IgA production [75, 76].

Effects of CPPs on In Vitro Bone Cells

The use of primary human osteoblast-like cells, established in culture from trabecular bone samples obtained from waste materials during orthopaedic surgery, has revealed interesting and new potentialities for CPPs when locally administered. Analogously to what monitored in intestinal in vitro cell lines, CPPs stimulate calcium uptake by primary human osteoblast-like cells and induce an increase in the expression and activity of alkaline phosphatase (ALP), a marker of human osteoblast differentiation. Like for intestinal cells, changes in the intracellular calcium concentration due to CPP administration return on modification in the cell proliferation rate and the apoptotic level. Moreover, the higher fraction of available calcium due to CPPs enhance calcification nodule formation by human SaOS-2 in vitro cell culture [77].

A successive study performed on mouse preosteoblast cells (MC3T3-E1 cell line) analyzed the influence of four selected CPPs that are theoretically present in the digestion products of different genetic variants of bovine β-casein and α_{s2}-casein on osteoblastic cell viability and differentiation, demonstrating that distinct CPPs can be either stimulatory or inhibitory [78].

Effects of CPPs on Fibroblasts

One of the most severe types of traumatic dental injury is avulsion, which comports the complete displacement of a tooth from the alveolar socket, and which can lead to necrosis of the pulp tissue [79]. When there is avulsion it is important to be able to carry out replantation by avoiding or minimizing the complications of periodontal ligament (PDL) tissue damage and pulpal infection. PDL cells can usually survive 15 min or less of dry time, but it is unlikely that they can survive dry times greater than 60 min [80, 81]. To maintain PDL viability a suitable transport medium is needed, and this also helps to prevent inflammatory and replacement resorption. Milk is known to be efficient in maintaining PDL cell viability, and this can be ascribed to factors like nutritional proteins and growth factors, physiologic osmolarity, pH buffering and the low bacterial content resulting from the pasteurization process [82, 83]. Demineralization can be prevented to a large degree by CPPs competing for calcium with dental plaque-forming bacteria, which explains a possible use of CPPs in the remineralization of tooth enamel [84]. To demonstrate biocompatibility, Cehreli [85] chose L929 fibroblast cells, a well established distinct line, in which the cells are genetically identical. A comparison has been made of the behavior of L929 cells with gingival fibroblasts, and it has been shown that their physiology and adhesive properties are similar [86]. It was shown that lower concentrations of CPP-amorphous calcium phosphate (CPP-ACP) can effectively maintain cell viability in vitro. Stronger dilution of CPP-ACP was effective for both the viability and the morphology of L929 cells; this was demonstrated in the 10^{-8} and 10^{-12} dilution groups that were similar in fibroblastic morphology and apoptotic response to the control group. A severe toxic and apoptotic response was found for fibroblasts at 10^{-3} and 10^{-4} dilution groups, from the very first day of incubation and thereafter. This could have been due to the CPP-ACP-dependent increase in calcium concentration in the culture medium. It could be that calcium triggers the mechanisms needed for cell proliferation; cytosolic enzymes executing apoptosis could also be stimulated. Because of its complex interaction with other signaling pathways, given ion channels could be involved in both cell proliferation and apoptosis [87]. This leads to the expectation of a decrease in possible toxic/apoptotic effects of ingredients with increase in the dilution factor. It was observed that the effect of CPP-ACP was dose-dependent.

Effects of CPPs on Lymphocytes

Mammalian neonates are immunologically immature because certain T and B lymphocyte functions are deficient. Cell function of an immature immune system is improved by the many host defense proteins and various proteins of milk [88-94].

Peptides liberated from casein by protease action are believed to have a physiological role, that particularly comes into play when digestion takes place. It was shown by Otani and Hata [95] that a bovine α_{s1}-casein tryptic digest significantly inhibites the response to proliferation of mouse spleen cells and Peyer's patch cells stimulated by concanavalin A (ConA). Thus immunomodulatory peptide(es) could be formed from bovine α_{s1}-casein, due to the action of trypsin because ConA mainly stimulates suppressor and killer T lymphocytes [96].

ConA-induced proliferation of mouse spleen cells and rabbit Peyer's patch cells was greatly inhibited by isolated peptides. Instead, cell proliferation was significantly enhanced by the peptides phytohaemagglutinin-P (PHA) and lipopolysaccharides (LPS), and there was also a mitogenic effect when commercial mitogen was absent. Calcium influx into lymphocytes is responsible for triggering lymphocyte stimulation as the first signal and the calcium ionophore A23187 is mitogenic towards lymphocytes, enhancing calcium uptake [97]. The effect of β-casein (1-25) on the proliferative response of mouse spleen cells and rabbit Peyer's patch cells stimulated by ConA, PHA and LPS was examined, and it was observed that this sequence also induces the proliferative response of mouse spleen cells and rabbit Peyer's patch cells in the absence of commercial mitogen. It enhances the PHA- and LPS-induced proliferation of both cells but suppresses the ConA-induced proliferation.

Effects of CPPs on CD19+ Cells

A recent study has shown that some CPPs have immune-enhancing activities. It was found by Hata et al. [98] that a commercial CPP preparation, CPP-III, consisting mainly of β-casein(1-28) and α_{s2}-casein (1-32), stimulates the mouse spleen cell proliferation. It was also shown that CPP-III enhances intestinal IgA levels by promoting IL-6 production, an oral administration of the CPP-III-supplemented diet being administered to mice [99]. Alongside these studies, it was found that β-casein (1-28), a major component of CPP-III, has mitogenic and immunoglobulin-enhancing effects on mouse spleen cell cultures [100], and there was clear evidence of a critical role of β-casein (1-28) in IgA production [101]. It was found that the SerP-X-SerP sequence was essential for the immune-enhancing activities of β-casein (1-28), chemically synthesized partial fragments of β-casein being used (1-28) [102]. A study by Tobita et al. [103] was done to explain which lymphocyte subsets predominantly interact with β-casein (1-28), and which receptors are involved in the β-casein (1-28)-induced immune-enhancing activity. Proliferation-enhancing substances, called mitogenic substances, are recognized by specific receptors, causing them to trigger naïve lymphocyte activation without prior sensitization. It is generally believed that mitogenic action by bacteria-derived substances can result from TLR-family-mediated signal transduction. Peptidoglycan (PG) and lipopolysaccharides (LPS) are known to be respective ligands for TLR2 [104] and TLR4 [105]. Among the CD4+, CD8+, and CD19+ cell subsets, β-casein (1-28) exerts a significant mitogenic effect on CD19+ cells from C3H/HeJ mice. Instead, in CD19+ cell proliferation from C3H/HeJ mice, β-casein (1-28) has little effect. C3H/HeJ mice are hyporesponsive to LPS because of the mutation in the exon of TLR4 gene [106]. Therefore the defective response of CD19+ cells from C3H/HeJ mouse indicates the need of β-casein(1-28) for TLR4 in order to achieve a mitogenic effect. Furthermore, the effect of β-casein (1-28) on CD19+ cells from C3H/HeJ mice was significantly inhibited in the presence of the TLR4-neutralizing antibody. Thus the hypothesis that β-casein (1-28) induces proliferation via TLR4 is strengthened. Furthermore, β-casein (1-28) enhances IL-6 expression by CD19+ cells. Also TLR4 signaling enhances the expression of cytokines like IL-6, IL-10, and the tumor necrosis factor (TNF- α) [107, 108]. It was found that β-casein (1-28) also enhances IL10 and TNF-α expression by CD19+ cells. This led to an examination to understand whether the IL6 expression of C3H/HeJ mouse derived CD19+ cells was not responsive to β-casein (1-28); indeed, as expected, the expression of IL6 was notably diminished. There was competitive inhibition in β-casein (1-28) -induced IL6 expression of CD19+ cells from C3H/HeJ mice; this was observed by treating β-casein (1-28) with anti-TLR4 antibody. The expression was not affected by treatment with PMB (Polymixin B). It was also suggested that IL6 expression of CD19+ cells by β-casein (1-28) was mediated by TLR4; indeed, it was revealed that the tripeptide sequence SerP-X-SerP is necessary for the mitogenic action of β-casein (1-28) [101]. Kitazawa [109] reported that some phosphorylated saccharides secreted from bacteria are known to have mitogenic activity to mouse spleen B cells, which were the primary population of CD19+ cells. Furthermore, for the mitogenic action of some of the saccharides a phosphate group is needed. The relationship between these phosphorylated structures and TLR is still unclear, but it could be that the phosphoserine residues of β-casein (1-28) act like phosphate groups in these phosphorylated mitogens. In conclusion, β-casein (1-28) enhances the proliferation and IL-6 production of CD19+ cells, and both enhancements were mediated by TLR4.

Table 1. Effects of CPPs on in vitro different cell lines

CPP sequence	Cell type	Biological Effect	Ref.
α_{s1}-casein (59-79)	Mouse spleen, rabbit Peyer's patch	Mitogenic activity and enhanced Ig production	[110]
β-casein (1-28)	Mouse spleen	Increased lymphocyte proliferation and IgA production	[102]
CPP purified commercial mixture	HT-29, Caco-2	Induced calcium uptake	[63]
β-casein (1-28) purified from commercial CPP-III	Human T, B and monocytes	Enhanced proliferation and stimulated IgA production	[72]
CPP-III: α_{s2}-casein (1-32) and β-casein (1-28)	Caco-2	IL-6 mRNA expression and TNF-α enhancement	[111]
CPP-III: α_{s2}-casein (1-32) and β-casein (1-28)	Int-407	Induced release of IL-6	[73]
β-casein (1-28) purified from commercial CPP-III	Mouse CD19+	Enhanced proliferation and stimulated IL-6 expression via TLR4	[103]
Commercial CPP-ACP	L929 fibroblasts	Cell viability preservation	[85]
CPP purified commercial mixture	Primary human osteoblast-like cells	Stimulated calcium uptake, increased ALP expression and activity	[77]
CPP purified commercial mixture	SaOS-2	Induced in vitro mineralization	[77]
Genetic variants of bovine β-casein and α_{s2}-casein	MC3T3-E1 mouse preosteoblasts	Cell growth, calcium uptake and deposition in the extracellular matrix	[78]

Conclusion

The search for an optimal health life style has involved nutrition as the science devoted to discover how nutrients and/or molecules associated to nutrients can interact with the gastro-intestinal system and cooperate to the body function, in view to maintain or acquire the health status.

During the last years, bioactive peptides with defined action on human health are increasingly added to food by modifying the usual manufacturing process, and these products are marketed as "functional foods" [112].

In this context, in milk and colostrum can be found proteins and peptides not only crucial to the neonate and adult alimentation, but with a potentiality as health-enhancing nutraceuticals [113]. Indeed, nutraceuticals, in the form of bioactive peptides, can be used to treat mineral malabsorption and dental diseases (phosphopeptides), immunodeficiency (immunopeptides), hypertension (casokinins), thrombosis (casoplatelins), diarrhea (lactoferrin) and other diseases like asthma and cancer. Furthermore, casein hydrolysates have been found to reduce tumor growth, lessen the symptoms of colic in babies and to protect animals that are diabetic [114].

In the present chapter we have proposed a review about the features and the properties of this family of casein-derived peptides, CPPs or casein phosphopeptides, which were the object of an EU research project (CT98-3077, Caseinphosphopeptide (CPP): Nutraceutical/functional food ingredients for food and pharmaceutical applications) aimed to demonstrate the potential role of casein phosphopeptides as functional foods/nutraceuticals.

Despite the controversial results coming from in vivo experiments, CPPs were demonstrated to be able to affect cell function mainly in intestinal vitro cell models, but also in bone cells, in fibroblasts and in lymphocytes. In all these cell types CPPs affect cell proliferation, cytochine production and, in the case of bone cells, the mineralization. The lack of a definite knowledge about the possibility for CPPs to be internalized by intestinal cells has sometimes stopped further projects of understanding how to employ them, even if their use as anti cariogenics is well documented. Similarly, the results obtained with in vitro bone cells open the way for a future employment of CPPs as a local anabolic factor for bone tissue engineering, in analogy to what demonstrated in the re-mineralization of carious lesions in dental enamel.

Anyway, the encouraging results published so far allowed to consider CPPs both potential functional food, due to their presence in dairy products and their physiological formation in vivo, and nutraceuticals, due to possibility to add to mouth rinse solutions, toothpaste and chewing gum, but also to confectionary products (breakfast foods, sweets, cakes, ice-cream, milk, powdered milk, yogurt, cheese, sport drinks, mayonnaise..) due to their good solubility in water, which will allow to use these peptides as an easy-to-formulate ingredient in food.

References

[1] Mellander, O; Olsson, N. The influence of peptide bound calcium and phosphorus on bone calcification in rickets. *Bol Med Hosp Infant Mex*, 1956 13, 243-6.

[2] Kitts, DD; Yuan, YV. Caseinophosphopeptides and calcium biovailability. *Trends in Food Science & Technology*, 1992 3, 31-5.

[3] Lee, YS; Noguchi, T; Naito, H. Phosphopeptides and soluble calcium in the small intestine of rats given a casein diet. *Br J Nutr*, 1980 43, 457-67.

[4] Sato, R; Noguchi, T; Naito, H. Casein phosphopeptide (CPP) enhances calcium absorption from the ligated segment of rat small intestine. *J Nutr Sci Vitaminol*, 1986 32, 67-76.

[5] Erba, D, Ciappellano, S; Testolin, G. Effect of caseinphosphopeptides on inhibition of calcium intestinal absorption due to phosphate. *Nutrition Research*, 2001 21, 649-56.

[6] Erba, D; Ciappellano, S; Testolin, G. Effect of the ratio of casein phosphopeptides to calcium (w/w) on passive calcium transport in the distal small intestine of rats. *Nutrition*, 2002 18, 743-6.

[7] Li, Y; Tome, D; Desjeux, JF. Indirect effect of casein phosphopeptides on calcium absorption in rat ileum in vitro. *Reprod Nutr Dev*, 1989 29, 227-33.

[8] Pointillart, A, Gueguen, L. Absence of effect of the incorporation of a milk phosphopeptide on the utilization of calcium and phosphorus in the young pig. *Reprod Nutr Dev*, 1989 29, 477-86.

[9] Kopra, N; Scholz-Ahrens, KE; Barth, CA. Effect of caseinphosphopeptides on utilisation of calcium in vitamin D-replete and vitamin D-deficient rats. *Milchwissenschaft*, 1992 47, 488-93.

[10] Brommage, R; Juillerat, MA; Jost, R. Influence of casein phosphopeptides and lactulose on intestinal calcium absorption in adult female rats. *Lait*, 1991 71, 173-80.

[11] Hansen, M; Sandstrom, B; Lonnerdal, B. The effect of casein phosphopeptides on zinc and calcium absorption from high phytate infant diets assessed in rat pups and Caco-2 cells. *Pediatr Res*, 1996 40, 547-52.

[12] Tsuchita, H; Goto, T; Shimizu, T; Yonehara, Y; Kuwata, T. Dietary casein phosphopeptides prevent bone loss in aged ovariectomized rats. *J Nutr*, 1996 126, 86-93.

[13] Reynolds, EC. Casein phosphopeptide-amorphous calcium phosphate: the scientific evidence. *Adv Dent Res*, 2009 21, 25-9.

[14] Heaney, R; Saito, Y; Orimo, H. Effect of caseinphosphopeptide on absorbability of co-ingested calcium in normal postmenopausal women. *Journal of Bone and Mineral Metabolism*, 1994 12, 77-81.

[15] Hansen, M; Sandstrom, B; Jensen, M; Sorensen, SS. Effect of casein phosphopeptides on zinc and calcium absorption from bread meals. *J Trace Elem Med Biol*, 1997 11, 143-9.

[16] Narva, M; Karkkainen, M; Poussa, T; Lamberg-Allardt, C; Korpela, R. Caseinphosphopeptides in milk and fermented milk do not affect calcium metabolism acutely in postmenopausal women. *J Am Coll Nutr*, 2003 22, 88-93.

[17] Teucher, B; Majsak-Newman, G; Dainty, JR; McDonagh, D; FitzGerald, RJ; Fairweather-Tait, SJ. Calcium absorption is not increased by caseinophosphopeptides. *Am J Clin Nutr*, 2006 84, 162-6.

[18] Kilara, A; Panyam, D. Peptides from milk proteins and their properties. *Crit Rev Food Sci Nutr*, 2003 43, 607-33.

[19] Korhonen, H; Pihlanto, A. Food-derived bioactive peptides--opportunities for designing future foods. *Curr Pharm Des*, 2003 9, 1297-308.

[20] Fuglsang, A; Nilsson, D; Nyborg, NC. Characterization of new milk-derived inhibitors of angiotensin converting enzyme in vitro and in vivo. *J Enzyme Inhib Med Chem*, 2003 18, 407-12.

[21] Gobbetti, M; Ferranti, P; Smacchi, E; Goffredi, F; Addeo, F. Production of angiotensin-I-converting-enzyme-inhibitory peptides in fermented milks started by Lactobacillus delbrueckii subsp. bulgaricus SS1 and Lactococcus lactis subsp. cremoris FT4. *Appl Environ Microbiol*, 2000 66, 3898-904.

[22] Pihlanto-Leppälä, A; Rokka, T; Korhonen, H. Angiotensin I Converting Enzyme Inhibitory Peptides Derived from Bovine Milk Proteins. *International Dairy Journal*, 1998 8, 325-31.

[23] Nurminen, ML, Sipola, M; Kaarto, H; Pihlanto-Leppala, A; Piilola, K; Korpela, R et al. Alpha-lactorphin lowers blood pressure measured by radiotelemetry in normotensive and spontaneously hypertensive rats. *Life Sci*, 2000 66, 1535-43.

[24] Yamamoto, N; Akino, A; Takano, T. Antihypertensive effect of the peptides derived from casein by an extracellular proteinase from Lactobacillus helveticus CP790. *J Dairy Sci*, 1994 77, 917-22.

[25] Juillard, V; Guillot, A; Le Bars, D; Gripon, JC. Specificity of milk peptide utilization by Lactococcus lactis. *Appl Environ Microbiol*, 1998 64, 1230-6.

[26] Fitzgerald, RJ; Murray, BA. Bioactive peptides and lactic fermentations. *International Journal of Dairy Technology*, 2006 59, 118-25.

[27] Leclerc, PL; Gauthier, SF; Bachelard, H; Santure, M; Roy, D. Antihypertensive activity of casein-enriched milk fermented by Lactobacillus helveticus. *International Dairy Journal*, 2002 12, 995-1004.

[28] Ashar, MN; Chand, R. Fermented milk containing ACE-inhibitory peptides reduces blood pressure in middle aged hypertensive subjects. *Milchwissenschaft*, 2004 59, 363–6.

[29] Ueno, K; Mizuno, S; Yamamoto, N. Purification and characterization of an endopeptidase that has an important role in the carboxyl terminal processing of antihypertensive peptides in Lactobacillus helveticus CM4. *Lett Appl Microbiol*, 2004 39, 313-8.

[30] Mao, XY; Yang, HY; Song, JP; Li, YH; Ren, FZ. Effect of yak milk casein hydrolysate on TH1/TH2 cytokines production by murine spleen lymphocytes in vitro. *J Agric Food Chem*, 2007 55, 638-42.

[31] Chabance, B, Marteau, P; Rambaud, JC; Migliore-Samour, D; Boynard, M; Perrotin, P et al. Casein peptide release and passage to the blood in humans during digestion of milk or yogurt. *Biochimie*, 1998 80, 155-65.

[32] Meisel, H, Frister, H. Chemical characterization of bioactive peptides from in vivo digests of casein. *J Dairy Res*, 1989 56, 343-9.

[33] Parrot, S; Degraeve, P, Curia, C; Martial-Gros, A. In vitro study on digestion of peptides in Emmental cheese: analytical evaluation and influence on angiotensin I converting enzyme inhibitory peptides. *Nahrung*, 2003 47, 87-94.

[34] Svedberg, J; de Haas, J; Leimenstoll, G; Paul, F; Teschemacher, H. Demonstration of beta-casomorphin immunoreactive materials in in vitro digests of bovine milk and in small intestine contents after bovine milk ingestion in adult humans. *Peptides*, 1985 6, 825-30.

[35] Shimizu, M. Food-derived peptides and intestinal functions. *Biofactors*, 2004 21, 43-7.

[36] Vermeirssen, V; Van Camp, J; Verstraete, W. Bioavailability of angiotensin I converting enzyme inhibitory peptides. *Br J Nutr*, 2004 92, 357-66.
[37] Dziuba, J; Minkiewicz, P; Puszka, K. Wheat gliadins as potential precursors of bioactive peptides. *Acta Acad Sci Tech Olst Technologia Alimentorum*, 1995 28, 3-16.
[38] Dziuba, J; Minkiewicz, P; Nalecz, D; Iwaniak, A. Database of biologically active peptide sequences. *Nahrung*, 1999 43, 190-5.
[39] Meisel, H; Frister, H. Chemical characterization of a caseinophosphopeptide isolated from in vivo digests of a casein diet. *Biol Chem Hoppe Seyler*, 1988 369, 1275-9.
[40] Juillerat, MA; Baechler, R; Berrocal, R; Chanton S; Scherz, JC; Jost, R. Tryptic phosphopeptides from whole casein. I. Preparation and analysis by fast protein liquid chromatography. *Journal of Dairy Research*, 1989 56, 603-11.
[41] Berrocal, R, Chanton, S; Juillerat, MA; Pavillard, B; Scherz, JC; Jost, R. Tryptic phosphopeptides from whole casein. II. Physicochemical properties related to the solubilization of calcium. *J Dairy Res*, 1989 56, 335-41.
[42] Kasai, T; Honda, T; Kiriyama, S. Caseinphosphopeptides (CPP) in Feces of Rats Fed Casein Diet. *Bioscience, Biotechnology and Biochemistry*, 1992 56, 1150-1.
[43] FitzGerald, RJ. Potential Uses of Caseinophosphopeptides. *International Dairy Journal*, 1998 8, 451-7.
[44] McDonagh, D; FitzGerald, RJ. Production of Caseinophosphopeptides (CPPs) from Sodium Caseinate Using a Range of Commercial Protease Preparations. *International Dairy Journal*, 1998 8, 39-45.
[45] Manson, W; Annan, WD. The structure of a phosphopeptide derived from -casein. *Arch Biochem Biophys*, 1971 45, 16-26.
[46] Adamson, NJ; Reynolds, EC. Characterization of tryptic casein phosphopeptides prepared under industrially relevant conditions. *Biotechnol Bioeng*, 1995 45, 196-204.
[47] Gerber, HW; Jost, R. Casein phosphopeptides: their effect on calcification of in vitro cultured embryonic rat bone. *Calcif Tissue Int*, 1986 38, 350-7.
[48] Kunst A. Process to isolate phosphopeptides. *Eur Patent Applic*, 1990 EP0476199A1.
[49] Ellegård, KH, Gammelgård-Larsen, C; Sørensen, ES; Fedosov, S. Process scale chromatographic isolation, characterization and identification of tryptic bioactive casein phosphopeptides. *International Dairy Journal*, 1999 9, 639-52.
[50] Meisel, H; Andersson, HB; Buhl, K; Erbersdobler, HF; Schlimme, E. Heat induced changes in casein-derived phosphopeptides. *Zeit Ernahrung*, 1991 30, 227-232.
[51] Meisel, H. Chemical characterization and opioid activity of an exorphin isolated from in vivo digests of casein. *FEBS Lett*, 1986 196, 223-7.
[52] Fiat, AM; Jolles, P. Caseins of various origins and biologically active casein peptides and oligosaccharides: structural and physiological aspects. *Mol Cell Biochem*, 1989 87, 5-30.
[53] Maruyama, S; Nakagomi, K; Tomizuka, N; Suzuki, H. Angiotensin I-converting enzyme inhibitor derived from an enzymatic hydrolysate of casein. *Agric Biol Chem*, 1985 51, 1405–9.
[54] Lottspeich, F; Henschen, A; Brantl, V; Teschemacher, H. Novel opioid peptides derived from casein (beta-casomorphins). III. Synthetic peptides corresponding to components from bovine casein peptone. *Hoppe Seylers Z Physiol Chem*, 1980 36, 1835-9.

[55] Meisel, H; Schlimme, E. Inhibitors of angiotensin converting enzyme derived from bovine casein (casokinins). In: Brantl V, Teschemacher H, editors. b-casomorphins and related peptides: Recent developments. Weinheim, New York: VCH; 1994. p. 27-33.

[56] Holt, C; de Kruif, CG; Tuinier, R; Timmins, PA. Substructure of bovine casein micelles by small-angle X-ray and neutron scattering. Colloids and Surfaces A: *Physicochemical and Engineering Aspects*, 2003 213, 275-84.

[57] Little, EM; Holt, C. An equilibrium thermodynamic model of the sequestration of calcium phosphate by casein phosphopeptides. *Eur Biophys J*, 2004 33, 435-47.

[58] Cross, KJ; Huq, NL; Palamara, JE; Perich, JW; Reynolds, EC. Physicochemical characterization of casein phosphopeptide-amorphous calcium phosphate nanocomplexes. *J Biol Chem*, 2005 280, 15362-9.

[59] Holt, C, Wahlgren, NM; Drakenberg, T. Ability of a beta-casein phosphopeptide to modulate the precipitation of calcium phosphate by forming amorphous dicalcium phosphate nanoclusters. *Biochem J*, 1996 314, 1035-9.

[60] Holt, C; Timmins, PA; Errington, N; Leaver, J. A core-shell model of calcium phosphate nanoclusters stabilized by beta-casein phosphopeptides, derived from sedimentation equilibrium and small-angle X-ray and neutron-scattering measurements. *Eur J Biochem*, 1998 252, 73-8.

[61] Gravaghi, C; Del Favero, E; Cantu, L, Donetti, E, Bedoni, M; Fiorilli, A et al. Casein phosphopeptide promotion of calcium uptake in HT-29 cells - relationship between biological activity and supramolecular structure. *FEBS J*, 2007 274, 4999-5011.

[62] Kibangou, IB, Bouhallab, S, Henry, G; Bureau, F; Allouche, S; Blais, A et al. Milk proteins and iron absorption: contrasting effects of different caseinophosphopeptides. *Pediatr Res*, 2005 58, 731-4.

[63] Ferraretto, A; Signorile, A; Gravaghi, C; Fiorilli, A; Tettamanti, G. Casein phosphopeptides influence calcium uptake by cultured human intestinal HT-29 tumor cells. *J Nutr*, 2001 131, 1655-61.

[64] Ferraretto, A; Gravaghi, C; Fiorilli, A; Tettamanti, G. Casein-derived bioactive phosphopeptides: role of phosphorylation and primary structure in promoting calcium uptake by HT-29 tumor cells. *FEBS Lett*, 2003 551, 92-8.

[65] Etcheverry, P; Wallingford ,JC, Miller, DD; Glahn, RP. The effect of calcium salts, ascorbic acid and peptic pH on calcium, zinc and iron bioavailabilities from fortified human milk using an in vitro digestion/Caco-2 cell model. *Int J Vitam Nutr Res*, 2005 75, 171-8.

[66] Perego, S; Cosentino, S; Fiorilli, A; Tettamanti, G; Ferraretto A. Casein phosphopeptides modulate proliferation and apoptosis in HT-29 cell line through their interaction with voltage operated L-type calcium channels. *Journal of Nutritional Biochemistry*, DOI 10.1016/j.jnutbio.2011.04.004.

[67] Cosentino, S; Donida, BM, Marasco, E; Del Favero; E, Cantù, L; Lombardi, G et al. Calcium ions enclosed in casein phosphopeptide aggregates are directly involved in the mineral uptake by differentiated HT-29 cells. *International Dairy Journal*, 2010 20, 770-6.

[68] Meisel, H; Olieman, C. Estimation of calcium-binding constants of casein phosphopeptides by capillary zone electrophoresis. *Analytica Chimica Acta*, 1998 372, 291-7.

[69] Cosentino, S; Gravaghi, C; Donetti, E; Donida BM, Lombardi, G; Bedoni, M et al. Caseinphosphopeptide-induced calcium uptake in human intestinal cell lines HT-29 and Caco2 is correlated to cellular differentiation. *J Nutr Biochem*, 2010 21, 247-54.
[70] MacDonald, RS; Thornton, WH Jr.; Marshall, RT. A cell culture model to identify biologically active peptides generated by bacterial hydrolysis of casein. *J Dairy Sci*, 1994 77, 1167-75.
[71] Ganjam, LS; Thornton, WH Jr.; Marshall, RT; MacDonald, RS. Antiproliferative effects of yogurt fractions obtained by membrane dialysis on cultured mammalian intestinal cells. *J Dairy Sci*, 1997 80, 2325-9.
[72] Kawahara, T; Katayama, D; Otani, H. Effect of beta-casein (1-28) on proliferative responses and secretory functions of human immunocompetent cell lines. *Biosci Biotechnol Biochem*, 2004 68, 2091-5.
[73] Kitts, DD; Nakamura, S. Calcium-enriched casein phosphopeptide stimulates release of IL-6 cytokine in human epithelial intestinal cell line. *J Dairy Res*, 2006 73, 44-8.
[74] Vitkus, SJ, Hanifin, SA; McGee, DW. Factors affecting Caco-2 intestinal epithelial cell interleukin-6 secretion. *In Vitro Cell Dev Biol Anim*, 1998 34, 660-4.
[75] Hosoi, T; Hirose, R; Saegusa, S; Ametani, A; Kiuchi, K; Kaminogawa, S. Cytokine responses of human intestinal epithelial-like Caco-2 cells to the nonpathogenic bacterium Bacillus subtilis (natto). *Int J Food Microbiol*, 2003 82, 255-64.
[76] Weinstein, DL; O'Neill, BL; Metcalf, ES. Salmonella typhi stimulation of human intestinal epithelial cells induces secretion of epithelial cell-derived interleukin-6. *Infect Immun*, 1997 65, 395-404.
[77] Donida, BM; Mrak, E; Gravaghi, C; Villa, I; Cosentino, S; Zacchi, E et al. Casein phosphopeptides promote calcium uptake and modulate the differentiation pathway in human primary osteoblast-like cells. *Peptides*, 2009 30, 2233-41.
[78] Tulipano, G; Bulgari, O, Chessa, S; Nardone, A; Cocchi, D; Caroli, A. Direct effects of casein phosphopeptides on growth and differentiation of in vitro cultured osteoblastic cells (MC3T3-E1). *Regul Pept*, 2010 160, 168-74.
[79] Andreasen, JO; Borum, MK; Jacobsen, HL; Andreasen, FM. Replantation of 400 avulsed permanent incisors. 4. Factors related to periodontal ligament healing. *Endod Dent Traumatol*, 1995 11, 76-89.
[80] Donaldson, M; Kinirons, MJ. Factors affecting the time of onset of resorption in avulsed and replanted incisor teeth in children. *Dent Traumatol*, 2001 17, 205-9.
[81] Trope, M. Clinical management of the avulsed tooth: present strategies and future directions. *Dent Traumatol*, 2002 18, 1-11.
[82] Blomlof, L. Milk and saliva as possible storage media for traumatically exarticulated teeth prior to replantation. *Swed Dent J Suppl*, 1981 8, 1-26.
[83] Patel, S; Dumsha, TC; Sydiskis, RJ. Determining periodontal ligament (PDL) cell vitality from exarticulated teeth stored in saline or milk using fluorescein diacetate. *Int Endod J*, 1994 27, 1-5.
[84] Rose, RK. Binding characteristics of Streptococcus mutans for calcium and casein phosphopeptide. *Caries Res*, 2000 34, 427-31.
[85] Cehreli, SB, Gurpinar, AO; Onur, AM; Dagli, FT. In vitro evaluation of casein phosphopeptide-amorphous calcium phosphate as a potential tooth transport medium: viability and apoptosis in L929 fibroblasts. *Dent Traumatol*, 2008 24, 314-9.

[86] Schmalz, G. Concepts in biocompatibility testing of dental restorative materials. *Clin Oral Investig*, 1997 1, 154-62.

[87] Berridge, MJ; Bootman, MD; Lipp, P. Calcium--a life and death signal. *Nature*, 1998 395, 645-8.

[88] Goldman, AS; Ham Pong, AJ; Goldblum, RM. Host defenses: development and maternal contributions. *Adv Pediatr*, 1985 32, 71-100.

[89] Hanson, LA; Ahlstedt, S; Andersson, B; Carlsson, B, Fallstrom, SP; Mellander, L et al. Protective factors in milk and the development of the immune system. *Pediatrics*, 1985 75, 172-6.

[90] Juto, P. Human milk stimulates B cell function. *Arch Dis Child*, 1985 60, 610-3.

[91] Lonnerdal, B. Biochemistry and physiological function of human milk proteins. *Am J Clin Nutr*, 1985 42, 1299-317.

[92] Shinmoto, H; Kawakami, H; Dosako, S; Sogo, Y. IgA specific helper factor (alpha HF) in human colostrum. *Clin Exp Immunol*, 1986 66, 223-30.

[93] Julius, MH, Janusz, M; Lisowski, J. A colostral protein that induces the growth and differentiation of resting B lymphocytes. *J Immunol*, 1988 140, 1366-71.

[94] Mincheva-Nilsson, L; Hammarstrom, ML; Juto, P; Hammarstrom, S. Human milk contains proteins that stimulate and suppress T lymphocyte proliferation. *Clin Exp Immunol*, 1990 79, 463-9.

[95] Otani, H; Hata I. Inhibition of proliferative responses of mouse spleen lymphocytes and rabbit Peyer's patch cells by bovine milk caseins and their digests. *J Dairy Res*, 1995 62, 339-48.

[96] Nakamura, A; Nagai, K; Ando, K; Tamura, G. Selective suppression by prodigiosin of the mitogenic response of murine splenocytes. *J Antibiot*, 1986 39, 1155-9.

[97] Toyoshima, S, Iwata, M; Osawa, T. Kinetics of lymphocyte stimulation by concanavalin A. *Nature*, 1976 264, 447-9.

[98] Hata, I; Ueda, J; Otani, H. Immunostimulatory action of a commercially available casein phosphopeptide preparation, CPP-III, in cell cultures. *Milchwissenschaft*, 1999 54, 3-7.

[99] Otani, H; Kihara, Y, Park, M. The immunoenhancing property of a dietary casein phosphopeptide preparation in mice. *Food Agric Immunol*, 2000 12, 165-73.

[100] Otani, H, Nakano, K; Kawahara, T. Stimulatory effect of a dietary casein phosphopeptide preparation on the mucosal IgA response of mice to orally ingested lipopolysaccharide from Salmonella typhimurium. *Biosci Biotechnol Biochem*, 2003 67, 729-35.

[101] Otani, H; Watanabe, T. Enhancement of immunoglobulin A formation by a casein phosphopeptide, b-casein (1-28) via modulation of cytokines produced by Th2 type of lymphocytes in mouse spleen cell cultures. *Milchwissenschaft*, 2006 61, 123-6.

[102] Otani, H; Watanabe, T; Tashiro, Y. Effects of bovine beta-casein (1-28) and its chemically synthesized partial fragments on proliferative responses and immunoglobulin production in mouse spleen cell cultures. *Biosci Biotechnol Biochem*, 2001 65, 2489-95.

[103] Tobita, K; Kawahara, T; Otani, H. Bovine beta-casein (1-28), a casein phosphopeptide, enhances proliferation and IL-6 expression of mouse CD19+ cells via Toll-like receptor 4. *J Agric Food Chem*, 2006 54, 8013-7.

[104] Schwandner, R; Dziarski, R; Wesche, H; Rothe, M; Kirschning, CJ. Peptidoglycan- and lipoteichoic acid-induced cell activation is mediated by toll-like receptor 2. *J Biol Chem*, 1999 274, 17406-9.

[105] Chow, JC, Young, DW; Golenbock, DT; Christ, WJ; Gusovsky, F. Toll-like receptor-4 mediates lipopolysaccharide-induced signal transduction. *J Biol Chem*, 1999 274, 10689-92.

[106] Poltorak, A; He, X; Smirnova, I; Liu, MY; Van Huffel, C; Du, X et al. Defective LPS signaling in C3H/HeJ and C57BL/10ScCr mice: mutations in Tlr4 gene. *Science*, 1998 282, 2085-8.

[107] Monodane, T; Kawabata, Y; Yang, S; Hase ,S; Takada, H. Induction of necrosis factor-alpha and interleukin-6 in mice in vivo and in murine peritoneal macrophages and human whole blood cells in vitro by Micrococcus luteus teichuronic acids. *J Med Microbiol*, 2001 50, 4-12.

[108] Schippers, EF, van 't Veer, C, van Voorden, S; Martina, CA; Huizinga, TW; le Cessie, S et al. IL-10 and toll-like receptor-4 polymorphisms and the in vivo and ex vivo response to endotoxin. *Cytokine*, 2005 29, 215-28.

[109] Kitazawa, H; Harata, T; Uemura, J; Saito, T; Kaneko, T; Itoh, T. Phosphate group requirement for mitogenic activation of lymphocytes by an extracellular phosphopolysaccharide from Lactobacillus delbrueckii ssp. bulgaricus. *Int J Food Microbiol*, 1998 40, 169-75.

[110] Hata, I, Higashiyama, S; Otani, H. Identification of a phosphopeptide in bovine alpha s1-casein digest as a factor influencing proliferation and immunoglobulin production in lymphocyte cultures. *J Dairy Res*, 1998 65, 569-78.

[111] Kawahara, T; Otani, H. Stimulatory effects of casein phosphopeptide (CPP-III) on mRNA expression of cytokines in Caco-2 cells. *Biosci Biotechnol Biochem*, 2004 68, 1779-81.

[112] Siro, I; Kapolna, E; Kapolna, B; Lugasi, A. Functional food. Product development, marketing and consumer acceptance--a review. *Appetite*, 2008,51, 456-67.

[113] Kanwar, JR, Kanwar, RK; Sun, X; Punj, V; Matta, H; Morley, SM et al. Molecular and biotechnological advances in milk proteins in relation to human health. *Curr Protein Pept Sci*, 2009 10, 308-38.

[114] Phelan, M; Aherne, A; FitzGerald, RJ; O'Brien, NM. Casein-derived bioactive peptides: Biological effects, industrial uses, safety aspects and regulatory status. *International Dairy Journal*, 2009 19, 643-54.

In: Casein: Production, Uses and Health Effects
Editors: A. M. Ventimiglia, et. al, pp. 101-121
ISBN: 978-1-62100-129-4

Chapter 5

CASEINS IN PORTUGUESE TRADITIONAL CHEESES: FROM SCIENCE TO APPLICATIONS

***Eliana Jerónimo*[1,2] *and F. Xavier Malcata*[1,2,3*]**

[1] Centro de Biotecnologia Agrícola e Agro-Alimentar do Baixo Alentejo e Litoral (CEBAL) / Instituto Politécnico de Beja (IPBeja), 7801-908 Beja, Portugal.
[2] Centre for Research in Ceramics & Composite Materials (CICECO), University of Aveiro, 3810-193 Aveiro, Portugal.
[3] ISMAI – Instituto Superior da Maia, Avenida Carlos Oliveira Campos, 4475-690 Avioso S. Pedro, Portugal.

ABSTRACT

Portugal holds a long tradition in cheesemaking, and several unique cheese specialities are indeed available; most such cheeses possess a Protected Denomination of Origin (PDO) status. Each artisanal cheese is intrinsically related to its region of origin, and the corresponding manufacture protocol entails a multicultural history and a complex set of empirical practices passed from generation to generation which confer unique features to each cheese variety, arising namely from distinct casein degradation patterns. The distinct patterns of casein breakdown are chiefly associated with the specific proteolytic activity of proteases present in the plant (*Cynara cardunculus*) or animal rennet used to bring about milk clotting. However, several other factors also affect the extent and quality of casein hydrolysis in said cheeses, e.g. ripening time, lactation season, proportion of sheep and caprine milk, and mode and amount of salting. Hence, cheesemaking conditions play a critical role upon the profile of casein-derived peptides generated during ripening; they determine not only the final sensory characteristics of Portuguese traditional cheeses, but also their nutritional value. Meanwhile, casein-derived bioactive peptides in cheeses manufactured following traditional procedures have been under extensive scrutiny; peptides possessing angiotensin-converting enzyme-inibitory and antioxidant activities were accordingly found in raw ovine cheese-like systems coagulated with *C. cardunculus*. Finally, adulteration of traditional cheeses bearing PDO labels is a current problem, so methods to monitor the authenticity of said cheeses based

* Corresponding author

on assessment of their casein profiles have been developed – and shown to be accurate and useful in detecting fakes. This chapter reviews the fate of caseins in Portuguese traditional cheeses, from a fundamental point of view – coupled with their being potential precursors of bioactive peptides, or else the applicability of the resulting peptides as markers of authenticity in quality assurance strategies.

1. INTRODUCTION

Caseins – the main proteinaceous components of milk, play a crucial role in cheesemaking. Casein molecules are hydrolysed throughout ripening, so the pattern and extent of their degradation influence strongly the sensory and nutritional features of the final cheeses (Sieber *et al.*, 2010; Sousa *et al.*, 2001). On the other hand, the various biochemical phenomena throughout cheesemaking are determined to a large extent by the chemical composition of the original milk – in particular, the concentrations of casein, fat and calcium, which depend directly on the milk source. However, beyond the factors known to affect the milk composition in caseins – viz. species, breed or stage of lactation (Park *et al.*, 2007), cheesemaking conditions also affect casein hydrolysis, and consequently the characteristics perceived in the final cheese (Sousa *et al.*, 2001).

Several typical cheeses are manufactured in Portugal – which are still produced mainly on a farmhouse-scale, and following traditional practices that have been empirically optimised from generation to generation (Costa *et al.*, 2008). The originality and uniqueness of most of them are protected by European legislation on Protected Denomination of Origin (PDO) status; this includes *Amarelo da Beira Baixa, Azeitão, Cabra Transmontano, Castelo Branco, Évora, Nisa, Picante da Beira Baixa, Pico, Rabaçal, São Jorge, Serpa, Serra da Estrela* and *Terrincho* cheeses (European Commission, 2011). All said cheeses are intrinsically related with their regions of origin ("terroir"), and thus receive the name of such regions; these are depicted in Figure 1 (with further details in Table 1). Each of these cheeses resorts to traditional resources and techniques prevailing in their region of origin, and the distinct artisanal mode of manufacture confers unique features to each cheese variety.

Portuguese traditional cheeses have attracted the attention of researchers in recent decades – covering from technological optimization, to microbial, biochemical and sensory characterization (e.g. Freitas and Malcata, 2000; Reis and Malcata, 2007). A major emphasis has been put on casein degradation, as well as on factors that affect it and their consequences upon the final organoleptic and nutritional features of said cheeses. The effects of the particular modes of manufacture of Portuguese traditional cheeses upon the qualitative and quantitative patterns of casein degradation, coupled with generation of casein-derived peptides possessing bioactivities will be presented and critically discussed along this chapter.

On the other hand, adulteration of traditional cheeses bearing PDO labels is a current problem, triggered by their premium price in the market when compared with that of regular cheeses. Hence, several methodologies have been designed and tuned to monitor the authenticity of Portuguese PDO cheeses, based chiefly on assessment of their casein profile. These methodologies will also be reviewed here, and discussed to some length.

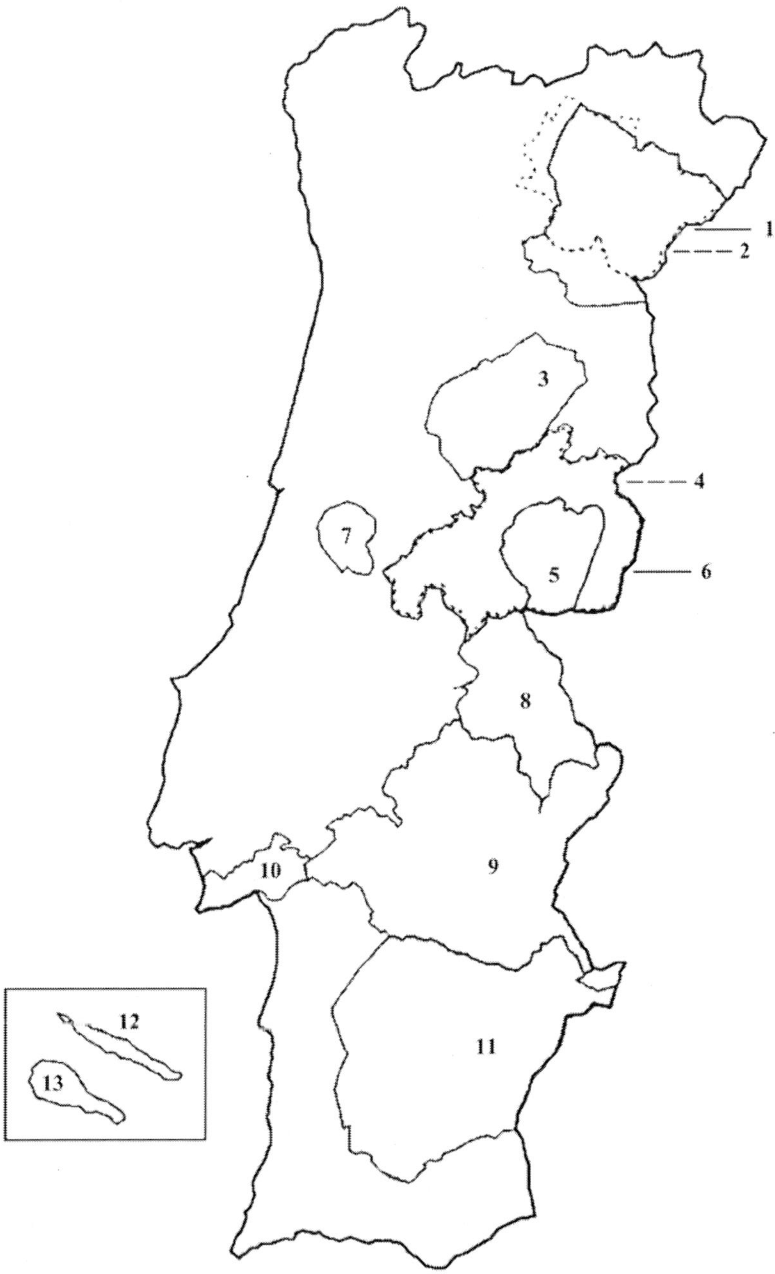

Figure 1. Geographical region of origin of cheeses with Protected Denomination of Origin status in Portugal. 1 – *Terrincho* cheese; 2 – *Cabra Transmontano* cheese; 3 – *Serra da Estrela* cheese; 4 – *Picante da Beira Baixa* cheese; 5 – *Castelo Branco* cheese; 6 – *Amarelo da Beira Baixa* cheese; 7 – *Rabaçal* cheese; 8 – *Nisa* cheese; 9 – *Évora* cheese; 10 – *Azeitão* cheese; 11 – *Serpa* cheese; 12 – *São Jorge* cheese; and 13 – *Pico* cheese. (Adapted from Freitas and Malcata, 2000).

Table 1. Legal requirements of cheeses with Protected Denomination of Origin status in Portugal[1]

Cheese variety	Region of origin	Milk source	Breed	Rennet type	Minimum ripening time (days)	Cheese type
Amarelo da Beira Baixa	Castelo Branco	Ewe & goat	-[2]	Animal	40	Semi-hard
Azeitão	Palmela, Sesimbra & Setúbal	Ewe	-[2]	Plant[3]	20	Semi-hard
Cabra Transmontano	Bragança & Vila Real	Goat	Serrana	Animal	60	Extra-hard
Castelo Branco	Castelo Branco	Ewe	-[2]	Plant[3]	40	Semi-hard
Évora	Évora	Ewe	-[2]	Plant[3]	30	Semi-hard or hard
Nisa	Nisa & Portalegre	Ewe	-[2]	Plant[3]	45	Semi-hard or hard
Picante da Beira Baixa	Castelo Branco	Ewe & goat	-[2]	Animal	120	Semi-hard or hard
Pico	Pico	Cow	-[2]	Animal	20	Soft
Rabaçal	Coimbra	Ewe & goat	-[2]	Animal	20	Semi-hard or hard
São Jorge	São Jorge	Cow	-[2]	Animal	90	Semi-hard or hard
Serpa	Serpa & Beja	Ewe	-[2]	Plant[3]	30	Soft
Serra da Estrela	Serra da Estrela	Ewe	Bordaleira da Serra da Estrela & Churra Mondegueira	Plant[3]	30	Soft
Terrincho	Bragança, Viseu & Guarda	Ewe	Churra da Terra Quente	Animal	30	Semi-hard

[1] – Based on manufacture specifications (available at European Commission website, 2011);
[2] – Milk from herds of ewes, goats or cows in the geographic region established for each cheese variety;
[3] – *Cynara cardunculus*.

2. Portuguese Traditional Cheeses and Cheesemaking

Portuguese traditional cheeses possess a high intrinsic value – arising not only from their unique sensory features, but also from their economic and social impacts in their rural regions of production, where they contribute actively to maintain the local employment and thus retain their inhabitants (Macedo and Malcata, 1994). Moreover, many of those cheeses are produced with milk from indigenous breeds – which are normally less productive than exotic breeds, yet are particularly well adapted to the local environmental conditions (e.g. geoclimatic characteristics and food availability); hence, they allow preservation of the genetic heritage of the country, especially that associated with small ruminant breeds. In Portugal, traditional cheeses are still manufactured mainly on a farmhouse-scale – and following traditional practices that are different from region to region, although an increasing fraction has meanwhile been taken up by small industrial dairies.

Traditional cheeses are produced with whole raw milk from ewes, goats or cows, or with a mixture of ewe's and goat's milk to various proportions, and coagulated with either plant or animal rennet. Hence, based on milk source and rennet type used, Portuguese PDO cheeses may be divided into 5 distinctive groups (see Table 1): (i) *Azeitão, Castelo Branco, Évora, Nisa, Serpa* and *Serra da Estrela* cheeses, which are manufactured from raw ovine milk and coagulated with plant rennet; (ii) *Terrincho* cheese, which is manufactured from raw ovine milk and coagulated with animal rennet; (iii) *Cabra transmontano* cheese, which is manufactured from raw caprine milk and coagulated with animal rennet; (iv) *Amarelo da Beira Baixa, Picante da Beira Baixa* and *Rabaçal* cheeses, which are manufactured from mixtures of raw ovine and caprine milks and coagulated with animal rennet; and (v) *São Jorge* and *Pico* cheeses, which are manufactured from raw bovine milk and coagulated with animal rennet (Freitas and Malcata, 2000).

It is noteworthy that most ovine cheeses (except *Terrincho* cheese) are coagulated with plant rennet; this is normally extracted from dried flowers of the wild thistle – *Cynara cardunculus*. Such a plant rennet has been classically obtained by soaking a handful of stylets and stigmae of its flowers in a bowl of tap water for some time, or mixing with a mortar and pestle in the case of pastes, followed by filtering through a piece of cotton cloth (Sousa and Malcata, 2002). The other cheeses are coagulated with animal rennet – which is normally extracted from the abomassa of recently slaughtered milk-fed calves, and is also known as calf rennet (Sousa *et al.*, 2001). On the other hand, all Portuguese traditional cheeses are manufactured without deliberate addition of a starter or nonstarter culture (Freitas and Malcata, 2000) – with the exception of *São Jorge* cheese, for which the whey from a previous batch is used as starter (Kongo *et al.*, 2009).

Although the manufacture protocols for the several varieties of Portuguese traditional cheeses differ in detail, they share a few essential steps: coagulation, dehydratation (via cutting the coagulum, working the curd and draining the whey), shaping (via moulding and pressing by hand), salting (via addition to milk, or rubbing on the curd surface) and ripening (Reis and Malcata, 2007). This final step takes place with or without control of room temperature and humidity, and for a period that is rather variable among cheese varieties – ranging from 20 days in *Azeitão, Pico* and *Rabaçal* cheeses, up to 120 days in *Picante da Beira Baixa* cheese (see Table 1). All manufacture steps are performed manually, and controlled by the cheesemaker based on her own intuition and experience (Reis and Malcata,

2007). Hence, the artisanal mode of manufacture confers unique features to each cheese variety, associated namely with distinct casein degradation patterns that are difficult to mimic elsewhere.

3. Casein Breakdown in Portuguese Traditional Cheeses

The typical texture of cheeses is directly determined by the proteolytic phenomena occurring during ripening – which also affect the flavour in a more indirect fashion (Upadhyay *et al.*, 2004). Proteolysis begins with coagulation: the peptide bond Phe_{105}-Met_{106} of κ-casein is specifically cleaved by rennet enzymes, followed by cooperative, physical aggregation of β- and α_S-caseins. During primary proteolysis, the casein molecules are hydrolysed into large- and medium-sized peptides by the coagulant and other indigenous milk enzymes. These peptides may be further degraded by exocellular and endocellular enzymes from lysed microorganisms – adventitious or deliberately added to milk, thus producing smaller peptides and eventually free amino acids; this process is termed secondary proteolysis. Said end-products of casein hydrolysis contribute indirectly to flavour as precursors for the synthesis of flavour compounds via secondary catabolic processes (McSweeney and Sousa, 2000). Moreover, the beakdown of the protein network contributes to noticeable textural changes of the cheese matrix (Sousa *et al.*, 2001). Hence, alteration in the patterns of proteolysis – e.g. extension or depth of degradation, affect the development of sensory features in the final cheese. These proteolytic phenomena are heavily influenced by the actual cheesemaking conditions (Sousa *et al.*, 2001; Upadhyay *et al.*, 2004); as expected, distinct proteolytic patterns are found among the various Portuguese traditional cheeses.

Proteolysis has been seen as the most important biochemical event during ripening – and rennet enzymes play a crucial role therein, especially at the first stages. As emphasized above, all Portuguese traditional cheeses are coagulated via the action of aspartic acid proteinases extracted from the abomasa of calves or, more often, from the flowers of *C. cardunculus*. Calf rennet contains two major proteolytic enzymes: chymosin, and pepsin to a lesser extent (Jacob *et al.*, 2011). Chymosin exhibits a high specificity toward hydrolysis of Phe_{105}-Met_{106} in κ-casein, where it triggers destabilisation of the casein micelles with consequent milk coagulation (Jacob *et al.*, 2011); the proteolytic activity of pepsin is much less specific (Jacob *et al.*, 2011; Visser, 1993). Extracts from *C. cardunculus* also possess two major enzymes: cardosin A and cardosin B, which are similar, in terms of specificity and activity, to chymosin and pepsin, respectively (Veríssimo *et al.*, 1995). In Portugal, ewe's milk cheeses are traditionally manufactured with *C. cardunculus*, whereas essentially all other traditional cheeses containing goat's or cow's milks are coagulated with calf rennet; *Terrincho* is the only traditional variety of ovine cheese manufactured exclusively with calf rennet.

Caseins in the aforementioned cheeses do undergo unique degradation patterns, chiefly associated with the rennet type employed. Sousa and Malcata (1997; 1998a) reported that plant rennet has a significantly less intense proteolytic activity upon β- and α_S-caseins than its animal counterpart, in ovine milk cheeses; recorded degradation levels of β- and α_S-caseins were 33 and 47%, respectively, in cheeses coagulated with *C. cardunculus* by 68 days of ripening, while in cheeses coagulated with calf rennet their degradation did not exceed 50 and 88%, respectively. Conversely, Roseiro *et al.* (2003b) observed higher extent of casein degradation in ewe's milk cheeses coagulated with *C. cardunculus* than with animal rennet; intact β- and α_{S1}-caseins were almost not detected by 60 days of ripening in cheeses manufactured with the former, while they were poorly degraded in cheeses manufactured with animal rennet. Similar results were claimed for *Picante da Beira Baixa* cheese (Freitas and Malcata, 1996).

The proteolytic activity of *C. cardunculus* and animal rennet in cheeses manufactured with ewe's milk following traditional procedures was comparatively studied by Sousa and Malcata (1997; 1998a). These authors concluded that proteolysis by *C. cardunculus* proceeds to a low degree, i.e. its proteinases cleave caseins into high-molecular weight peptides that are not subsequently broken down to low-molecular weight peptides – whereas proteolysis effected by animal rennet proceeds to a higher degree. Moreover, they showed that the peptide profile is considerably different (both qualitatively and quantitatively) – thus reflecting the different specificities of the rennet enzymes, because the manufacture conditions and the native microflora were similar for both cheeses. The reverse-phase chromatographic peptide profile of the water-soluble extract obtained from cheeses manufactured with *C. cardunculus* exhibited a higher number of peaks than that from cheeses manufactured with animal rennet – thus suggesting a broader specificity of the proteinases from *C. cardunculus* than of those from animal rennet (Sousa and Malcata, 1998a).

Cardosins are indeed able to cleave more peptide bonds than chymosin (Macedo *et al.*, 1996; Sousa and Malcata, 1998b); the major cleavage sites by the former in ovine β-casein are Leu_{127}-Thr_{128} and Leu_{190}-Tyr_{191}, and in α_{S1}-casein is Phe_{23}-Val_{24}, followed by Trp_{156}-Tyr_{157} and Tyr_{165}-Thr_{166} – whereas in α_{S2}-casein it is Phe_{88}-Tyr_{89} (Sousa and Malcata, 1998b). Regarding chymosin, the major cleavage sites in ovine β-casein are Leu_{190}-Tyr_{191} and Ala_{187}-Phe_{188}, and Phe_{23}-Val_{24} in α_{S1}-casein (Whyte, 1995). Macedo *et al.* (1996) claimed that proteinases from *C. cardunculus* display a stronger preference for peptide bonds between bulky hydrophobic residues than does chymosin; in bovine caseins, proteinases from *C. cardunculus* cleave all bonds in the hydrophobic regions of β-casein (f189-193) and α_{S1}-casein (f163-167), whereas chymosin cleaves only Ala_{189}-Phe_{190} and Leu_{192}-Tyr_{193} in that region of β-casein, and Trp_{164}-Tyr_{165} in that region of α_{S1}-casein. Based on the most susceptible cleavage sites of ovine caseins by proteinases from either *C. cardunculus* or animal rennet, the primary products of degradation of ovine β- and α_S-caseins are β-(f1-190) and α_{S1}-(f24-191), respectively (Sousa and Malcata, 1998a). The complementary peptides α_{S1}-(f1-23) and α_{S1}-(f24-191) were detected with either rennet (Sousa and Malcata, 1998a); however, in cheeses manufactured with *C. cardunculus*, the bond Phe_{23}-Val_{24} was hydrolysed earlier (i.e. by 7 days of ripening) than with animal rennet (only by 28 days) – which may explain the smoother texture of the former ovine cheeses coagulated with *C. cardunculus* than when animal rennet is used. The peptide fragments identified in the water-soluble extracts obtained from ovine cheeses, manufactured with *C. cardunculus* or animal rennet, by 68 days of ripening are shown in Table 2. Peptides α_{S1}-(f24-165), β-(f128-*), β-(f166-*) and β-(f191-*) are released only by *C. cardunculus*, whereas peptides α_{S1}-(f120-191), β-(f164-*) and β-(f191-*) are produced only by animal rennet (Sousa and Malcata, 1998a). These results suggest that proteolysis in traditional cheeses is associated mainly with the action of rennet enzymes, rather than with the prevailing microflora or milk adventitious enzymes; and some peptides identified appear to be specific for each rennet type.

Table 2. Peptide fragments in water-soluble extracts, by 68 days of ripening, of traditional ovine cheeses coagulated with two alternative rennets. (Adapted from Sousa and Malcata, 1998a)

Cynara cardunculus	Animal rennet
α_{S1}-(f1-*)	α_{S1}-(f1-*)
α_{S1}-(f24-*)	α_{S1}-(f25-*)
α_{S1}-(f166-*)	α_{S1}-(f120-*)
α_{S2}-(f74-*)	α_{S2}-(f14-*)
α_{S2}-(86-*)	α_{S2}-(f17-*)
α_{S2}-(f194-*)	α_{S2}-(f24-*)
	α_{S2}-(f26-*)
	α_{S2}-(f89-*)
	α_{S2}-(f131-*)
β-(f1-*)	β-(f1-*)
β-(f7-*)	β-(f51-*)
β-(f29-*)	β-(f69-*)
β-(f45-*)	β-(f164-*)
β-(f83-*)	β-(f191-*)
β-(f94-*)	
β-(f128-*)	
β-(f129-*)	
β-(f141-*)	
β-(f143-*)	
β-(f166-*)	
β-(f177-*)	
β-(f183-*)	
β-(f191-*)	
β-(f197-*)	
β-(f201-*)	

The processing conditions of the *C. cardunculus* feedstock – such as time of grinding, pH of aqueous buffer, time of homogenization of the aqueous suspension and time of storage thereafter, have proven important upon the proteolytic activity of its aqueous extracts (Sousa and Malcata, 1996; Tavaria *et al.*, 2001). Sousa and Malcata (1996) found that increases in pH and time of grinding of stigmae and stylets affect positively the proteolytic activity of *C. cardunculus* extract, whereas time of homogenization affects it otherwise. The salt content of the aqueous buffer had a negligible effect upon the proteolytic activity exhibited. The maximum caseinolytic activity of extracts of *C. cardunculus* was obtained via grinding the stylets for ca. 36 s, using an extraction buffer at pH 5.9 without salt, and homogenizing the suspension for 15 min (Sousa and Malcata, 1996). The effect of storage of fresh *C. cardunculus* extract at 4 °C upon proteolytic breakdown of ovine Na-caseinates was meanwhile studied by Tavaria *et al.* (2001) – who showed that degradation of ovine β- and α_S-caseins by fresh *C. cardunculus* extract increased until 1 week of storage, but decreased by 6.2 and 13.7% between 1 and 2 weeks, respectively.

In addition to type of rennet, several other factors can affect the extent and pattern of casein degradation in Portuguese traditional cheeses: ripening time and position within the cheesemaking season (Macedo and Malcata, 1997), proportion of ewe's and goat's milks (Freitas *et al.*, 1997), and mode and amount of salting (Freitas and Malcata, 1996). Macedo and Malcata (1997) showed that hydrolysis of caseins in *Serra da Estrela* cheese is strongly affected by ripening time: two variants of α_S-casein (i.e. α_{S2}- and α_{S3}-caseins) were degraded only after 7 days of ripening, while the two variants of β-caseins (i.e. -β_1- and β_2-caseins) were degraded equivalently at all stages of ripening (Figure 2). Although α_S-caseins tend to be degraded more slowly than β-caseins during the initial stages of ripening, the degree of degradation of α_S-caseins was higher than that of β-caseins by the end of ripening time (35 days) – 82 and 76%, respectively. Urea-polyacrylamide gel electrophoretograms of samples of *Serra da Estrela* cheese taken throughout ripening unfolded the presence of several primary products of casein breakdown; the compounds with higher electrophoretic mobility than α_S-caseins were associated to degradation of α_S-caseins, while products with mobility between β-caseins and α_S-caseins seemed to result from hydrolysis of β-caseins – with no bands accounting for by γ-caseins showing up in this region. Note that γ-casein is originated mainly from hydrolysis of β-caseins effected by plasmin, a milk indigenous enzyme that is activated by thermal treatment (Upadhyay *et al.*, 2004); hence, low plasmin activity is expected in *Serra da Estrela* cheese, since its manufacture does not entail any form of heat processing or cooking (Macedo *et al.*, 1993).

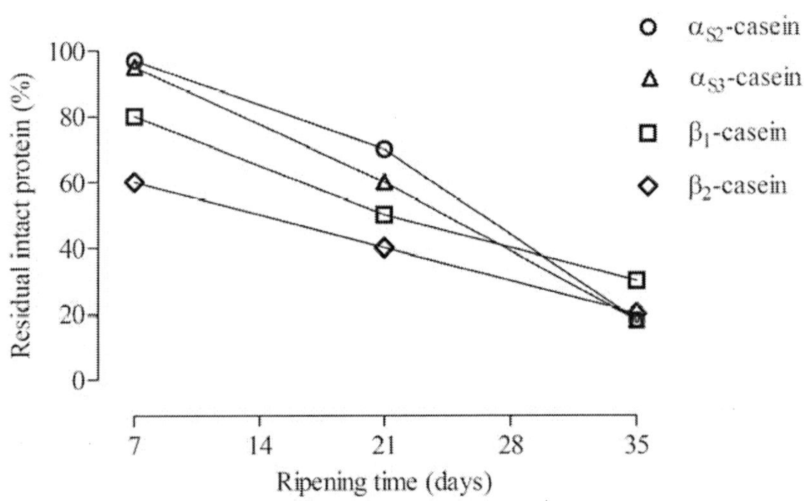

Figure 2. Extent of degradation of α_S- and β-caseins throughout ripening in *Serra da Estrela* cheese. (Adapted from Macedo and Malcata, 1997).

Moreover, the period within the cheesemaking season affects significantly the hydrolysis of β_2- and α_{S3}-casein in *Serra da Estrela* cheese – but not the degradation of β_1- and α_{S2}-casein (Macedo and Malcata, 1997). The level of hydrolysis of α_{S3}-casein was indeed lower in cheeses manufactured in February (75% protein remained intact) than in November or May (52% protein still intact), by 21 days of ripening; and hydrolysis of β_2-casein by 7 or 21 days of ripening was lower in May (69 and 49% protein still intact, respectively) than in November or February (57 and 30% of protein remaining, respectively). This distinct rate of casein hydrolysis between cheesemaking seasons is apparently due to variations in chemical composition of cheeses, viz. in terms of moisture and ash; but also to the weather prevailing outdoors, as temperature changes widely throughout the 3 months considered (Macedo and Malcata, 1997).

Casein degradation in *Picante da Beira Baixa* cheese manufactured with different combinations of ewe's and goat's milks (viz. 0, 25, 50, 75 and 100 %(v/v) caprine milk) was ascertained by Freitas *et al.* (1997) during 180 days of ripening. Likewise to *Serra da Estrela* cheese, β-casein exhibited a higher resistance to enzymatic hydrolysis than α_S-casein in *Picante da Beira Baixa* cheese. As expected, the extents of α_S- and β-casein hydrolysis varied with the composition of the milk used – and increased with incorporation of goat's milk (Figure 3). The extents of degradation of α_S-caseins, by 180 days of ripening, were 36 and 69% for ovine and caprine cheeses, respectively – while the extents of degradation of β-caseins were 19 and 37%, respectively. These results were ascribed to the distinct structure of the casein micelles in goat's and ewe's milks – which probably implicate dissimilar degradation mechanisms (Freitas and Malcata, 1999). High molecular weight peptides – with electrophoretic mobilities similar to that of γ-casein, were detected in all cheeses, and up to four bands were observed. γ-Caseins increased in content during ripening in all cheeses, with caprine milk leading to higher levels of γ-caseins – in agreement with the larger extent of degradation of β-caseins in caprine milk.

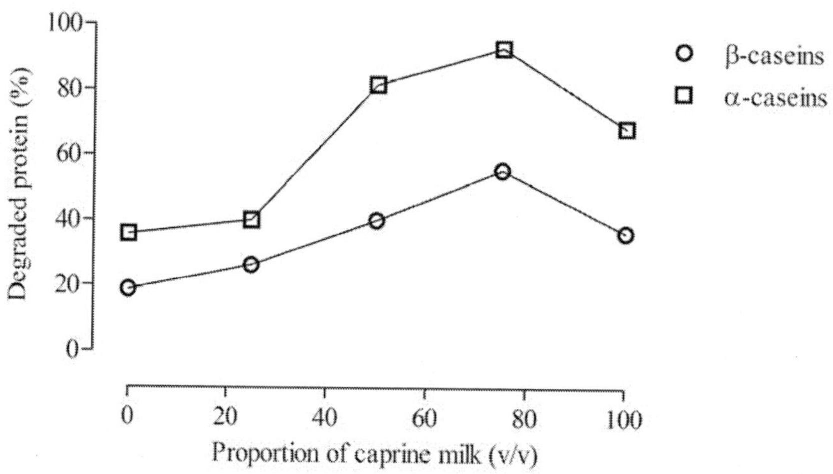

Figure 3. Extent of degradation of α_S- and β-caseins, by 180 days of ripening, in *Picante da Beira Baixa* cheeses manufactured with several combinations of ewe's and goat´s milks. (Adapted from Freitas *et al.*, 1997).

The degradation of caseins in *Picante da Beira Baixa* cheese is also affected by the mode and amount of salting (Freitas and Malcata, 1996): higher levels of degradation of β- and α_S-caseins were observed in cheeses salted only once than twice. In the latter, the degradation of both caseins was more gradual up to 120 days of ripening – which is in agreement with the inhibitory effect of high levels of NaCl upon the proteolytic activity of chymosin (Guinee and Fox, 2004).

In *Terrincho* cheese, β-caseins also showed higher resistance to degradation than α_S-caseins (Ferreira *et al.*, 2006; Veloso *et al.*, 2004). Conversely, hydrolysis of α_{S1}- and α_{S2}-casein was significant, mainly at earlier stages of ripening; by 30 days of storage, only 20 and 25% of α_{S1}- and α_{S2}-casein, respectively, remained intact (Ferreira *et al.*, 2006). Roseiro *et al.* (2003a; b) assessed proteolysis in *Serpa* cheese, during 60 days of ripening, by capillary zone electrophoresis; they observed as well a rapid degradation of α_{S1}-casein, which essentially vanished by 30 days of ripening. The biochemical profile of *São Jorge* cheese was studied more recently (Kongo *et al.*, 2009); in this cheese, there is extensive primary proteolysis of α_{S1}- and β-caseins, with said substrate protein being essentially absent by 130 days of ripening. Concomitantly, electrophoretic bands appeared, which were characterized by higher mobility than that of α-casein, and lower than that of β-casein – with the rate of hydrolysis of α_{S1}-casein being slightly greater than that of β-casein.

Although proteolysis at large has been monitored in most Portuguese traditional cheeses, casein degradation specifically was only evaluated in some of those (viz. *Serra da Estrela, Picante da Beira Baixa, Terrincho, Serpa* and *São Jorge* cheeses). Distinct casein degradation patterns were thus observed, but in general α_S-casein was degraded more extensively than β-casein (Sousa and Malcata, 2001). Casein degradation patterns also revealed that the contribution of several proteases to primary proteolysis is different, depending on the cheese type in question: in *Serra da Estrela*, rennet proteases seem to be the primary proteolytic agent, whereas in *Terrincho, Picante da Beira Baixa* and *São Jorge* cheeses, the milk indigenous enzyme – plasmin, also contributes to proteolysis.

4. Casein-Derived Bioactive Peptides in Portuguese Traditional Cheeses

Consumers are becoming more aware of the relationship between diet and health – and this has increased the interest on the nutritional value of foods. Cheese has suffered from a negative health image, mainly due to association of saturated fatty acid, cholesterol and salt content to an increased risk of cardiovascular diseases; this has led to a decline in consumption of this dairy product. However, cheeses are also a good dietary source of healthy nutrients, such as proteins (e.g. bioactive peptides), lipids (e.g. conjugated linoleic acid), vitamins and minerals (Walther *et al.*, 2008).

Casein-derived peptides have been under active scrutiny, owing to their potential health effects. Several *in vitro* and *in vivo* studies have indeed shown that some of such peptides possess interesting biological activities – including antithrombotic, antihypertensive, antimicrobial, immunomodulatory and opioid effects (Silva and Malcata, 2005). Throughout cheese ripening, a great many peptides are formed as a result of casein breakdown, and many of them possess physiological roles (Sieber *et al.*, 2010); hence, ripened cheeses have been proposed as potential functional foods to prevent (or, at least, delay) the incidence of specific diseases.

Unfortunately, not much is known at present on the peptide profile of Portuguese traditional cheeses in terms of bioactivity. Of the few peptides with recognized beneficial physiological effects identified in traditional cheeses, Silva *et al.* (2006) studied the angiotensin-converting enzyme (ACE)-inhibitory activity and antioxidant activity of peptides in water-soluble extracts obtained from raw ovine cheese-like systems coagulated with *C. cardunculus*. By 45 days of ripening, those authors found four peptides able to inhibit *in vitro* the activity of ACE; two were derived from β-casein, viz. Try-Gln-Glu-Pro and Val-Pro-Lys-Val-Lys – and the latter showed higher ACE-inhibitory activity (IC_{50} = 93.8 μg/mL) than the other bioactive peptides found. Among such peptides, one was identified the as N-terminal sequence of α_{S1}-casein – Arg-Pro-Lys, and possessed a high ACE-inhibitory activity (IC_{50} = 36.7 μg/mL). The water-soluble extracts of the ovine cheeses tested showed an ACE-inhibitory activity of ca. 62% – which is within the range reported in the literature for other ripened cheeses (Sieber *et al.*, 2010).

Moreover, some ACE-inhibitory peptides found in ovine cheeses – e.g. Val-Pro-Lys-Val-Lys and Try-Gln-Glu-Pro, also possess antioxidant activity. The radical-scavenging activity of these peptides, expressed as Trolox-equivalent antioxidant capacity (TEAC), ranged from 0.079 m*M* in the former to 0.102 m*M* in the latter; hence, it is comparable to that associated with some carotenoids (Miller *et al.*, 1996) and polyphenols, such as monosubtituted benzoic acid derivatives (Rice-Evans *et al.*, 1996).

Despite the little information on casein-derived bioactivity in Portuguese traditional cheeses, the results conveyed by Silva *et al.* (2006) are particularly promising; traditional ovine cheeses coagulated with *C. cardunculus* appear to be good sources of casein-derived bioactive peptides for eventual incorporation in the human diet. Besides improving the nutritional image of cheeses, an extra added value will consequently be brought to cheeses produced via traditional procedures – with obvious economic and social impacts on their rural regions of production.

5. Caseins as Markers of Authenticity of Portuguese Traditional Cheeses

PDO cheeses are manufactured following predefined procedures, within a unique geographical area of production – encompassing milk species, coagulant type and ripening time. However, the popularity of such cheeses and the increasing demand thereof have prompted a higher market value, which may encourage less ethical manufacturers to adulterate the feedstocks and violate the traditional mode of manufacture, so as to increase the associated profit margin.

Inclusion of bovine milk in (supposedly) play ovine and caprine cheeses, as well as change of ripening time or substitution of plant coagulant by other coagulants, have been identified as fraudulent practices in Portuguese traditional cheesemaking. Hence, analytical ability to detect said adulterations is urged to ensure the authenticity of PDO cheeses – which is particularly important if said cheeses are to remain in the global market. Several methodologies suitable to detect cheese adulterations have been proposed, such as protein-based methods (de la Fuente and Juárez, 2007); although other methods may also ascertain fakes in Portuguese PDO cheeses, e.g. polymerase chain reactions – PCR (Mafra *et al.*, 2004) and enzyme-linked imunosorbent assays – ELISA (Costa *et al.*, 2008), procedures based on assessment of the casein profile have to date received most attention, and are chiefly carried out via chromatography and electrophoresis.

Unfortunately, adulteration of traditional ovine and caprine cheeses with bovine milk is relatively common in the dairy industry (Veloso *et al.*, 2004); the seasonal fluctuations of supply and the lower milk yields of ewes and goats have led to this fraudulent practice (Mafra *et al.*, 2004). Recall that most Portuguese PDO cheeses are manufactured exclusively with ewe's or goat's milk, so Veloso *et al.* (2002) explored reversed-phase liquid chromatography (HPLC-RP) and urea-polyacrylamide gel electrophoresis (urea-PAGE) to detect bovine milk that contaminates goat's or ewe's milks. Those authors claimed that addition of 5 %(v/v) cow's milk to goat's milk produces a rather distinct chromatographic profile – associated chiefly with bovine α-casein. It is also possible to quantitate the amount of cow's milk added to goat's milk, by integrating the bovine α-casein peak area *vis a vis* with a standard curve prepared with samples of known composition. However, HPLC-RP could not detect addition of cow's milk to ewe's milk. Although chromatographic methods are more efficient in quantitative terms, urea-PAGE is qualitatively more sensitive to detect milk adulteration – with sensitivities down to 2 %(v/v) cow's milk in ewe's or goat's milk, based on presence of a band in the region of the bovine α_{s1}-casein (Veloso *et al.*, 2002).

The proteolytic activity during ripening induces changes in the casein profile – with formation of several peptides that make it harder to claim authenticity of the cheese product (Ferreira *et al.*, 2006). In attempts to develop analytical methods suitable for detection of adulteration in ripened *Terrincho* cheese, Veloso *et al.* (2004) evaluated casein proteolysis of plain ovine cheeses, ovine cheeses containing 10% and 20 %(v/v) cow's milk, and plain bovine cheeses – all manufactured according to the traditional *Terrincho* protocol and ripened for 30 days (i.e. the minimum period required), using HPLC-RP and urea-PAGE. Typical chromatographic profiles of the casein fraction are depicted in Figure 4. During ripening, the chromatographic profile of caseins from ovine cheese was more variable than that from its bovine counterpart – especially with regard to the peaks accounted for by α-caseins, due to the intense proteolysis of α-casein and probable formation of α_{s1}-I peptide (Veloso *et al.*, 2004). Later during ripening, reversal of these two peaks was observed – with the α_{s1}-I peptide peak becoming greater than that of α-casein: α-caseins and α_{s1}-I peptide accounted for 63 and 37% of the total α-casein-related peaks by 5 days of ripening, but 28 and 73% by 30 days. A similar α-casein degradation pattern was found in cheeses adulterated with 10 % cow's milk. Unlike ovine cheeses, the chromatographic profile of caseins in bovine cheeses remained fairly constant, with similar relative area of α-casein peaks throughout the whole ripening time tested (α-caseins and α_{s1}-I peptide accounted for ca. 20% and 80% of the total α-casein-related peaks during ripening, respectively). A similar trend was observed in ovine cheeses added with 20% cow's milk, thus allowing detection of the latter in *Terrincho* cheeses via HPLC-RP – provided that at least 20% cow's milk had been added. The electrophoretic results for all cheeses were consistent with the chromatographic ones. Once again, electrophoresis appeared to be more sensitive to cheese adulteration than HPLC-RP, since cow's milk could be detected at 10 %(v/v) (Figure 5).

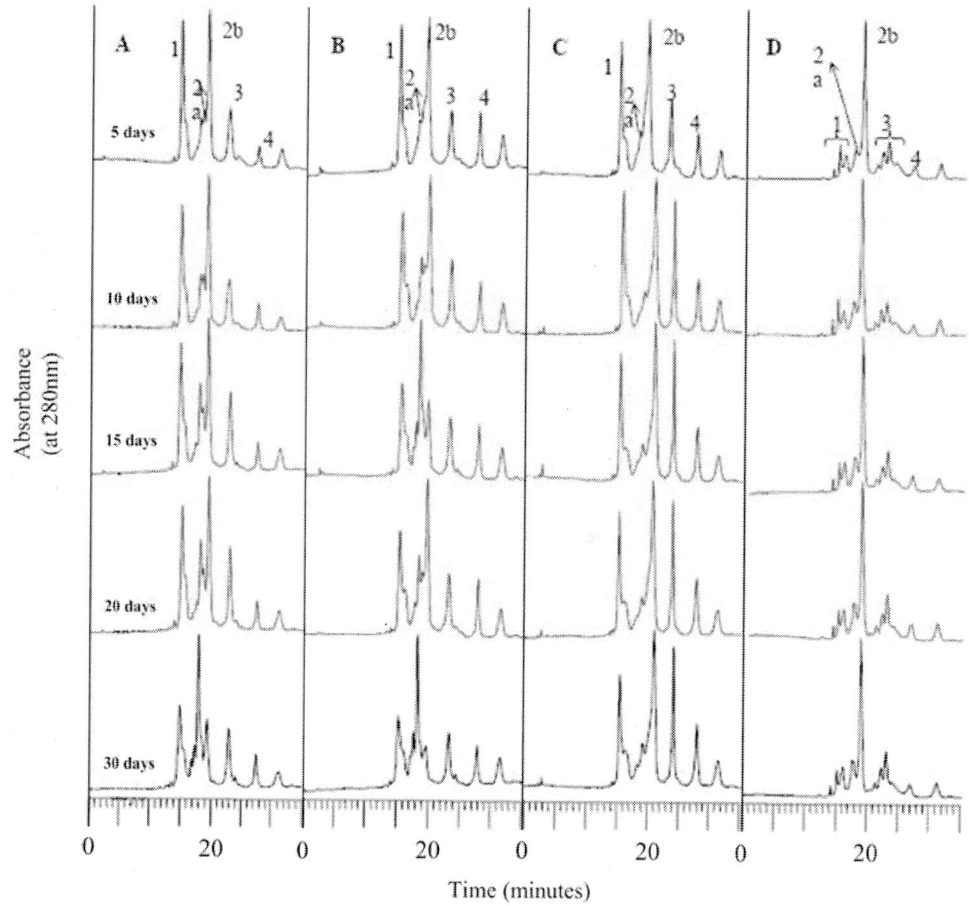

Figure 4. Chromatograms of α_S- and β-caseins throughout ripening in Portuguese traditional cheeses. A – ovine cheese; B – ovine cheese adulterated with 10 %(v/v) cow milk; C – ovine cheese adulterated with 20 %(v/v) cow milk; D – bovine cheese; 1 – κ-casein, 2a – α_{S1}-I peptide; 2b – α_S-casein; 3 and 4 – β-casein. (Adapted from Veloso *et al.*, 2004).

However, differences in proteolysis development derived from specific artisanal protocols of cheesemaking – e.g. cheese size, native microflora, salt amount or ripening conditions, further complicate detection of adulteration in ripened cheeses. For instance, Ferreira *et al.* (2006) were not able to detected adulteration of *Terrincho* cheeses with 10 or 20 %(v/v) cow's milk by 30 days of ripening, by either HPLC-RP or urea-PAGE of caseins. At early stages of ripening (say, up to 7 days), the chromatographic profile of *Terrincho* cheeses is clearly distinct from that of adulterated cheeses – due to the different retention times of α_{S1}-caseins from ewe's and cow's milks, which produce an overlap of bovine α_{S1}-casein with ovine β_1- and α_{S1}-caseins in adulterated cheeses. However, *Terrincho* and adultered cheeses showed essentially similar chromatographic profiles of caseins by 30 days. A similar problem was encountered using urea-PAGE: based on the different migration speeds of bovine and ovine α_{S1}-caseins, it was possible to identify adulteration by 20 days, but not by 30 days – owing to the intense hydrolysis of α_{S1}-casein associated with appearance of several degradation products with electrophoretic mobility similar to that of bovine α_{S1}-

casein. Hence, in ripened cheeses characterized by a considerable degree of proteolysis, the techniques discussed seem not to be appropriate for detection of cow's milk.

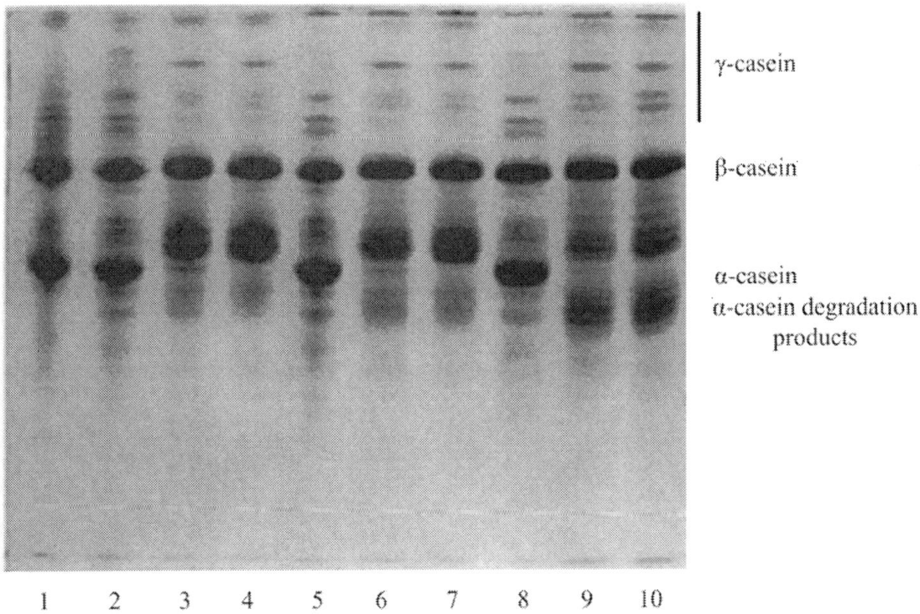

Figure 5. Electrophoretograms of α_S- and β-caseins throughout ripening in Portuguese traditional cheeses. Lane 1 – bovine casein standard; Lane 2 – bovine cheese by 10 days of ripening; Lane 3 – ovine cheese adulterated with 10 %(v/v) cow milk by 10 days of ripening; Lane 4 – ovine cheese by 10 days of ripening; Lane 5 – bovine cheese by 20 days of ripening; Lane 6 – ovine cheese adulterated with 10 %(v/v) cow milk by 20 days of ripening; Lane 7 – ovine cheese by 20 days of ripening; Lane 8 – bovine cheese by 30 days of ripening; Lane 9 – ovine cheese adulterated with 10 %(v/v) cow milk by 30 days of ripening; Lane 10 – ovine cheese by 30 days of ripening. (Adapted from Veloso *et al.*, 2004).

Multivariate statistical analysis has proven successful in evaluating the proteolytic process, as well as in differentiating cheese types and predicting ripening time (Coker *et al.*, 2005). Ferreira *et al.* (2006) applied a canonical discriminate analysis to HPLC-RP data obtained for *Terrincho* cheese, mixtures with 10 and 20 %(v/v) cow's milk, and bovine cheeses, by 30 days of ripening; only the peaks corresponding to κ-, α_{S2}- and β_1-caseins, and α_{S1}-derived peptides were considered – with exclusion of the peaks that co-elute in chromatography. This procedure was able to discriminate correctly between the 4 types of cheese (Figure 6). Moreover, such a chemometric approach showed that only 3 variables (viz. κ- and β_1-caseins, and α_{S1}-derived peptide) contributed to explain most variability among cheese types. The κ-casein and α_{S1}-derived peptide led to the highest discriminant power to function 1, and β_1-casein to function 2. Considering the former, *Terrincho* and mixture cheeses were located on the left side of the plot, whereas the bovine cheese was located on the right side. According to function 2, *Terrincho* cheese received negative scores, while mixed milk cheeses had positive scores, and bovine cheese remained close to the origin; this suggests that differences between *Terrincho* and adulterated cheeses are due chiefly to β_1-casein content. This casein fraction is higher in bovine milk and more resistant to enzyme hydrolysis until 30 days of ripening, thus resulting in higher β_I-casein content in mixture cheeses. Therefore, β_I-casein was selected as suitable marker for *Terrincho* cheese authenticity (Ferreira *et al.*, 2006).

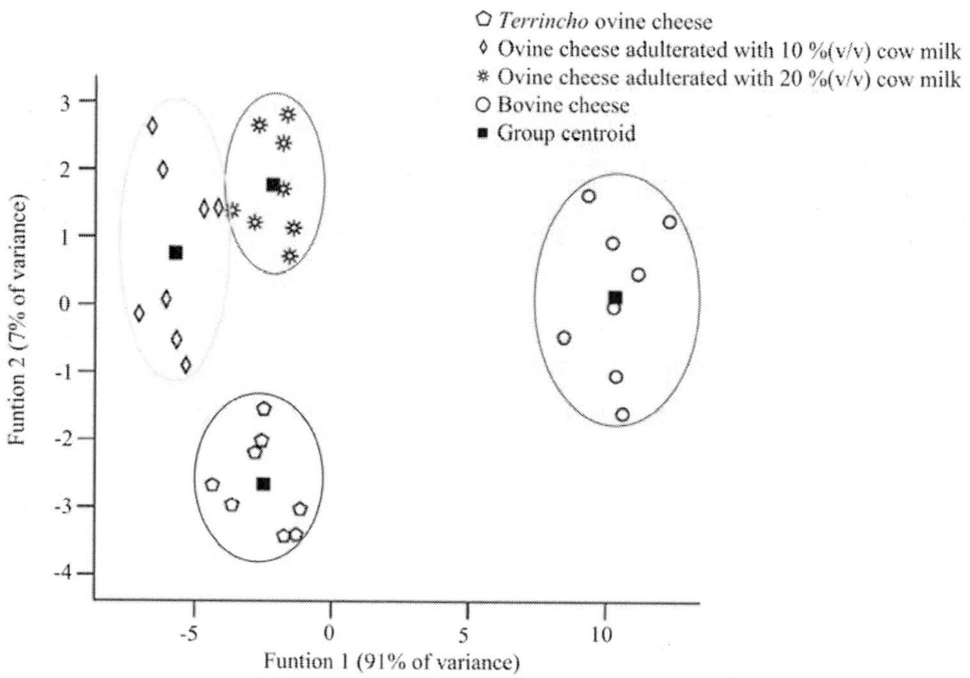

Figure 6. Results of principal component analysis, plotted as major discriminant functions for classification of Portuguese traditional cheeses. (Adapted from Ferreira *et al.*, 2004).

As mentioned above, all Portuguese PDO cheeses are manufactured from raw milk – which has been shown to be a potential source of contamination with pathogenic bacteria (Kausta *et al.*, 2010). During ripening, several biochemical, physical and microbial alterations occur – in which such factors as pH, a_w and salt content interact with the extent of inhibition of growth of pathogenic bacteria (Donnelly, 2004). To assure microbiological safety, a minimum ripening time is necessary – thus minimizing the risk of disease outbreaks caused by cheese intake. This is why Portuguese PDO cheeses can only be sold after a predetermined ripening period, which varies from cheese to cheese (see Table 1). On the other hard, it is during this period that the characteristic flavour and texture of each cheese variety develop (McSweeney, 2004) – so an adequate maturation extent is important to the final cheese quality. Since control of the extent of cheese ripening is crucial to ensure safety and quality, methods based on rheological, physical and chemical data have been explored in attempts to estimate the ripening time of Portuguese PDO cheese varieties – such as *Serpa* and *Terrincho* cheeses (Alvarenga *et al.*, 2008; Pinho *et al.*, 2004). Ferreira *et al.* (2006) proposed, in particular, utilization of the chromatographic peak areas of the major casein fractions (κ-, α_{S1}-, α_{S2}-, β_I- and β_2-caseins, and α_{S1}-I peptide) to predict the ripening time of *Terrincho* cheese. Resorting to multiple linear regression analysis, those authors showed that *Terrincho* ripening time can be predicted using only the normalized peak area of α_{S1}-casein and α_{S1}-I peptide, according to the following equation:

Terrincho ripening time = 13.56 + 10.7 [α_{S1}-I peptide] – 11.8 [α_{S1}-casein] (R = 0.92) ; the estimation error of this method was ± 2.5 days, which is lower than that obtained for prediction based on texture and colour parameters, i.e. ± 4.2 (Pinho *et al.*, 2004). Therefore, utilization of HPLC-RP casein profiles leads to a more reliable estimate of *Terrincho* ripening time than only physical properties of cheeses. Moreover, methods to estimate cheese ripening time based on analysis of casein fractions are faster and easier to perform than those based on rheological, physical and chemical data, which involve numerous determinations (Alvarenga *et al.*, 2008; Pinho *et al.*, 2004).

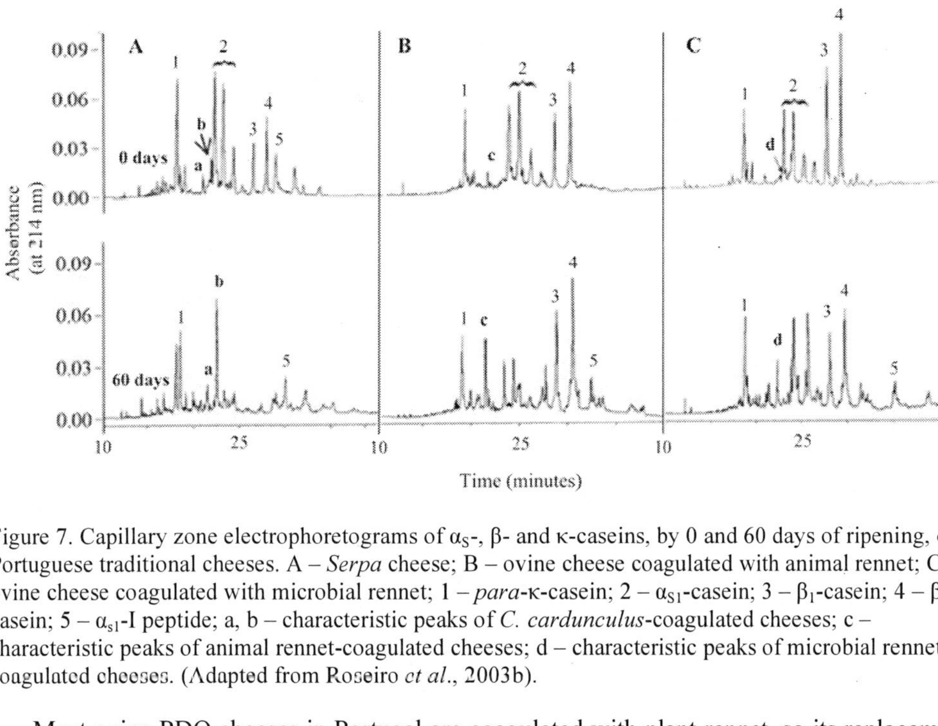

Figure 7. Capillary zone electrophoretograms of α_S-, β- and κ-caseins, by 0 and 60 days of ripening, of Portuguese traditional cheeses. A – *Serpa* cheese; B – ovine cheese coagulated with animal rennet; C – ovine cheese coagulated with microbial rennet; 1 – *para*-κ-casein; 2 – α_{S1}-casein; 3 – β_1-casein; 4 – β_2-casein; 5 – α_{s1}-I peptide; a, b – characteristic peaks of *C. cardunculus*-coagulated cheeses; c – characteristic peaks of animal rennet-coagulated cheeses; d – characteristic peaks of microbial rennet-coagulated cheeses. (Adapted from Roseiro *et al.*, 2003b).

Most ovine PDO cheeses in Portugal are coagulated with plant rennet, so its replacement by less expensive coagulants – e.g. commercial animal or microbial rennet, is another fraudulent practice. To detect the coagulant origin in ripened *Serpa* cheeses, Roseiro *et al.* (2003b) evaluated, by capillary zone electrophoresis and during 60 days of ripening, the casein profile of ovine milk cheeses manufactured with *C. cardunculus*, animal and microbial coagulants – all following traditional protocols. The capillary electrophoretograms of each cheese type showed characteristic peaks (denoted as *a, b, c* and *d* in Figure 7), which seem to be specific for each coagulant type. At all stages of ripening, a peak (denoted as *b*) was found in *Serpa* cheeses made with *C. cardunculus*, which was not detected in cheeses manufactured with other rennets. This peak – which appears to be the result of proteolytic activity of *C. cardunculus* upon β-casein, can thus be used as a marker of cheese coagulation brought about by *C. cardunculus* (Roseiro *et al.*, 2003b).

In view of the above, methodologies based on assessment of the casein profile proved accurate and useful to ensure authenticity of some Portuguese PDO cheeses; this entails detection of cow's milk added to ovine or caprine cheeses, prediction of cheese ripening time and identification of rennet origin. However, further validation is needed to conclude on whether such methodologies can be extended to detect fakes in other Portuguese traditional cheese varieties.

6. Conclusion

Traditional cheeses in Portugal exhibit distinct patterns of casein degradation, arising mainly from their traditional and artisanal modes of manufacture – which confer unique features to each cheese variety. Those distinct patterns are chiefly associated with the specific proteolytic activity of proteases present in the plant or animal rennet employed. Moreover, a number of other factors affect that pattern – e.g. ripening time, lactation season, proportion of ewe's and goat's milk, and mode and amount of salting. Therefore, cheesemaking conditions critically affect casein degradation – thus constraining not only the final sensory characteristics of cheeses, but also their nutritional value.

Regarding bioactive peptides, not much is known at present, yet the data available already suggest that traditional ewe's milk cheeses coagulated with *C. cardunculus* are a pool of casein-derived peptides possessing angiotensin-converting enzyme-inibitory and antioxidant activities; this realisation has emphasized the new interest generated by these traditional cheeses in terms of favourable contribution to health.

On the other hand, assessment of the cheese casein profile has proven useful in attempts to detect fakes of traditional cheeses; application of chemometric analysis to such chemical data as chromatographic casein profile is indeed a powerful tool to detect adulteration in ripened cheeses.

Acknowledgements

A postdoctoral fellowship (ref. SFRH/ BPD / 74527 / 2010, supervised by author F.X.M.) was granted to author E.J. by Fundação para a Ciência e a Tecnologia (Portugal), within Programa Operacional Potencial Humano – under the auspices of the European Social Fund and the Portuguese State.

References

Alvarenga, N., Silva, P., Garcia, J. R. & Sousa, I. (2008). Estimation of *Serpa* cheese ripening time using multiple linear regression (MLR) considering rheological, physical and linear chemical data. *Journal of Dairy Research*, *75*, 233-239.

Coker, C. J., Crawford, R. A., Johnston, K. A., Singh, H. & Creamer, L. K. (2005). Towards the classification of cheese variety and maturity on the basis of statistical analysis of proteolysis data – a review. *International Dairy Journal*, *15*, 631-643.

Costa, N., Ravasco, F., Miranda, R., Duthoit, M. & Roseiro, L. B. (2008). Evaluation of a commercial ELISA method for the quantitative detection of goat and cow milk in ewe milk and cheese. *Small Ruminant Research*, *79*, 73-79.

de la Fuente, M. A. & Juárez, M. (2007). Authenticity of dairy products. *Critical Reviews in Food Science and Nutrition*, *45*, 563-585.

European Commission. Geographical indications and traditional specialities, Door database. 2011, May, 13. Available from: http://ec.europa.eu/agriculture/quality/door/list.html; jsessionid=pL0hLqqLXhNmFQyFl1b24mY3t9dJQPflg3xbL2YphGT4k6zdWn34!37087 9141?&recordStart=0&filter.dossierNumber=&filter.comboName=&filterMin.milestone __mask=&filterMin.milestone=&filterMax.milestone__mask=&filterMax.milestone=&fi lter.country=PT&filter.category=PDOPGI_CLASS_13&filter.type=PDO&filter.status=R EGISTERED&locale=en

Donnelly, C. W. (2004). Growth and survival of microbial pathogens in cheese. In Fox, P. F., McSweeney, P. L. H., Cogan, T. M., & Guinee, T. P (Eds.), *Cheese: Chemistry, Physics and Microbiology – General Aspects 1* (vol. 3, pp. 541-559). Elsevier Academic Press, London.

Ferreira, I. M. P. L. V. O., Veiros, C., Pinho, O., Veloso, A. C. A., Peres, A. M. & Mendonça, A. (2006). Casein breakdown in *Terrincho* ovine cheese: comparison with bovine cheese and with bovine/ovine cheeses. *Journal of Dairy Science*, *89*, 2397-2407.

Freitas, C. & Malcata, F. X. (2000). Microbiology and biochemistry of cheeses with *Appélation d'Origine Protegée* and manufactured in the Iberian Peninsula from ovine and caprine milks. *Journal of Dairy Science*, *83*, 584-602.

Freitas, C. A., Fresno, M. J., Prieto, B., Malcata, F. X. & Carballo, J. (1997). Effects of ripening time and combination of ovine and caprine milks on proteolysis of *Picante* cheese. *Food Chemistry*, *66*, 219-229.

Freitas, C. A. & Malcata, F. X. (1996). Influence of milk type, coagulant, salting procedure and ripening time on the final characteristics of *Picante* cheese. *International Dairy Journal*, *6*, 1099-1116.

Freitas, C. A. & Malcata, F. X. (1999). Technological optimisation of *Picante* cheese using microbiological, chemical and physical criteria. *Journal of Food Enginnering*, *41*, 163-175.

Guinee, T. P. & Fox, P. F. (2004) Salt in cheese: physical, chemical and biological aspects. In Fox, P. F., McSweeney, P. L. H., Cogan, T. M., & Guinee, T. P (Eds.), *Cheese: Chemistry, Physics and Microbiology – General Aspects 1* (vol. 3, pp. 207-260). Elsevier Academic Press, London.

Jacob, M., Jaros, D. & Rohm, H. (2011). Recent advances in milk clotting enzymes. *International Journal of Dairy Technology*, *64*, 14-33.

Kausta, M., Mataragas, M., Skandamis, P. & Drosinos, E. H. (2010). Prevalence and sources of cheese contamination with pathogens at farm and processing levels. *Food Control*, *21*, 805-815.

Kongo, J. M., Gomes, A. M., Malcata, F. X. & McSweeney, P. L. H. (2009). Microbiological, biochemical and compositional changes during ripening of *São Jorge* – a raw milk cheese from the Azores (Portugal). *Food Chemistry*, *112*, 131-138.

Macedo, A. C. & Malcata, F. X. (1994). Lacticínios tradicionais e inovações tecnológicas. *Via Láctea*, *5*, 63-65.

Macedo, A. C. & Malcata, F. X. (1997). Hydrolysis of α_s- and β-caseins during ripening of *Serra* cheese. *Food Chemistry, 58*, 43-48.

Macedo, A. C., Malcata, F. X. & Oliveira, J. C. (1993). The technology, chemistry, and microbiology of *Serra* cheese: a review. *Journal of Dairy Science, 76*, 1725-1739.

Macedo, I. Q., Faro, C. J. & Pires, E. M. (1996). Caseinolytic specificity of cardosin, an aspartic protease from the cardoon *Cynara cardunculus* L.: action on bovine α_s- and β-casein, and comparison with chymosin. *Journal of Agricultural and Food Chemistry, 42*, 42-47.

Mafra, I., Ferreira, I. M. P. L. V. O., Faria, M. A. & Oliveira, B. P. P. (2004). A novel approach to the quantification of bovine milk in ovine cheeses using a duplex polymerase chain reaction method. *Journal of Agricultural and Food Chemistry, 52*, 4943-4947.

McSweeney, P. L. H. (2004). Biochemistry of cheese ripening: introduction and overview. In Fox, P. F., McSweeney, P. L. H., Cogan, T. M., & Guinee, T. P (Eds.), *Cheese: Chemistry, Physics and Microbiology – General Aspects 1* (vol. 3, pp. 347-360). Elsevier Academic Press, London.

McSweeney, P. L. H. & Sousa, M. J. (2000). Biochemical pathways for the production of flavour compounds in cheeses during ripening: a review. *Le Lait, 80*, 293-324.

Miller, N. J., Sampson, J., Candeias, L. P., Bramley, P. M. & Rice-Evans, C. A. (1996). Antioxidant activities of carotenes and xanthophylls. *FEBS Letters, 384*, 240-242.

Park, Y. W., Juárez, M., Ramos, M. & Haenlein, G. F. W. (2007). Physico-chemical characteristics of goat and sheep milks. *Small Ruminant Research, 68*, 88-113.

Pinho, O., Mendes, E., Alves, M. M. & Ferreira, I. M. P. L. V. O. (2004). Chemical, physical, and sensorial characteristics of *Terrincho* ewe cheese: changes during ripening and intravarietal comparison. *Journal of Dairy Science, 87*, 249-257.

Reis, P. J. M. & Malcata, F. X. (2007). Improvements in small scale artisanal cheesemaking via a novel mechanized apparatus. *Journal of Food Enginnering, 82*, 11-16.

Rice-Evans, C. A., Miller, N. J. & Paganga, G. (1996). Structure-antioxidant activity relationships of flavonoids and phenolic acids. *Free Radical Biology and Medicine, 20*, 933-956.

Roseiro, L. B., García-Risco, M., Barbosa, M., Ames, J. M. & Wilbey, A. (2003a). Evaluation of *Serpa* cheese proteolysis by nitrogen content and capillary zone electrophoresis. *International Journal of Dairy Technology, 56*, 99-104.

Roseiro, L. B., Gómes-Ruiz, J. A., García-Risco, M. & Molina, E. (2003b). Vegetable coagulant (*Cynara cardunculus*) use evidenced by capillary electrophoresis permits PDO *Serpa* cheese authentication. *Le Lait*, 83, 343-350.

Sieber, R., Bütikofer, U., Egger, C., Portmann, R., Walther, B. & Wechsler, D. (2010). ACE-inhibitory activity and ACE-inhibiting peptides in different cheese varieties. *Dairy Science and Techology, 90*, 47-73.

Silva, S. V., Pihlanto, A. & Malcata, F. X. (2006). Bioactive peptides in ovine and caprine cheeselike systems prepared with proteases from *Cynara cardunculus*. *Journal of Dairy Science, 89*, 3336-3344.

Silva, S. V. & Malcata, F. X. (2005). Caseins as source of bioactive peptides. *International Dairy Journal, 15*, 1-15.

Sousa, M. J., Ardö, Y. & McSweeney, P. L. H. (2001). Advances in the study of proteolysis during cheese ripening. *International Dairy Journal, 11*, 327-345.

Sousa, M. J. & Malcata, F. X. (1996). Effects of processing conditions on the caseinolytic activity of crude extracts of *Cynara cardunculus* L. *Food Science and Technology International*, *2*, 255-263.

Sousa, M. J. & Malcata, F. X. (1997). Comparison of plant and animal rennets in terms of microbiological, chemical, and proteolysis characteristics of ovine cheese. *Journal of Agricultural and Food Chemistry*, *45*, 74-81.

Sousa, M. J. & Malcata, F. X. (1998a). Identification of peptides from ovine milk cheese manufactured with animal rennet or extracts of *Cynara cardunculus* as coagulant. *Journal of Agricultural and Food Chemistry*, *46*, 4034-4041.

Sousa, M. J. & Malcata, F. X. (1998b). Proteolysis of ovine and caprine caseins in solution by enzymatic extracts from flowers of *Cynara cardunculus*. *Enzyme and Microbial Technology*, *22*, 305-314.

Sousa, M. J. & Malcata, F. X. (2002). Advances in the role of the plant coagulant (*Cynara cardunculus*) *in vitro* and during ripening of cheeses from several milk species. *Le Lait*, *82*, 151-170.

Tavaria, F. K., Sousa, M. J. & Malcata, F. X. (2001). Storage and lyophilization effects of extracts of *Cynara cardunculus* on the degradation of ovine and caprine caseins. *Food Chemistry*, *72*, 79-88.

Upadhyay, V. K., McSweeney, P. L. H., Magboul, A. A. A., & Fox, P. F. (2004). Proteolysis in cheese during ripening. In Fox, P. F., McSweeney, P. L. H., Cogan, T. M., & Guinee, T. P (Eds.), *Cheese: Chemistry, Physics and Microbiology – General Aspects 1* (vol. 3, pp. 391-433). Elsevier Academic Press, London.

Veloso, A. C. A., Teixeira, N. & Ferreira, I. M. P. L. V. O. (2002). Separation and quantification of the major casein fractions by reverse-phase high-performance liquid chromatography and urea-polyacrylamide gel electrophoresis. Detection of milk adulterations. *Journal of Chromatography A*, *967*, 209-218.

Veloso, A. C. A., Teixeira, N., Peres, A. M., Mendonça, A. & Ferreira, I. M. P. L. V. O. (2004). Evaluation of cheese autenticity and proteolysis by HPLC and urea-polyacrylamide gel electrophoresis. *Analytical, Nutritional and Clinical Methods*, *87*, 289-295.

Veríssimo, P. C., Esteves, C. L. C., Faro, C. J. F. & Pires, E. M. V. (1995). The vegetable rennet of *Cynara cardunculus* L. contains two proteinases with chymosin and pepsin-like specificities. *Biotechnology Letters*, *17*, 621-626.

Visser, S. (1993). Proteolytic enzymes and cheese ripening. *Journal of Dairy Science*, *76*, 329-350.

Walther, B., Schmid, A., Sieber, R. & Wehrmüller, K. (2008). Cheese in nutrition and health. *Dairy Science and Techology*, *88*, 389-405.

Whyte, N. H. (1995). Proteolysis of bovine and ovine caseins by lamb and calf chymosin. M.Sc. Thesis, National University of Ireland, Cork.

In: Casein: Production, Uses and Health Effects
Editors: A. M. Ventimiglia, et. al, pp. 123-139
ISBN: 978-1-62100-129-4

Chapter 6

OPTIMIZING OF STERILIZATION OF CASEIN FOR NOVEL USE

Kouass Sahbani Saloua[a,*b], ***Aloui Amine***[a], ***Kouass Salah***[c], ***Zourgui Lazhar***[d] ***and Landoulsi Ahmed***[a]

[a] Unité de recherche de Biochimie des lipides et interactions avec les macromolécules. 03/UR/0902. Laboratoire de Biochimie et de Biologie moléculaire, Faculté des sciences de Bizerte, 7021, Zarzouna, Tunisia.
[b] Groupe en Sciences des Radiations, Département de Médecine Nucléaire et de Radiobiology, Faculté de Médecine, Université de Sherbrooke, Québec, Canada J1H 5N4.
[c] Laboratoire de Cristallochimie et Matériaux, Faculté des Sciences de Tunis, 1092 El Manar II Tunisia.
[d] Research Unit of Macromolecular Biochemistry & Genetics, Faculty of Sciences of Gafsa, Gafsa, Tunisia.

ABSTRACT

Milk proteins are natural vehicles, which evolved to deliver essential micronutrients and building blocks, as well as immune system components, from mother to the newborn. Casein is the principal protein found in cow's milk from which it has been extracted commercially for most of the 20th century. It is responsible for the white, opaque

[*] Author to whom correspondence should be addressed:
Kouass Sahbani Saloua,
Unité de recherche de Biochimie des lipides et interactions avec les macromolécules. 03/UR/0902. Laboratoire de Biochimie et de Biologie moléculaire Faculté des sciences de Bizerte Zarzouna7021, Tunisia. Groupe en Sciences des Radiations, Département de Médecine Nucléaire et de Radiobiology, Faculté de Médecine, Université de Sherbrooke, Québec, Canada J1H 5N4.
Tel.: (001)819 346 1110 ext 12103 / (216) 23 347283
Fax: (001) 819-564-5442 /(216)-72-590566
E-mail: kouasssahbanisaloua@rocketmail.com

appearance of milk in which it is combined with calcium and phosphorus as clusters of casein molecules, called .micelles.

The major uses of casein until the 1960s were in technical, non-food applications such as adhesives for wood, in paper coating, leather finishing and in synthetic fibres, as well as plastics for buttons, buckles... During the past 30 years, however, the principal use of casein products has been as an ingredient in foods to enhance their physical (so-called .functional.) properties, such as whipping and foaming, water binding and thickening, emulsification and texture, and to improve their nutrition. The caseins are proline-rich, open-structured rheomorphic proteins which have distinct hydrophobic and hydrophilic domains. The major advances of the past year in harnessing milk proteins for novel health-promoting delivery applications were mainly in nanosizing, conjugation, crosslinking and targeting. Novel milk-protein nanoparticles were introduced for solubilizing and protecting hydrophobic nutraceuticals in clear systems, or for targeting gastric tumors, utilizing the natural digestibility of caseins. Several interesting advancements were reported in the application of milk proteins for drug targeting. Casein like all raw material of animal origin requires sterilization before the processing into consumable products or drugs. Many methods can be used to sterilize casein. Ionizing irradiation can be a very powerful method to sterilize casein and can also reduce the allergenicity of milk proteins by the destruction of the human IgE-binding epitopes in protein allergens. Ionizing irradiation can damage protein. One approach to achieve this goal is the addition of compounds that act a radioprotectors. Milk proteins have a common feature related to their cross-linking affinity, and its effect on texture, as well as nutritional properties. The reactions of milk proteins with some functional groups of other compounds in food have gained increasing attention in food research. It was found that *Camellia sinensis* extract can protect casein against gamma irradiation. Also a decrease of the dose of bacterial decontamination D_{10} was observed after addition of *Camellia sinensis* extract to casein.

INTRODUCTION

Proteins are important constituents of the human diet, since they are a principal source of nitrogen and essential amino acids. Milk proteins have a high nutritional value compared to other proteins because of their relative high content of essential amino acids and their good digestibility [1]. Moreover, milk proteins may contribute to structural properties of end products, such as emulsion, foam and gel properties. Milk proteins are used in many different food products, ranging from dairy products to beverages, dietary and medical products [2, 3]. For some food applications proteins are hydrolysed, amongst others for hypoallergenic infant nutrition, for nutrition for patients with digestion disorders and for sports nutrition [4, 5]. Caseins and whey proteins are the two main protein groups in milk. Caseins, representing about 80% of the protein content in bovine milk, are isolated from milk by acid or by rennet precipitation. The acid, or isoelectric, precipitation is performed at pH 4.6. Acid precipitated caseins can be resolubilised by raising the pH through alkali addition, usually sodium hydroxide, yielding soluble caseinates. In rennet precipitation, one of the casein molecules, i.e. k-casein, is selectively hydrolysed at the pH of milk (pH 6.7). As a result, the casein micelles in milk become susceptible towards precipitation with calcium. Casein consists of four primary proteins, i.e. αs_1-, αs_2-, β- and κ-casein [6], which all have genetic variants that differ in their molecular weight and isoionic pH. Caseins are flexible, heat stable, proteins. Typical for casein proteins is the clustering of amino acids in hydrophobic and hydrophilic domains, giving rise to the amphipathic character of the proteins [2, 7]. The caseins are proline-rich, open-structured rheomorphic proteins (i.e. they assume any one of several energetically favourable conformations in solution), which have distinct hydrophobic and hydrophilic domains. αs1-CN, αs2-CN and β-CN have serine–phosphate residue centers for calcium sequestration. κ-CN is a glycoprotein, with 2 cysteines forming intermolecular disulfide bridges. 95% of the caseins are naturally self-assembled into casein micelles which are spherical colloidal particles, 50–500 nm (average 150 nm) in diameter and molecular mass between 10^6 and 3×10^9 Da. They comprise about 94% protein and 6% low molecular weight compounds collectively called colloidal calcium phosphate. The micelles contain ~ 2 g water/g protein, and there is 10^{14}–10^{16} micelles/ml milk. They are extremely well adapted to their evolutionary tasks of concentrating, stabilizing and delivering calcium, phosphate and protein from the mammalian mother to its neonate [8]. Moreover, the open structure of the caseins, due to their high proline content, makes them readily accessible for proteolytic cleavage. This along with the acid-soluble calcium-phosphate bridging, make an excellent target-activated release mechanism for unloading the cargo in the stomach. The caseins are known to contain in their structure potent bioactive peptides, some with opioid properties [6, 9]. An interesting blood cell regulatory properties of complete casein molecules, or bioactive peptides contained was found. The casein micelle is indeed a remarkable example of natural nanovehicle for nutrient delivery. Many regulatory properties on lymphoid cells were observed with a couple of those casein subunits. For example, k-casein can block splenocyte mitogenesis [9, 10], while β-casein can help in the induction of proliferation of B and T lymphocytes [11]. Also, Milk proteins were used for novel health-promoting delivery applications were mainly in nanosizing, conjugation, crosslinking and targeting. Novel milk-protein nanoparticles were introduced for solubilizing and protecting hydrophobic nutraceuticals in clear systems, or for targeting gastric tumors, utilizing the natural digestibility of caseins [12]. Application of milk proteins for drug targeting has got more interesting attention [13]. Casein like all raw material of animal origin requires decontamination before using in the processing into consumable products or drug. The International Dairy Federation (IDF) Standard [14], specifies the following upper limits for bacterial contamination of edible casein: total count not more than 3.10^4 g^{-1}, coliforms negative in 0.1 g, mould and yeasts not more than 50 g^{-1} and thermophilic organisms not more than 5.10^3 g^{-1}. As for the microbial load of industrial caseins there is no legislation imposed yet [15]. The typical high microbial load of industrial raw products clearly limits the applications of contaminated casein. Sterilization of milk protein without heating is of great interest. Gamma irradiation is a very powerful method to decontaminated casein. Irradiation technology has positive effects in preventing a decay of food products by sterilizing the microorganisms as well as improving the safety and shelf-stability without compromising the nutritional or sensorial quality [16]. At least 12 and 22 countries are using irradiation in raw red meat and poultry, respectively, to control parasites and food-borne pathogens and extend the shelf-life of the meat [17]. However, limited countries are approved for irradiation of milk products as sanitation purpose. Dried milk products for disinfestations (max. 3 kGy) and microbial control (max. 30 kGy) are approved for irradiation in Croatia and casein or caseinates are approved for microbial control (max. 6kGy) in Belgium [18]. FAO also approved 6kGy a maximum dose to sterilize casein. We were shown that this dose can be less than 2kGy if casein was irradiated in the presence of *Camellia sinensis* extract [19]. Besides the sanitary purposes, however, irradiation has been studied to reduce or eliminate undesirable or toxic materials including food allergens [20, 21], carcinogenic volatile N-nitrosamines [22], biogenic amines [23], embryotoxicity of gossypol [24], and enhancement of bioactivity [25, 26]. Modification of a protein structure may cause it to lose its functional activity by physical, chemical, and biological damages [27]. Irradiation of proteins produces a structural denaturation and creates changes in the binding ability of IgE against allergens. Ionizing radiation can reduce the allergenicity of milk proteins by the destruction of the human IgE-binding epitopes in food allergens [20, 28, 29]. Especially, Lee et al. studied using milk-hypersensitive patients' IgE and rabbit IgGs individually produced α-casein and investigated the changes of allergenicity and antigenicity of irradiated proteins by competitive indirect enzyme-linked immunosorbent assay [20]. The authors reported that epitopes on milk allergens were structurally altered by gamma irradiation. Gamma irradiation is a very powerful method to decontaminated casein but it is known to cause irreversible alterations of protein conformation at the molecular level, such as fragmentation or aggregation [29]. Lee and Sung, showed, via Sodium dodecyl sulphate-polyacrylamide gel electrophoresis (SDS-PAGE) study, that gamma-irradiation of myoglobin in solution, first caused the disruption and then aggregation due to the cross-linking of myoglobin fragments and that degradation and aggregation could be prevented by antioxidants such as ascorbic acid [30]. Protein damages due to indirect irradiation effects could be minimized by optimization of the irradiation parameters. One approach to achieve this goal is the addition of compounds that act as radioprotectors [31-33] e.g., *Camellia sinensis* green tea extract (up to 5%) to protect the protein from the ROS generated during the irradiation process [19]. Try to make the review of knowledge acquired about casein: novel applications and novel methods of sterilisation of casein establish one of the aims of this present chapter.

NOVEL USES OF CASEIN: NATURAL NANODELIVERY VEHICLES FOR CHEMOTHERAPEUTIC DRUGS

Casein is the principal protein found in cow's milk from which it has been extracted commercially for most of the 20th century. Casein has been used commercially in non-food technical applications since the mid-19th century, initially in adhesives and water-based paints. In food technical applications casein is generally not consumed as a food on its own. Modifying the physical properties of food product or providing nutritional supplementation can explain the mainly uses of casein as ingredients in foods. The nutritional quality of a protein is determined primarily by its essential amino acid content. For adult man, eight amino acids are essential, i.e. they must be supplied in the diet. These are isoleucine, leucine, lysine, methionine, phenylalanine, threonine, tryptophan and valine; the infant requires histidine as well. In comparison with an ideal reference protein composition that was developed by the FAO in 1973, casein contains an adequate supply of all the essential amino acids with the possible exception of the sulphur-containing amino acids methionine and cysteine. They have recently shown that casein micelles can be utilized as natural nanodelivery vehicles for lipid-soluble vitamins (e.g., vitamin D) [34]. It is known that many of the current chemotherapeutic drugs for the treatment of multiple human malignancies are administered intravenously. Studies suggest it is a major source of cost, discomfort, and stress to patients, and multiple hospitalizations are required to complete the relatively long intravenous sessions of combination chemotherapeutic regimens [35]. The availability of suitable and effective oral therapeutic agents would make a significant contribution to patients' quality of life, may significantly reduce cost, and may prove even more effective than current treatment modalities. More recently they introduced β-CN nanoparticles as potential nanovehicles for entrapment and oral delivery of hydrophobic bioactive molecules, such as 9, 10-anthracenedione, better known as mitoxantrone (MX), an anticancer drug [36]. It has been suggested in the literature that β-CN is analogous to a diblock copolymer [37]. As such, its micelles are expected to be extremely more stable than those of low-molecular-weight surfactant micelles, and the kinetics of release of entrapped hydrophobic molecules are expected to be several orders of magnitude slower than release from low molecular-weight surfactants [38]. This is a major advantage for using the block-copolymer-like β-CN for entrapment of hydrophobic bioactive compounds, compared with using low-molecular-weight surfactants. The contamination of casein can reduce it novel application.

Casein Contamination

Casein is used as an ingredient in many food products; there are strong needs to meet some standards given in obligatory legal acts. Total bacteria count (TBC), number of thermophylic bacteria, coliforms and such pathogenic microorganisms as *Listeria monocytogenes* and *Salmonella spp.* are the most often used like indicators of biological impurity which characterise edible acid casein [39, 40, 41]. Micro-organisms correspond to a multitude of living organisms. Pathogenic microorganisms are harmful to humans [42]. Their destruction is completely required to adapt the processes to inhibit their proliferation. Many microorganisms has two forms: a "vegetative" form in which the microorganisms is present when the environmental conditions are favourable for its development and a "spore" form which appears when these conditions become unfavourable [42]. The "spore" forms are thus much less sensitive to potential unfavourable environmental situation and they are more harmful to humans. The destruction of these germs is generally need two stages: the first aims to trigger the spores' germination and the second takes place destroying the germinated spores [42]. They were found if germination is incomplete germs inactivation will then be incomplete and will lead to final product partially treated but still infected. These microorganisms are present in our environment and can be found in our diet [42]. It was also found, the microorganisms that suffered sub-lethal stress factor can develop new resistance mechanisms [43, 44, 45, 46]. This evolution is usually not abrupt but more gradual [45]. Safety standards become a shared concern increasingly limiting food market access. Reducing the contamination of raw materials and applying final appropriate technologies to limit microbial load two solutions of the food industry to limiting this constraint. Indeed, the precautionary measures settled in various manufacturing processes to acquire healthy foods are in no way sufficient to fully guard against various sources of microbiological contamination [42]. The food industry must provide products with high clean (microbiologically speaking) and also high nutritional, gustative... quality while remaining "natural" [42]. During the edible acid casein production processes it is very important to control the temperature of pasteurization [47]. The number of thermophylic bacteria, which survive the pasteurization process can affect the quality of the pasteurized product [48]. The need for a specific optimization for each case is hence highly needed [42]. There are so many obstacle related to the need of a good decontamination operation in terms of technical performances (microorganism's destruction efficiency, energy consumption…) and great preservation of product attributes and quality [42]. Increase of the number of microorganisms in edible acid casein can be observed if the hygiene is not observed during packing or storing the products in inadequate conditions (high humidity) [49].

EFFECT OF GAMMA IRRADIATION ON CASEIN

Temperature, early techniques of microorganisms destruction was used. However, products that treated by temperature require special storage conditions. In fact, these products still contain germs capable of multiplying. If they are placed under favourable environmental conditions to their multiplication, the products will then quickly become unfit for human consumption [42]. Sterilization has been used for a long time in order to inactivate microorganisms [50, 51], but these heat treatments have very negative effects on nutritional, gustative, etc. qualities [52]. Furthermore, the use of thermal shock (positive or negative) poorly controlled may lead to increase resistance to subsequent decontamination techniques. They have studied the case of *Penicillium glabrum* [53]; *Cronobacter sakazakii* [54]; and *Listeria monocytogenes* [55], have studied application to increase the germs preservation *(Lactococcus lactis)* used as processing assessment [56]. Ethylene oxide or propylene oxide has long been practiced at room temperature. Nevertheless, standards are more and more stringent mainly regarding residual molecules introduced into the treated product [57]. Many studies provide that the ionization by nuclear gamma irradiation has for a very long time and for many solid and/or powders foodstuffs emerged as the best decontamination treatment [58, 59, 60, 61]. This treatment has proven to be highly effective and very convenient since the product can be treated in its airtight packaging away from any recontamination. γ Irradiation has been used for the decontamination and/or sterilization of dehydrated vegetables, fruits, seasonings and animal foods, and to prolong the storage period [62, 63, 64]. The dose to deliver to food depends on the desired effect: "low" dose irradiations (50-150 Gy) don't decontaminate and are used to only inhibit sprouting of potatoes and onions. Food sterilization requires much higher doses (10-50 kGy) [65] irradiation dose of 7 kGy can be effective to control microbial growth in black tea and in extending their shelf life without any significant deterioration of quality constituents. It was found that high dose irradiation kills bacteria but does not affect the toxins previously produced. However, these toxins are often responsible for many foodborne illnesses. Ionizing radiation suffers from a very bad public opinion due to the confusion between radiation and radioactivity. The World Health Organization (W.H.O.) expert committee on the wholesomeness of food irradiation agreed with the food and drug administration [66] that foods subjected to low dosages (10 kGy) of γ irradiation are safe and do not require toxicological testing [60, 67, 68]. γ Irradiation can extend the shelf life of treated foods without inducing the formation of any radionuclide in food products [69]. It is known gamma irradiation can inducing fragmentation and/or aggregation of the protein [29]. We were shown in our previous study [19] that electrophoretic profiles of casein irradiated with gamma rays in the absence of free radical scavengers in the presence of oxygen show that even for low doses; irradiation induces the rupture of the peptide chain (Figure 1). The result of this breakdown is the formation of low molecular weight peptides from the initially irradiated. Older studies with other proteins led to similar results [70-72], Schuessler and Schilling, have proposed that proline residues are the main targets that cause fragmentation of the peptide chain [71]. Wolff, Garner and Dean, reported that cleavage of the peptide chain could be the result of the direct oxidation of proline residues [73]. The electrophoretic profiles of casein exposed to gamma rays show changes due to an increase in molecular mass of the protein. It seems that the caseins undergo aggregation when irradiated with high dose [74]. It was found when caseins were irradiated with a dose greater than 6 kGy (Figure 1), the two bands characteristic of total caseins were completely absent this disappearance is aggregation which prevents the proteins from migrating through the gel. Similar results were observed with another proteins like Ceruloplasmin [72-74] and hemoglobin [75]. Indeed, irradiation induces fragmentation and aggregation of ovalbumin [72]. Stuart, Houee-Levin and Potier, suggested that when the protein is irradiated in solution most of the damage observed in chain peptide are due to the action of the hydroxyl radical OH generated by radiolysis of water [76]. This species is responsible for polymerization and also the fragmentation observed [77]. Increasing the molecular weight of proteins after aggregation is due to the formation of proteins on the reactions of cross-links and the formation of Disulfide Bridge [71-78], but in our study for probably increasing the molecular weight of casein is due to the reaction of cross-links. Aromatic amino acids are modified by oxidative action of free radicals (species generated by radiolysis of water). They are reactive and form covalent bonds (cross-linking) due to the formation of intermolecular bityrosine [78-80]. This represents a basic mechanism of the formation of cross-linking in proteins irradiated with gamma irradiation [81], which generates a high molecular mass after aggregation [78]. This process has also been observed with ovalbumin [72] and even to some interpretations are reported previously with egg white lysozyme and BSA when the damage caused by irradiation.

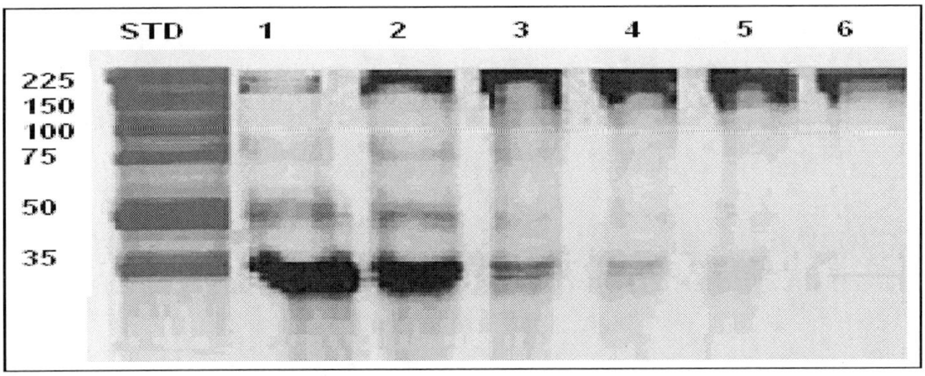

Figure 1. SDS/PAGE electrophoretic patterns of caseins (0,22mM) non-irradiated and irradiated at increasing doses: STD: stands for Standard Molecular Weights in kDa, (1) control whole casein prepared non-irradiated (0,22mM), (2) whole casein cow's milk (0,22mM) irradiated to 2 kGy, (3) 4 kGy, (4) 6 kGy, (5) 8 kGy and (6) 10 kGy [19].

Camellia Sinensis Green Tea Extract Protected Casein Gamma Irradiated

Gamma irradiation can damage casein, to protect our protein against aggregation and fragmentation we were tested the presence of *Camellia sinensis* green tea extract with the protein in our previous study. Camellia sinensis green tea extract (5 and 10%) as shown is able to protect casein against aggression of gamma irradiation [19]. Electrophoretic profile of caseins that were irradiated with a dose of 10 kGy has migrated through the gel separation and intensities of two bands characteristic of these proteins are identical for all doses (Figure 2). Similar results were observed in other studies where the formation of organic radicals was decreased by addition of ascorbic acid [82]. According to the electrophoretic profile, green tea extract protected against caseins fragmentation and aggregation during the radiotraitement, but the protective role of green tea depends on its concentration. Indeed, green tea extract (2.5%) present during the duration of exposure of casein solution gamma radiation appears able to prevent aggregation and degradation of protein especially for a 6 kGy dose ≤ if we consider the intensity of the bands of casein. This shows that the extract of green tea (2.5%) effectively prevents the aggregation of casein in solution at low doses of gamma irradiation (2 to 6 kGy), but for high doses the effect of this is much less effective. For against, the green tea extract 5% added to the solution of casein appears more effective in preventing the aggregation of these proteins prepared. Indeed, the intensities of two bands characteristic of casein is almost identical to those of commercial casein and prepared unirradiated even high doses 6 kGy, 8 kGy and 10 kGy. This shows that the caseins have migrated through the gel separation. Green tea has played a very important role in scavenging free radicals (species generated by radiolysis of water). This antioxidant effect is probably due to the wealth of green tea polyphenols and flavonoids and their ability to trap free radicals. Green tea is a powerful antioxidant that has neutralized during exposure the casein to gamma-irradiation the effect of hydroxyl radical. This antioxidant effect is certainly the wealth of green tea polyphenols and flavonoids and their ability to trap free radicals. Hasimoto and al., have evaluated the in vitro antioxidant effect of green tea polyphenols. Indeed, they have shown that polyphenols have antioxidant activity much greater than that of vitamin E [83].

Microbial Decontamination of Total Cow's Milk Caseins by Gamma Irradiation in the Presence of *Camellia Sinensis* Green Tea Extract

Camellia sinensis green tea extract was found can protected casein against gamma irradiation and also it was shown in a previous study, the reduction of the bacterial microflora in the casein depended on the concentration of *Camellia sinensis* green tea extract. Indeed, the concentration of 10% *Camellia sinensis* green tea extract (Figure 3) was more effective in reducing the number of viable bacteria in the caseins without irradiation [19]. Many studies have shown the antimicrobial effect of tea compounds: antibacterial, antiviral and antifungal [84-90]. Phenolic compounds known as catechins, [91] are the major active ingredients of organic green tea. Catechins represent approximately 10% of the dry weight of tea [92]. The concentration 10% (10 g leaves of green tea/100ml) has proven most effective in inhibiting the growth of bacteria [19]. The radiotreatment of casein in the presence of *Camellia sinensis* green tea extract 10% was found to be able to lower D_{10} down to a value of 2 kGy decontaminate the proteins even with a dose of 2 kGy (Figure 3) [19]. This dose is lower than that set by the Joint F.A.O. 6 kGy.

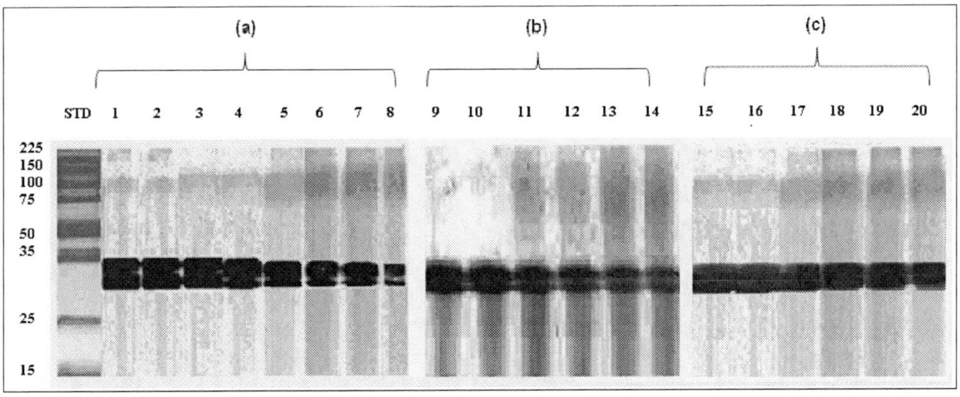

Figure 2. SDS/PAGE electrophoretic patterns of caseins (0,22mM) non-irradiated and irradiated at increasing doses in the absence and in the presence of *Camellia sinensis* green tea extract. (a) Casein irradiated in the presence of *Camellia sinensis* green tea extract 2,5%, (b) casein irradiated in the presence of *Camellia sinensis* green tea extract 5%, (c) casein irradiated in the presence of *Camellia sinensis* green tea extract 10%. STD: stands for Standard Molecular Weights in kDa, (1) control whole casein without green tea non-irradiated, (2) control whole standard casein non-irradiated, (3, 9 and 15) respectively control whole casein prepared non-irradiated with *Camellia sinensis* green tea extract: 2,5%, 5% and 10%, (4, 10 and 16) casein irradiated 2 kGy, (5, 11 and 17) 4 kGy, (6, 12 and 18) 6 kGy, (7, 13 and 19) 8 kGy and (8, 14 and 20) 10 kGy [19].

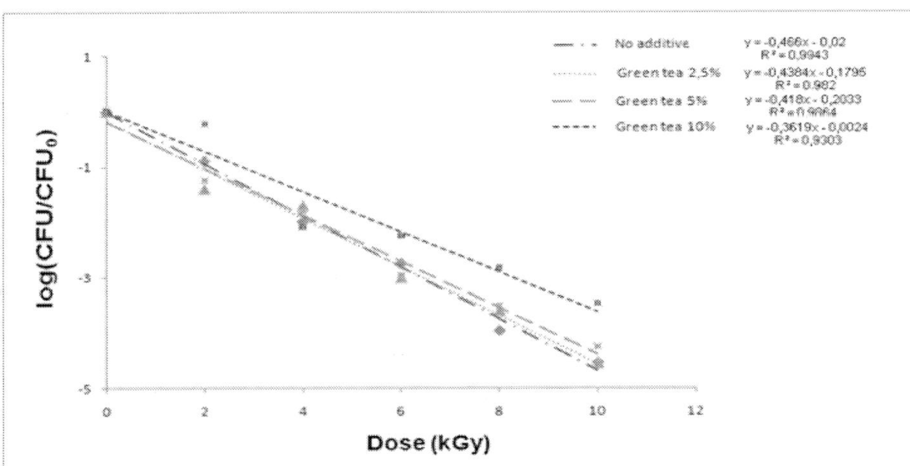

Figure 3. Combined effect of *Camellia sinensis* green tea extract (2.5% (2,5g/100ml); 5% (5g/100ml) and 10% (10g/100ml)) and gamma irradiation at 2, 4, 6, 8 and 10kGy on total aerobic plate counts of casein cow milk (n=3; P<0.05) the results were expressed as $\log(CFU/CFU_0)$ [19].

GAMMA IRRADIATION REDUCE THE ALLERGENICITY OF MILK PROTEIN

Gamma irradiation can be a powerful method to decontaminate casein and also it can reduce the allerginicity and antigenicity of this protein [20]. Many studies were carried out to evaluate the application of food irradiation technology as a method for reducing milk allergies. Bovine alpha-casein [20] and β-lactoglobulin [93] were used as model food allergens for experiments on allergenic and molecular properties by gamma irradiation. Others authors were observed a change of allergenicity and antigenicity of Bovine alpha-casein after irradiation. They were used milk-hypersensitive patients' immunoglobulin E (IgE) and rabbit IgGs individually produced to Bovine alpha-casein [20]. They confirm the changes of allergenicity and antigenicity of irradiated proteins were observed by competitive indirect enzyme-linked immunosorbent assay. Allergenicity and antigenicity of the irradiated proteins were changed with different slopes of the inhibition curves [20]. They were found that an increase of the turbidity showed that solubility of the proteins decreased by radiation, and this decrease probably caused by agglomeration of the proteins. They were concluding the epitopes on milk allergens were structurally altered by gamma irradiation [20].

CONCLUSION

Casein is used in different domains. It is more used to day as an ingredient in many food products and also as natural vehicles for bioactive. The structure and physiochemical properties of casein facilitate their natural or employed functionality in delivery tasks. Casein need to be sterilizing if it used in the medical application for example for a transport of chemotherapeutic drug. Gamma irradiation can be a powerful method to sterilize medical and surgical equipment and food materials but it known induces fragmentation and aggregation of protein. Green tea extract can reduce the impact of irradiation on the protein and it can prevent these fragmentation and aggregation probably polyphenol fraction in green tea extract play a most of action in this processes [19]. Polyphenol have a high antioxidant activity. We also were found in our previous study, that D_{10} was lowered to 2 kGy for all concentrations of green tea extract, which gives us the right to suggest that a decrease of the dose of bacterial decontamination D_{10} 2 kGy in the absence of *Camellia sinensis* green tea extract at a dose lower than 2 kGy in the presence of *Camellia sinensis* green tea extract is due to the additive effect of green tea to the effect of gamma irradiation [19]. This effect is comparable to synergistic combinations of compounds of green tea (catechins) and antibiotics [94-97]. It was found also in another study that tea polyphenols inhibited the development and growth of bacterial spores and reduced the heat resistance of thermophilic spores [98]. It is an advantage to reduce the probability of contamination of foodborn also to reduce the contamination of casein by vegetative or spore form of microorganism when was irradiated in the presence of green tea. Irradiation can also reduce antigenity and allergenicity of protein. When quantifying allergens in irradiated solutions, the amount of intact allergens in an irradiated solution was reduced by gamma irradiation depending upon the dose [20]. These results indicate that epitopes on allergens were structurally altered by gamma irradiation and immunoglobulin did not recognize well antigen-determinant sites on allergens [20].

REFERENCES

[1] Hambraeus, L., (1992). Nutritional aspects of milk proteins, In Advanced Dairy Chemistry, Vol 1: Proteins; Fox, P.F., Ed. Elsevier Science Publishers: Essex, 457-490.

[2] Mulvihill, D.M., (1992). Production, functional properties and utilization of milk protein products. In Advanced Dairy Chemistry, Vol 1: Proteins; Fox, P.F., Ed.; Elsevier Science Publishers: Essex, 369- 404.

[3] Walstra, P., Geurts, T.J., Noomen, A., Jellema, A., & Van Boekel, M.A.J.S., (1999). Dairy technology: principles of milk properties and processes; Marcel Dekker: New York.

[4] Mahmoud, M.I., (1994). Physicochemical and functional properties of protein hydrolysates in nutritional products. Food Technol., 48, 89-95.

[5] Swartz, M.L., (1995). Milk proteins and hydrolysates in nutritional foods. *Food Market Technol.*, 9, 4-12.

[6] Meisel, H., (2005). Biochemical properties of peptides encrypted in bovine milk protein. *Curr. Med. Chem.*, 12, 1905–1919.

[7] Swaisgood, H.E., (1992). Chemistry of the caseins. In Advanced Dairy Chemistry, Vol 1; Proteins; Fox, P.F., Ed.; Elsevier Science Publishers: Essex, 63-110.

[8] Fox, P.F., (2003). Mc Sweeney PLH Advanced Dairy Chemistry, Proteins Part A, vol. 1, 3 edn. New York: Kluwer Academic/Plenum Publishers.

[9] Otani, H., & Hata, I., (1995). Inhibition of proliferative responses of mouse spleen lymphocytes and rabbit Peyer's patch cells by bovine milk caseins and their digests. *J. Dairy Res.*, 2, 339–348.

[10] Otani, H., Monnai, M., & Hosono, A., (1992). Bovine k-casein as inhibitor of the proliferation of mouse splenocytes induced by lipopolysaccharide stimulation. *Milchwissenschaft*. 47, 512–515.

[11] Wong, C.W., Seow, H.H.S., Liu, A.H., Husband, A.J., Smithers, G.W., & Watson, D.L., (1996). Modulation of immune responses by bovine b-casein. *Immunol. Cell Biol.*, 74, 323–329.

[12] Livney, Y.D., (2010). Milk proteins as vehicles for bioactives. *Current Opinion in Colloid & Interface Science* 2009, 15, 73–83.

[13] Hu, KL; Li, JW; Shen, YH; Lu, W; Gao, XL; Zhang QZ; & Jiang, XG., (2009). Lactoferrin-conjugated PEG-PLA nanoparticles with improved brain delivery: in vitro and in vivo evaluations. J Controll Rel., 134, 55–61.

[14] International Dairy Federation Standard, (1969). Compositional standards for casein (edible and industrial). IDF Standard, 45.

[15] Egota, H.Z., & Ma!olepszy, B., (2008). The decontamination of industrial casein and milk powder by irradiation. *Radiation Physics and Chemistry*, 77, 1108– 1111.

[16] Abu-Tarboush, H.M., Al-Kahtani, H.A., Abou-Arab, A.A., Baijaber, A.S., & El-Mojadid, M.A., (1996). Sensory and microbial quality of chicken as affected by irradiation and post-irradiation on storage at 4 °C. *J. Food Prot.*, 60,761–770.

[17] Waje, C.,Kim, M.Y.,Nam, K.C., Jo, C.,Kim, D.H., Lee, J.W., & Kwon, J.H., (2008). Effects of irradiation on the color, microbiological quality, and sensory attributes of frozen ground beef, pork, and chicken after 6 months at 6 °C. *Food Sci. Biotechnol.*, 17, 212–215.

[18] International Atomic Energy Agency, (2008). Clearance of irradiated food database. Available from: /http://www.iaea.org/NUCLEUS/nucleus/content/application/ FICdb/Food Irradiation Clearance. jspS.

[19] Saloua K.S., Salah, K., Nasreddine B., Samia, A., Mouldi, S., & Ahmed, L., (2011). Sterilization and protection of protein in combinations of *Camellia sinensis* green tea extract and gamma irradiation. *International Journal of Biological Macromolecules*, 48, 452–458.

[20] Lee, J.W., Kim, J.H., Yook, H.S., Kang, K.O., Lee, S.Y., Hwang, H.J., Byun, M.W., (2001a). Effects of gamma radiation on the allergenic and antigenic properties of milk proteins. *J. Food Prot.*, 64, 272–276.

[21] Lee, J.W., Yook, H.S., Cho, K.H., Lee, S.Y., & Byun, M.W., (2001b). The changes of allergenic and antigenic properties of egg white albumin (Gal d 1) by gamma irradiation. *J. Korean Soc. Food Sci. Nutr.*, 30, 500–504.

[22] Ahn, H.J., Yook, H.S., Rhee, M.S., Lee, C.H., Cho, Y.J., & Byun, M.W., (2002). Application of gamma irradiation on breakdown of hazardous volatile N nitrosamines. *J. Food Sci.*, 67,596–599.

[23] Kim, J.H., Ahn, H.J., Kim, D.H., Jo, C., Yook, H.S., Park, H.J., & Byun, M.W., (2003). Irradiation effects on biogenic amines in Korean fermented soybean paste during fermentation. *J. Food Sci.*, 68, 80–84.

[24] Jo, C., Yook, H.S., Lee, M.S., Kim, J.H., Song, H.P., Kwon, J.S., & Byun, M.W., (2003). Irradiation effects on embryotoxicity and oxidative properties of gossypol dissolved in methanol. *Food Chem. Toxicol.*, 41, 1329–1336.

[25] Ahn, H.J., Kim, J.H., Jo, C., Kim, M.J., & Byun, M.W., (2004). Comparison of irradiated phytic acid and other antioxidants for antioxidative activity. *Food Chem.* 88, 173–178.

[26] Perez, M.B., Calderon, N.L., & Croci, C.A., (2007). Radiation-induced enhancement of antioxidant activity in extracts of rosemary (*Rosmarinus officinalis* L.). *Food Chem.*, 104, 585–592.

[27] Byun, M.W., Lee, J.W., Yook, H.S., Jo, C., & Kim, H.Y., (2002). Application of gamma irradiation for inhibition of food allergy. *Radiat. Phys. Chem.*, 63, 369–370.

[28] Byun, M.W., Kim, J.H., Lee, J.W., Park, J.W., Hong, C.S., & Kang, I.J., (2000). Effects of gamma irradiation on the conformational and antigenic properties of a heat- stable major allergen in brown shrimp. *J. Food Prot.*, 63, 940–9440.

[29] Filali-Mouhim, A., Audette, M., St-Louis, M., Thauvette, L., Denoroy, L., Penin, F., Chen, X., Rouleau, N., Le Vacr, J.P., Rossier, J., Potier, M., & Le Maire, M., (1997). Lysozyme fragmentation induced by gamma radiolysis. *Int. J. Radiat. Biol.*, 72, 63–70.

[30] Lee Y., & Song K.B., (2002). Effect of gamma-irradiation on the molecular properties of myoglobin. *J. Biochem. Mol. Biol.*, 35, 590–594.

[31]Grieb T., Forng R.Y., Brown R., Owolabi, T., Maddox, E., McBain, A., Drohan, W.N., Mann, D.M., Burgess, W.H., (2002). Effective use of gamma irradiation for pathogen inactivation of monoclonal antibody preparations. *Biologicals*, 30, 207–216.

[32]Amareld, R.W., Wersocki, M., Drohan, M., Drohan, W.N., Burgess, W.H., Mann, D.M., & Forng, R.Y., (2003). Controlled gamma-irradiation mediated pathogen inactivation of human urokinase preparations with significant recovery of enzymatic activity. *Biologicals*, 31, 261–264.

[33]Zbikowska, H.M., Nowak P., & Wachowicz B., (2006). Protein modification caused by a high dose of gamma irradiation in cryo-sterilized plasma: Protective effects of ascorbate. *Free Radical Biol. Med.*, 40, 536–542.

[34] Semo E., Kesselman E., Danino D., Livney Y.D., (2007). Casein micelle as a natural nano-capsular vehicle for nutraceuticals. *Food Hydrocolloids*, 21:936-42.

[35] Liu G., Franssen E., Fitch M.I., & Warner E., (1997). Patient preferences for oral versus intravenous palliative chemotherapy. *J Clin Oncol*, 15, 110-5.

[36] Shapira A., Assaraf Y.G., & Livney Y.D., (2010). Beta-casein nanovehicles for oral delivery of chemotherapeutic drugs. *Nanomed Nanotechnol Biol. Med.*, doi:10.1016/j.nano.2009.06.006.

[37] Horne D.S., (2002). Casein structure, self-assembly and gelation. *Curr Opin Colloid Interface Sci.*, 7: 456-61.

[38] Zana R., (2005). Dynamics in micellar solutions of amphiphilic block copolymers. In: Hubbard A.T., editor. Dynamics of surfactant selfassemblies. New York: CRC Press, Taylor & Francis Group, 161-231.

[39] Al-Haddad, K.S. H.S., Robinson, R.K., (2003). Salmonella survival. *Dairy Ind. Int.*, Vol. 68. P. 20–22.

[40] Olsen, S.J., Ying M., Davis, M. F., Deasy, M., Holland, B., Iampierto, L., Baysinger, C.M., Sassano, F., Polk, K., Gormoley, B., Hung, M., J., (2004). Multidrung-resistant *Salmonella typhimurium* infection from milk contaminated after pasteurization. *Emerg. Infect. Dis*. Vol. 10. P. 932–935.

[41] Šarkinas, A., 2000. Žalio pieno užteršimas listerijomis. *Maisto chemijair technologija*, T. 34. P. 40.

[42] Tamara A., Colette B., Ismail M., Laurent L. and & Karim A., (2011). Decontamination of solid and powder foodstuffs using DIC technology. *Advances in Computer Science and Engineering.*, 260-282.

[43] Hill, C., Cotter, P.D., Sleator, R.D. & Gahan, C.G.M. (2002). Bacterial stress response in *Listeria monocytogenes*: jumping the hurdles imposed by minimal processing. *International Dairy Journal*, Vol., 12, 273–283.

[44] Lou, Y.Q. & Yousef, A.E. (1997). Adaptation to sublethal environmental stresses protects *Listeria monocytogenes* against lethal preservation factors. *Applied and Environmental Microbiology*, Vol., 63, 1252–1255.

[45] Rajkovic, A., Smigic, N., Uyttendaele, M., Medic, H., De Zutter, L., Devlieghere, F. (2009). Resistance of *Listeria monocytogenes*, *Escherichia coli* O157 :H7 and *Campylobacter jejuni* after exposure to repetitive cycles of mild bactericidal treatments. *Food microbiology*, Vol., 26, 889-895.

[46] Rowan, N.J., (1999). Evidence that inimical food-preservation barriers alter microbial resistance, cell morphology and virulence. *Trends in Food Science & Technology*, Vol., 10, 261–270.

[47] Hope, E. W., (1901). Sterilisation and pasteurisation v. tubercle-free herds, & c. *The Lancet*, Vol. 158, Issue 4065, (July 1901), 197-198.

[48] Novella-Rodriguez, S., Veciana-Nogues, M., Roig-Sagues, A. X.,Trujillo-Mesa, A., J., Vidal-Carou, M. C. (2004). Evaluation of biogenic amines and microbial counts throughout the ripening of goat cheese from pasteurized and raw milk. *J. Dairy Res.*, Vol. 71. P. 245–252.

[49] Nair, M. K. M., Vasudevan, P., Hoagland, T., Venkitanarayanan, K. (2004). Inactivation of *Escherichia coli* O 157: H7 and *Listeria monocytogenes* in milk by caprylic acid and monocaprylin. *Food Microbiol.*, Vol. 21. P. 611–616.

[50] Ramanauskas, R., Alenčikienė, G. (2002). Aukštos koncentracijos parakazeino komplekso aktyviojo rūgštingojo įtaka jo reologiniams rodikliams. *Maisto chemija ir technologija*, T. 36. P. 129.

[51] Kim, S. R., Rhee, M. S., Kim, B. C., Lee, H. & Kim, K. H. (2007). Modeling of the inactivation of *Salmonella typhimurium* by supercritical carbon dioxide in physiological saline and phosphate-buffered saline. *Journal of Microbiological Methods*, Vol., 70, 131–141.

[52] Jo, C., Kim, D. H., Shin, M. G., Kang, I. J. & Byun, M. W., (2003). Irradiation effect on bulgogi sauce for making commercial Korean traditional meat product, bulgogi. *Radiation Physics and Chemistry*, Vol., 68, 851–856.

[53] Nevarez, L., Vasseur, V., Debaets, S., Barbier, & G. (2010). Use of response surface methodology to optimise environmental stress conditions on *Penicillium glabrum*, a food spoilage mould. *Fungal biology*, Vol. 114, 490-497.

[54] Chang, C.-H., Chiang, M.-L. & Chou, C.C. (2009).The effect of temperature and length of heat shock treatment on the thermal tolerance and cell leakage of *Cronobacter sakazakii* BCRC 13988. *International Journal of Food Microbiology*, 134, 184–189.

[55] Lin, Y.D. & Chou C.H. (2004). Effect of heat shock on thermal tolerance and susceptibility of *Listeria monocytogenes* to other environmental stresses. *Food Microbiology*, 21, 605–610.

[56] Broadbent, J. R. & Lin, C. (1999). Effect of Heat Shock or Cold Shock Treatment on the Resistance of *Lactococcus lactis* to Freezing and Lyophilization. *Cryobiology*, 39, 88–102.

[57] Lewis, M.J. & Heppell, N. (2000). Continuous thermal processing of foods Pasteurization and UHT Sterilization , Aspen Food Engineering Series, Maryland, USA, ISBN 0-8342-1259-5.

[58] Fan, X., Niemira, B. A., & Sokorai, K. J. B. (2003). Sensorial, nutritional and microbiological quality of fresh cilantro leaves as influenced by ionizing radiation and storage. *Food Research International*, 36, 713–719.

[59] Molins, R. A. (Ed.)., (2001). Food irradiation principles and applications. New York: Wiley- Inter science.

[60] WHO, (1988). Food irradiation a technique for preserving and improving the safety of food. Geneva, Switzerland: World Health Organization.

[61] Molins, R.A., Motarjemi, Y. & Käferstein, F.K. (2001). Irradiation : a critical control point in ensuring the microbiological safety of raw foods. *Food Control*, 12, 347-356.

[62] Chwla, S. P., Kim, D. H., Jo, C., Lee, J. W., Song, H. P., & Byun, M. W. (2003). Effect of γ irradiation on the survival of pathogens in Kwamegi, a traditional Korean semidried sea food. *Journal of Food Protection*, 66, No 11, 2093–2096.

[63] Fu, J., Shen, W., Bao, J., & Chen, Q. (2000). The decontamination effects of γ irradiation on the edible gelatin. *Radiation Physics and Chemistry*, 57, 345–348.

[64] Mahrour, A., Caillet, S., Nketsa-Tabiri, J., & Lacroix, M. (2003). Microbial and sensory quality of marinated and irradiated chicken. *Journal of Food Protection*, 66, No 11, 2156– 2159.

[65] Thomas, J., Senthilkumar, R.S., Raj Kumar, R., Mandal, A.K.A. & Muraleedharan, N., (2008). Induction of gamma irradiation for decontamination and to increase the storage stability of black teas. *Food Chemistry*, 106, Issue 1, 180-184.
[66] C.A.C., (2003). Codex general standard for irradiated foods. CODEX STAN 106–1983, Rev. 1.
[67] W.H.O., (1981). Wholesomeness of irradiated food: report of a Joint FAO/ I.A.E.A./W.H.O. expert committee. In World Health Organization technical report series 659, 36. Geneva, Switzerland: World Health Organization.
[68] F.D.A., (2005). Irradiation in the production, processing and handling of food. *Federal Register*, 70, (157), 48057–48073.
[69] Lacroix, M., & Quattara, B., (2000). Combined industrial processes with irradiation to assure innocuity and preservation of food products a review. *Food Research International*, 33, (719–724).
[70] Schuessler, H., & Schilling, K., (1984). Oxygen effect in the radiolysis of proteins. *International Journal of Radiation and Bioliogy*, 45, 267–281.
[71] Le Maire, M., Thauvette, L., De Foresta, B.,Viel,A., Beauregard,G., & Potier, M., (1990). Effects of ionizing radiations on proteins. *Biochem. J.*, 267, 431–439.
[72] Sunae, M., & Kyung, B.S., (2001). Effect of g-irradiation on the molecular properties of ovalbumin and ovomucoid and protection by ascorbic acid. *Food Chemistry*, 74, 479-483.
[73] Wolff, S.P., Garner, A., & Dean, R.T., (1986). Free radicals, lipids and protein degradation. *Trends Biochemistry and Science*, 11, 27–31.
[74] Assemand, E., Lacroix, M., & Mateescu, M. A., (2004). Protective role of l-tyrosine in the sterilization of Ceruloplasmin therapeutic protein by gamma-irradiation. *Radiation Physics and Chemistry*, 71, 403–407.
[75] Puchala, M., & Schessler, H., (1993). Oxygen effect in the radiolysis of proteins. *International Journal of Radiation Biology*, 64, 149–156.
[76] Stuart, A., Houee-Levin, C., & Potier, M., (2005). Radiation-induced protein fragmentation and inactivation in liquid and solid aqueous solutions. Role of OH and electrons. *Radiat. Phys. Chem.*, 72, 301–306.
[77] Garrison, W.M., (1987). Reaction mechanisms in the radiolysis of peptides, polypeptides, and proteins. *Chem. Rev.*, 87, 381–398.
[78] Davies, K.J.A., & Delsignore, M.E., (1987). Protein damage and degradation by oxygen radicals III. Modification of secondary structure and tertiary structure. *Journal of Biological Chemistry*, 262, 9908–9913.
[79] Lacroix, M., Jobin, M., Mezgheni, E., Srour, M., & Boileau, S., (1998). Polymerization of calcium caseinates solutions induced by gamma irradiation. *Radiat. Phys. Chem.*, 52, 1-6, 223-227.
[80] Getoff, N., Platzer, I., & Winkelbauer, C., (1999). Transients and cooperative action of beta carotene, Vitamin E and C in biological systems in vitro under irradiation. *Rad. Phys. Chem.*, 55 (5–6), 699–704.
[81] Boguta, G., & Dancewicz, A.M., (1983). Radiolytic and enzymatic dimerization of tyrosyl residues in insulin ribonuclease papain and collagen. *Int. J. Radiat. Biol.*, 43, 249–26.

[82] Stevens, C.O., Sauberlich, H.E., & Bergstrom, G.R., (1967). Radiation-produced aggregation and inactivation in egg white lysozyme. *Journal of Biology Chemistry*, 242, 1821–1826.

[83] Hasimoto, F., Ono, M., Masuoka, C., Ito, Y., Sakata, Y., Shimizu, K., Nonaka, G., Nishioka, I., & Nohara, T., (2003). Evaluation of the antioxidative effect in vitro of tea polyphenols. *Biosci. Biotechnol. Biochem.*, 67, 396-401.

[84] Diker, K.S., Akan, M., Hascelik, G., & Yurdakök, M., (1991). The bactericidal activity of tea against *Campylobacter jejuni* and *Campylobacter coli. Letters Appl. Microbiol.*, 12, 34–35.

[85] Toda, M., Okubo, S., Ikigai, H., Suzuki, T., Suzuki, Y., & Shimamura, T., (1991). The protective activity of the tea against infection by *Vibrio cholerae. J. Appl. Bacterial.*, 70,109–112.

[86] Chou, C.C., Lin, L.L., & Chung, K.T., (1999). Antimicrobial activity of tea as affected by the degree of fermentation and manufacturing season. *Int. J. Food Microbiol.*, 48, 125-130.

[87] Sakanaka, S., Raj, J.L., & Taniguchi, M., (2000). Antimicrobial effects of green tea polyphenols on thermophilic spore-forming bacteria. *J. Biosci. Bioengin.*, 90, 81-85.

[88] Caturla, N., Vera-Samper, E., Villalaín, J., Reyes, M. C., Micol,V., (2003). The relationship between the antioxidant and the antibacterial properties of galloylated catechins and the structure of phospholipid model membranes. *Free Radic. Biol. Med.*, 34, 648–662.

[89] Cushnie, T.P.T., & Lamb, A.J., (2005). Antimicrobial activity of flavonoids. International Journal of Antimicrobial Agents, 26(5), 343-356.

[90] Friedman, M., Mackey, B.E., Kim, H.J., Lee, I.S., Lee, K.R., Lee, S.U., Kozukue, E., & Kozukue, N., (2007). Structure–activity relationships of tea compounds against human cancer cells; *J. Agric. Food Chem.*, 55, 243–53.

[91] Juneja, V.K., & Friedman, M., (2007). Carvacrol, cinnamaldehyde, oregano oil, and thymol inhibit *Clostridium perfringens* spore germination and outgrowth in ground turkey during chilling. *J. Food Prot.*, 70, 218–222.

[92] Balentine, D.A.S.A., Wiseman, L.C., & Bouwens, L.C., (1997). The chemistry of tea flavonoids. *Crit. Rev. Food Sci. Nutr.*, 37, 693–704.

[93] Byun, M.W., Lee, J.W., Yook, H.S. , Jo, C. & Kim, H.Y.(2002). Application of gamma irradiation for inhibition of food allergy. *Radiation Physics and Chemistry*, 63, Issues 3-6, 369-370.

[94] Yam, T.S., Hamilton-Miller, J.M., & Shah, S., (1998). The effect of a component of tea (*Camellia sinensis*) on methicillin resistance, PBP2'synthesis, and beta-lactamase production in *Staphylococcus aureus. J. Antimicrob. Chemother.*, 42, 211–216.

[95] Hatano, T., Kusuda, M., Hori, M., Shiota, S, Tsuchiya, T, Yoshida, T., (2001). A tea polyphenol formed from epigallocatechin gallate, suppresses antibiotic resistance of methicillin-resistant *Staphylococcus aureus*, *Planta Med.*, 69, 984–989.

[96] Hu, Z.Q., Zhao, W.H., Yoda, Y., Asano, N., Hara, Y., & Shimamura, T., (2003). Additive, indifferent and antagonistic effects in combinations of epigallocatechin gallate with 12 non-b-lactam antibiotics against methicillin-resistant *Staphylococcus aureus*, *J. Antimicrob. Chemother.*, 50, 1051–1054.

[97] Lee, Y.S., Han, C.H., Kang, S.H., Lee, S.J., Kim, S.W., Shin, O.R., Sim, Y.C., Lee, S.J., & Cho, Y.H., (2005). Synergistic effect between catechin and ciprofloxacin on chronic bacterial prostatitis rat model. *Int. J. Urol.* 12, 383 –389.

[98] Senji, S., Lekh, R. J., Makoto, T., (2000). Antimicrobial effects of green tea polyphenols on thermophilic spore-forming bacteria. *Journal of Bioscience and Bioengineering*, 90, Issues 1, 81-85.

In: Casein: Production, Uses and Health Effects
Editors: A. M. Ventimiglia, et. al, pp. 141-158
ISBN: 978-1-62100-129-4

Chapter 7

USING CASEIN PROTEIN FOR MICROSPHERES' SYNTHESIS, ENCAPSULATION DRUG AND TEXTILE COATINGS

Ulyana Shimanovich[1], Suyeon Kim[2], Artur Ribeiro[2], Artur Cavaco-Paulo[2] and Aharon Gedanken[1]

[1] Department of Chemistry and Kanbar Laboratory for Nanomaterials
Bar-Ilan University Center for Advanced Materials and Nanotechnology,
Bar-Ilan University, Ramat-Gan 52900, Israel.
[2] University of Minho, Textile Engineering Department,
P-4800058 Guimaraes, Portugal.

ABSTRACT

Casein is a protein that is found in milk and used independently in many foods as a binding agent. Presently, there are numerous applications of casein such as coatings, adhesive, as a paper packaging material, in the leather, textile, and food industries. In this review we describe the use of casein as a building block for the fabrication of proteinaceous microspheres (PM) and film materials. Casein can be used as an interface material holding stable microsphere emulsions made from water and hydrophobic solvents, as well it can be used as matrix material for producing stable films for medical applications. The current manuscript will attempt to review the experiments that have been conducted for the production of microsphere emulsions and the encapsulation of drugs inside the casein spheres. Moreover, the paper will describe the chemical modifications of the casein protein made with the bifunctional compound, glutaraldehyde.

Keywords: casein; sonochemistry; drug encapsulation; microspheres; textile coating.

The current manuscript will attempt to describe possible applications of the casein protein. The following applications of casein protein will be discussed: use of casein for microcapsule formation, and surface coatings. We have used ultrasonic waves in our research for the formation of the microcapsules named "proteinaceous microspheres" (PM) and surface coating. Applications of ultrasound can be found in many areas, but sonochemical processes are most widely developed for heterogeneous reactions.

In addition, the protein (casein)-flavonoid conjugate, besides non-enzymatic covalent bonding, was produced by enzymatic radical generation using laccase (EC 1.10.3.2). A multicopper polyphenol oxidases found to catalyze several amino acids such as tyrosine and cysteine that result in the peptide polymerization and certain proteins' crosslink, as well as fragmentation since the radical generation in proteins by enzymatic oxidation (Chung et al., 2003; Selinheimo et al., 2008; Steffensen et al., 2008). The medium size random coil protein; alpha-casein was studied as a modification model for intermolecular crafting into proteins of flavonoids such as catechin and quercetin by laccase, introducing new antioxidant property to the conjugates. The attachment of the casein-flavonoid conjugates onto flax fabrics was carried out under mild conditions.

THE USE OF ULTRASOUND FOR SYNTHESIS OF PROTEINACEOUS MICROSPHERES

The use of ultrasound in chemical reactions in solution provides specific activation based on a physical phenomenon: acoustic cavitation. Cavitation is a process in which mechanical activation destroys the attractive forces of molecules in the liquid phase. In the case of proteins it destroys electrostatic interactions between positively and negatively charged amino acids. Moreover, high intensity ultrasonic waves might cause cleavage of disulfide bonds in protein chain. Applying ultrasound, the compression of the liquid is followed by expansion, in which a sudden pressure drop forms small, oscillating bubbles of gaseous substances. These bubbles expand with each cycle of the applied ultrasonic energy until they reach an unstable size; they can then collide and/or violently collapse. There are two types of effects mediated by ultrasound: chemical and physical. When the quantity of bubbles is low - using standard laboratory equipment - it is mainly physical rate acceleration of reactions that plays a role. For example, a specific effect is the asymmetric collapse near a solid surface, which forms microjets. This effect is the reason why ultrasound is very effective in cleaning, and is also responsible for rate acceleration in multiphasic reactions, since surface cleaning and erosion lead to improved mass transport.

In our work we applied ultrasonic waves to casein. Using the ultrasonic emulsification method, casein proteinaceous microspheres were synthesized in a 3 minute reaction.

In 1990, Suslick and co-workers developed a method in which they used high-intensity ultrasound to make aqueous suspensions of proteinaceous microcapsules filled with water-insoluble liquids, and demonstrated the chemical mechanism of their formation. [1] Micrometer-sized gas- or liquid-filled microspheres can be produced from various kinds of proteins such as BSA,[2–4] human serum albumin (HAS),[5] hemoglobin (Hb),[6] and a mixture of two proteins.[7] The first liquid-filled proteinaceous microspheres (PM), prepared by Suslick, were made of BSA (bovine serum albumin) and were filled with n-dodecane, n-decane, n-hexane, cyclohexane, or toluene. The synthesis was conducted under high-intensity ultrasonic probe and $1.5x10^9$ microcapsules per mL were obtained upon sonicating the precursor solution under air or O2. The average diameter of the PM was 2.5 mm with a narrow size distribution (Gaussian distribution=_1.0 mm). Ultrasonic irradiation of human serum albumin (HSA) generates similar microcapsules to those of the BSA; the same is true for the PM of hemoglobin (Hb).[1] We applied ultrasonic waves to casein for the synthesis of proteinaceous microspheres made of casein. The difference between BSA-PMs and casein PMs was found in the average size of the obtained spheres. The microspheres made of casein protein were larger in diameter (2.6 μm) than BSA spheres (2.34 μm) prepared under the same conditions. The formation of larger spheres is due to the larger size of casein as compared with BSA. Electrical charge of -35 mV was measured for both types of microspheres (BSA and casein PMs).

The first question raised by Suslick and co-workers was how are the microcapsules are formed and what holds them together? [1] The answer to this question relates also to the importance of the nonoaqueous liquid. Emulsification must occur during the microscopic dispersion of the nonaqueous phase into the aqueous protein solution.

Ultrasonic emulsification is a well-known process and does occur in this biphasic system. Emulsification is necessary for microcapsule formation. However, if vortex mixing emulsification is used instead, microcapsules are not formed. Consequently, emulsification by itself is not sufficient for microsphere formation. Thermal or solvent denaturation (for which O2, N2, and Ar should give similar results) cannot explain the microcapsule permanence. Because Suslick found that PMs are created only in O2 or air, the answer must involve the chemistry associated with ultrasound radiation. Aqueous sonochemistry caused by the implosive collapse of bubbles produces OH• and H•.[8] The radicals so-produced form H2, H2O2, and, in the presence of O2, superoxide HO2•.[9-10] Hydroxyl, superoxide, and peroxide radicals are all potential protein cross-linking agents. Using various trapping agents, Suslick concluded [1] that the important oxidant involved in microcapsule formation is the superoxide radical. He proposed that the cysteine, which is present in BSA, HSA, Hb, and casein is oxidized by the superoxide radical. The microcapsules are held together by protein cross-linking through disulfide linkages from cysteine oxidation. Myoglobin (Mb), which has no cysteine, revealed a substantial decrease in microcapsule yield relative to Hb. The employment of a protein disulfide cleavage reagent, such as ithioerythritol, destroyed the PM. In a similar way, inhibiting the oxidation of cysteine by alkylation with N-ethylmaleimide reduced the formation of PM considerably. Yields and size distribution of human and BSA microbubbles were determined as a function of various experimental parameters. The ultrasound irradiation was conducted at 50 ^{0}C and lasted three minutes.[3] This irradiation time is typical for the optimal formation of the PM. If longer irradiation times are used, a reduction in the number of PM is observed. It is worth mentioning [4] that the difference in the formation of the liquid- and air-filled bubbles is in the position of the sonicator. In a typical synthesis of liquid-filled PM, the organic liquid is layered over a 5% w/v protein solution and the horn is positioned at the organic–water inter interface. For air-filled microbubbles, the horn is placed at the air–water interface.[4] The chemical nature of these microbubbles and the origin of their remarkably long lifetimes have been explored. The microbubbles are held together primarily by interprotein cross-linking of cysteine residues. The principal cross-linking agent is superoxide radical, created by the extremely high temperatures produced during acoustic cavitation.

ENCAPSULATING A DRUG IN A CASEIN PROTEINACEOUS MICROSPHERE

The creation of the microbubbles is a short process that lasts a few minutes. In Gedanken's laboratory the process was stopped after 3 min. It was demonstrated that if during this period a drug is found in the precursor mixture, it will be encapsulated in the PM. Tetracycline (TTCL) was successfully encapsulated in BSA and casein microspheres.[11] TTCL is an antibiotic drug with a broad spectrum of activity, it is a relatively safe drug that can be used by many routes of administration, and is widely used. The encapsulation was achieved in a one-step, 3 min, sonochemical process, starting with the native aqueous solution of the BSA and TTCL over layered by an organic solvent, dodecane. The product was analyzed and characterized by scanning electron microscopy (SEM) and dynamic light scattering (DLS) measurements. The amount of TTCL loaded in the casein microspheres was also determined. TTCL loaded in casein microspheres was prepared by using the sonochemical method. Dodecane (6.7 ml, 98.0% Fluka) was layered over of 10 mL aqueous casein solution (casein sodium salt, Sigma-Aldrich) mL, 5% w/v). A separation flask was used to separate the product from the mother solution. The separation was accomplished within a few minutes due to the lower density of the microspheres relative to water. We waited 24 h to ensure a complete phase separation.

The preparation of TTCL-loaded microspheres was repeated by using a different concentration of the drug in the precursor solution. The amount of TTCL loaded in the microspheres was determined by subtracting the amount of the drug in the residual microsphere phase (the lower phase in the separation flask) from its total amount in the precursor solution. The sonochemistry did not destroy the TTCL, as evidenced by the very small changes in the TTCL concentration (less than 5%) that occur during sonication in the absence of protein. No residue of TTCL was found in the excess dodecane (the upper phase).

The distribution of the total amount of tetracycline (encapsulated, not encapsulated and the residue in the solution) was determined by absorption measurements as explained below. The UV studies of the amount of TTCL were based on measurements conducted on a Carry 100 spectrophotometer at 366 nm. The calculations were based on the Beer-Lambert Law: $A=\varepsilon \cdot c \cdot l$ (the TTCL molar extinction coefficient, ε, is $14.15 \cdot 10^{-3}$ L/mol·cm). The TTCL loading studies showed that the maximum TTCL loading capacity was found to be 65%. The percentage of the loaded drug in the BSA and casein microspheres increased with the increase in the concentration of the TTCL in the original solution. However, this behavior changed when the concentration of TTCL reached 3.6×10^{-3}M (the percentage of laded spheres under this concentration is 65%). At this concentration saturation is obtained, and the percentage of the drug in the microsphere does not grow with an increase of its concentration in the precursor solution. The increase in the amount of TTCL that was loaded in the microspheres can be explained by understanding the sonochemical method. The microspheres are formed by chemically cross-linking cysteine residues of the protein with an HO_2 radical formed around a micron-sized gas bubble or a nonaqueous droplet. The TTCL is hardly dissolved in the dodecane. However, it was found that TTCL is partially dissolves in dodecane after the sonochemical reaction. The droplet of the solution trapped upon the collapse of the bubble encapsulates the dodecane and the TTCL molecules. The higher the TTCL concentration in the solution, the higher the amount of TTCL loaded in the microspheres, until it reaches maximum. However, since the solubility of TTCL in dodecane is limited, saturation is obtained when this limit is reached. The saturation reached at a certain percentage of entrapped TTCL is due to molecules of TTCL leaving the microspheres through the walls, and the equilibrium that is attained between leaving and entering the molecules. An additional investigation was carried out to find whether the measured amount of tetracycline is due to molecules adsorbed on the outer surface of the sphere or due to the TTCL molecules encapsulated inside the microspheres. The first experiment involved heating (50 °C) the as-prepared solution in a separation flask for 5 h. TTCL that is adsorbed on the surface of the microspheres is known to dissolve in hot water. After a complete separation, the amount of TTCL in the residue phase (water) was measured. The average of a few experiments yielded 4% of TTCL adsorbed on the surface of the microspheres. The second experiment was carried out by measuring the absorbance of a sample of TTCL-containing casein microspheres before and after washing with ethyl acetate/diethyl ether (6:4). The results yielded a value of 1– 3% of TTCL adsorbed on the microsphere surface. The differences between the results of these two experiments can be explained by assuming that even during gentle heating (the first experiment), some microspheres can be destroyed. Nevertheless, these results indicate that most of the TTCL molecules are found inside the microspheres, and only a very small amount on its surface. The high capacity of the microspheres, produced by the sonochemical method, can be used in the future for antibiotic treatment.

Sonochemical Coating of Cotton and Polyester Fabrics with Tetracycline Loaded Casein Microspheres

The transmission of infectious disease agents in health-care facilities is an increasingly important concern to medical providers and the public.[12-15] The survival of microorganisms has occurred on textiles materials and plastics [14] and their transmission has been demonstrated as a result of surface contact with hands, fabrics, and hospital devices such as surgeons' gowns and nurses' clothing.[5-6] We have applied an ultrasonic waves for imparting cotton and polyester fabrics with durable antimicrobial functions.[16-17] The same method when applied to microspheres instead of NPs revealed that sonochemical radiation is an excellent technique for attaching drug-loaded proteinaceous microspheres to cotton and polyester fabrics. [18] Casein was used for coating all the fabrics experimented with in this research. "Antibacterial fabrics" were prepared by coating sonochemically cotton and polyester fabrics with TTCL-loaded proteinaceous microspheres. An organic solvent, such as dodecane, was used as the co-solvent, and the sonochemical process was continued for only 3 minutes. The creation and the attachment of TTCL-loaded PMs (proteinaceous microspheres) were achieved in a one-step process.

The morphology of the cotton and polyester bandages coated with TTCL-loaded casein microspheres was determined by light microscopy. Figure 1(a) presents the image of an uncoated yarn of cotton. The fibrous nature of the yarn is clearly observed. In Figure 1(b) we observe many microspheres attached to the cotton fiber. Figure 1(c) and 1(d) show an uncoated polyester fiber and a coated polyester fiber, respectively. Similar morphological results are observed for coating cotton and polyester bandages with casein microspheres encapsulated with a drug.

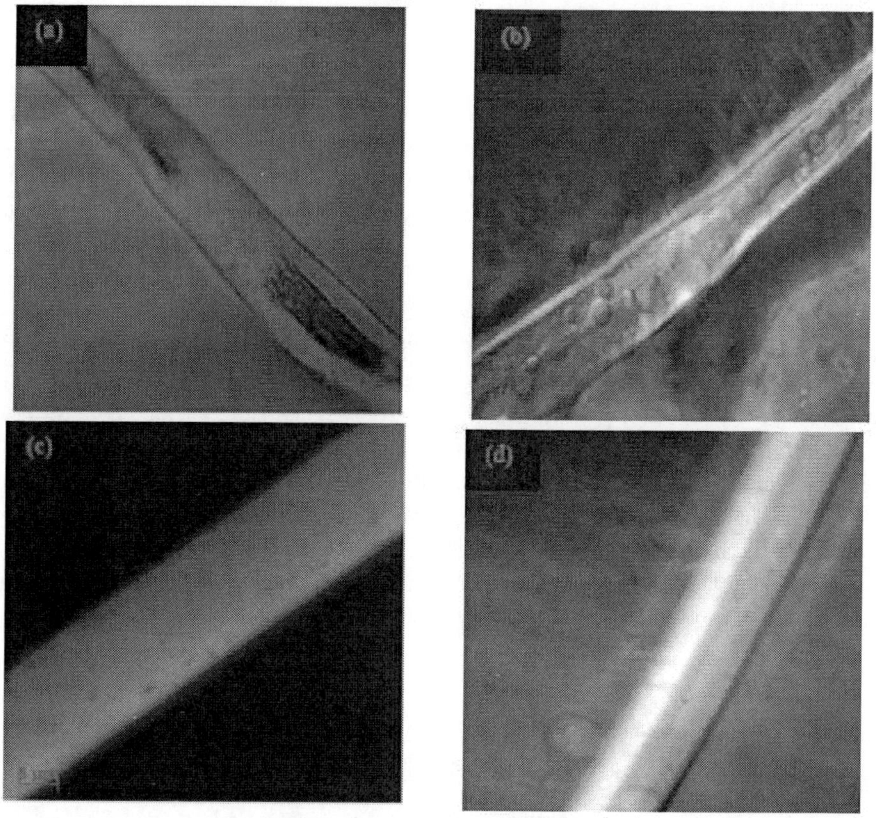

Figure 1. (a) An Apo-Tome image of a pristine wet cotton fiber; (b) Apo-Tome image of casein spheres on wet cotton surface; (c) Apo-Tome image of a pristine wet polyester fiber; (d) Apo-Tome image of casein spheres on polyester surface. For optical microscopy analyses the cross-linking agent gluteraldehyde was not used. Scale bar=5 micrometers.

The efficiency of the sonochemical method in the creation and attachment of drug-loaded microspheres to cotton and polyester fabrics was studied using UV spectrophotometric measurements. The sonochemistry did not destroy the TTCL, as evidenced by the very small changes in the TTCL concentration (2-3%) that occur during the sonication of an identical precursor solution sonicated for the same time in the absence of the casein protein. No residues of the antibiotic were found in the excess dodecane (the upper phase).

In order to find the optimal concentration of the TTCL in the PMs, the preparation of coated fabrics was repeated by using different concentrations of the drug in the precursor solution while the protein concentration remains constant. The TTCL-loading studies showed that the optimal TTCL concentration in the precursor solution for microsphere attachment to cotton and polyester fabrics is 5.4×10^{-3} M. Using this concentration led to the PMs attachment to the fabrics, containing the maximum amount of TTCL. Although the percentage of the drug (TTCL in PMs) increased with the increase in the concentration of the TTCL in the original solution (Figure 2), it has reached a maximum when the concentration of the TTCL was $5.4\text{x}10^{-3}$.

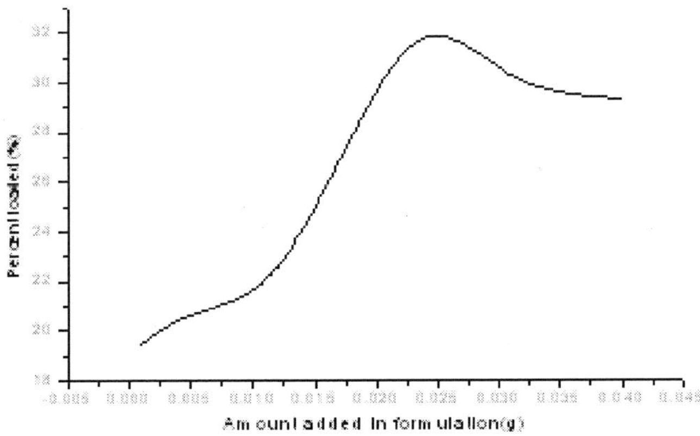

(A)

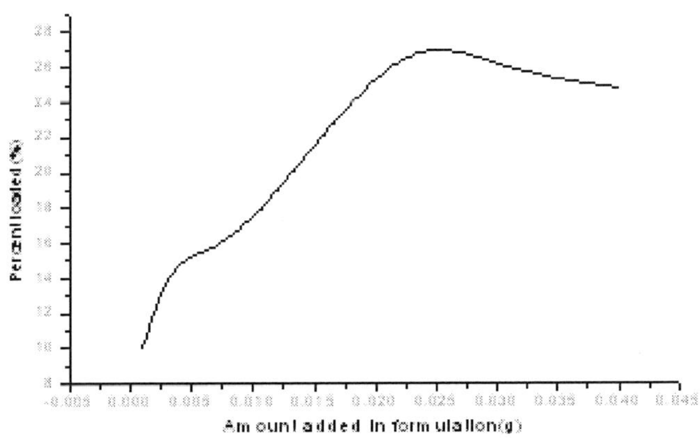

(B)

Figure 2. The amount of TTCL (in percent) that was encapsulated in casein microspheres attached to: (A) polyester and (B) cotton fabrics.

The percentage of the drug in the microspheres attached to the fabric's surface decrease with further increase of its concentration in the precursor solution. Reaching such a maximum can be explained as follows. The organic solvent is the major liquid to be found inside the newly formed microsphere. The solubility of TTCL in dodecane is limited, saturation is obtained when this limit is reached. Although some water can also be found in the microsphere its amount is small and can't add to the amount of encapsulated TTCL. After saturation was reached the curve of % TTCL lading (See Figure 2 A and b) goes down. This behavior could be explained by destruction of some part TTCL laded spheres with the time and releasing TTCL to the solution.

The way each of the steps taken to find the concentration of TTCL was described above the results is as follows. The results obtained from the long process heating (4 days at 50 ^{0}C) were: 26.9% of TTCL was found in/on the cotton fabric and 31.8% of TTCL was found in/on the polyester fabric. These amounts include not only the TTCL encapsulated in the microspheres but also the amount of free TTCL directly adsorbed to the cotton and polyester bandages. The direct adsorption of TTCL to cotton and polyester bandages was determined according to the above-mentioned process. Only 7.5% of TTCL was directly adsorbed to the cotton bandage and 11.4% of TTCL was adsorbed to polyester bandage.

To summarize the results and to find the exact amount of TTCL inside the microspheres attached to fabrics, we subtract the amount of TTCL directly adsorbed to fabrics from the total amount of TTCL which was found on the surface of fabrics. The same calculation was repeated after laundering the coated fabrics at the above-mentioned conditions. The results of calculation are presented in table 1.

Table 1. Calculation of amount (%) TTCL encapsulated in PMs attached to the surface of fabrics. Loss of TTCL during the "laundering process" was determined by measuring the amount of TTCL which remain in the laundering solution after the laundering process

Type of fabric	Total amount of TTCL on the surface of fabric (%)	TTCL directly adsorbed to fabric (%)	TTCL encapsulated in PMs attached to fabric(%)	TTCL directly adsorbed to fabric after "laundering test(%) "	TTCL encapsulated in PMs attached to fabric after "laundering test(%) "	Loss of TTCL in this "laundering" process(%)
Cotton	26.9	7.5	**19.4±2**	3.7	**17.6±1.8**	5.6
Polyester	31.8	11.4	**20.4±2**	5.1	**18.9±1.8**	7.8

These results point out that the loss of TTCL in this "laundering" process is 5.6 % and 7.8 % for the cotton and polyester, respectively. If we take into account that most of the removed TTCL is pristine TTCL, we can conclude that the microspheres are strongly bonded to these fabrics and perhaps can sustain a few machines washing cycles. Moreover, we have conducted the following control experiment. The adsorption of the TTCL loaded spheres (PMs with TTCL drug inside) to the surface of fabrics was studied without sonication. The pieces of cotton and polyester fabrics were incubated with the solution of TTCL loaded in PMs for 24 hours, and then washed several times with distilled water. The drug loaded PMs weren't found attached to the fabrics' surfaces and only ~2.6-3% of TTCL was directly adsorbed to the surface of fabrics. The results indicate that drug loaded PMs can be attached to the fabrics only when ultrasonic radiation was applied.

We have also examined the amount of protein attached to the textiles by spectrophotometric analysis (at 280 nm.). First, the amount of protein (casein) left in the aqueous solution after sonication was subtracted from the total amount of protein introduced into the aqueous solution. The results indicate that the total amount of protein converted to microspheres is ~87% for cotton and 89% in the case of polyester fabric in the reaction cell, which is composed of PM bonded and non-bonded to the fabric. To determine the amount of protein bonded to the coated fabrics, we placed them in water in a glass vial and heated the solution for 4 days at 45-50 °C. Heating the solution destroyed completely the microspheres in the solution. The results showed that 47.3% of PM of casein protein was anchored to the polyester fabric, and 31.6% PM of casein was attached to the cotton fabric.

It is worth noting that when four pieces of 5×5 cm cotton or four pieces of polyester bandages were sonicated with the precursor solution, all four pieces were coated with PMs and the amount of protein on the surface of each piece of fabric was found to be ~20-22%.

Thus, we found that a maximum amount of four bandages could be coated simultaneously in our 50 mL sonication cell.

The average sizes and the electrical charges of the PMs formed in the presence of cotton and polyester fabrics in the reaction cell were examined by DLS measurements. The average size and the electrical charge of TTCL loaded in casein are 1282 nm and -15.7 mV, respectively. When pristine casein protein was sonicated the DLS results yielded spheres with an average size of 2.6 μm for casein PMs, and an electrical charge of -35 mV was measured for microspheres. The size and the electrical charge difference between pristine PMs and drug loaded PMs could be explained by influence of the TTCL on microspheres formation. The electrical charge was decreased due to the presence of TTCL molecules on the outer surface of the microspheres. The next step was to check the influence of each type of fabric on the size and electrical charge of the created microsphers. The average size distribution of the spheres formed in the reaction cell with a piece of cotton and polyester fabrics are 1328 nm and 680 nm, respectively. The electrical charges of the PMs formed in the reaction cell with a piece of cotton and polyester fabrics are -13.1mV and -8.16 mV, respectively. While the cotton coating exhibits PMs that are very similar in size and electrical charge to regular PM prepared without a fabric in the sonication cell, the size and the electrical charge are drastically reduced for the PMs anchored to the polyester fabric. These dramatic changes in size on the two fabrics were detected for the caseine spheres. A possible explanation is that a higher concentration of PMs was found on the polyester fabric due to the better bonding of the protein to the polyester, which forms many centers around which the polymerization continues. The many seeds distributed upon the fabric continue to grow, and since the amount of protein is spread over a larger amount of seeds, smaller PMs are formed. The decrease in electrical charge from -15.7mV (TTCL loaded PMs formed without presence of polyester in the reaction cell) to -8.16mV could be explained by the more important role of the TTCL molecules on the smaller spheres formed on the polyester fabric leading to a larger decrease of the -35mV detected for the bare PMs.

The antimicrobial activity of the TTCL loaded in casein microspheres attached to cotton and polyester bandages was tested on two bacterial strains (*Staphylococcus aureus and Escherichia coli*) that are sensitive to tetracycline. Coated bandages with microspheres loaded with TTCL have shown (see point 3 in the various images) an inhibition zone similar to inhibition zone of a commercial tablet of TTCL (about 27 mm for *S. aureus bacterial strains* and 19 mm for *E. coli*). The inhibition sone for the commercial tablet was 27mm for coated cotton versus *S. aureus,* the same size sone was observed also for the PM coated on cotton. For the *E. coli* the commercial tablet on cotton revealed an inhibition zone of 19 mm, while the PM coated on cotton showed an inhibition zone of 18mm. The zhones of inhibition for coated polyester: 26.5 and 27 mm for *S. aureus* (27mm for commercial tablet of TTCL and 26.5 mm for PM coated polyester); and 19 and 17 for *E. coli.* (17 mm for PM coated polyester and 19 mm for commercial TTCL tablet). The difference in inhibiting zones, for commercial tablet of TTCL and for PM coated cotton and polyester fabrics, is due to the difference in the amounts of TTCL in the commercial tablet and inside the microspheres attached to the surface of fabrics which are 30 μg and ~12 μg, respectively. If we take into account these quantities, the coated PM shows a stronger killing affect towards these two bacterial strains. The results of antibacterial tests are presented in Figure 3.

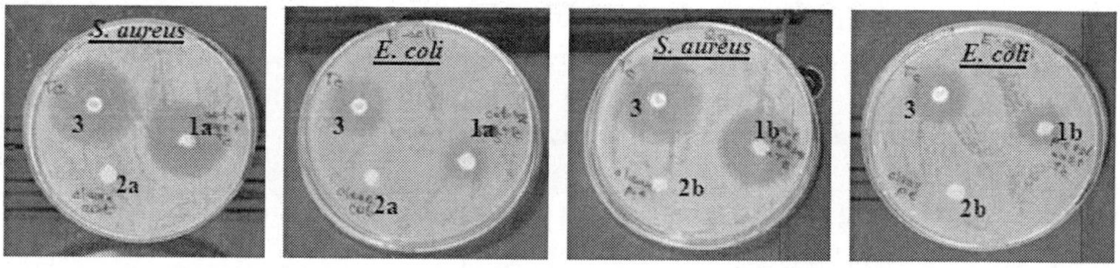

Figure 3. Zones of inhibition of *S. aureus* and *E. coli* on agar plates as a result of: (1a) TTCL freed from micospheres attached to cotton fabric; 1(b) TTCL freed from micospheres attached to polyester fabric; (2a) clean cotton fabric; (2b) clean polyester fabric; (3) commercial tablet of TTCL.

It seems that the TTCL trapped in the microspheres attached to the surface of cotton and polyester fabrics and released to the medium is as active as the TTCL freed from the commercial tablet. Nevertheless, the "inhibition zone" results indicate that the encapsulated TTCL "antibacterial" effect is superior to that of the tablet since a 1/3 of the amount of TTCL in the Tablet is needed to reach the same antimicrobial activity as that of the commercial TTCL. It is worth mentioning that the TTCL is released in a fast mode due to the proteases activity. These proteases destroy the PMs and the drug is release at once.

The sonochemically treated bandages of cotton and polyester coated with microspheres loaded with TTCLs were found to be active as antimicrobial agents.

LACCASE CATALYZED COUPLING OF CASEIN PROTEIN WITH FLAVONOIDS PRODUCING STABLE FILMS FOR MEDICAL APPLICATIONS

Proteins such as collagen, gelatin, casein, zein, elastin, and albumin are widely used in medical textiles for wound healing, tissue culture, tissue repair, and so on.[19] Numerous chemicals have been attempted to afford antimicrobial and antioxidant activity in spite of their toxicity and harmfulness to human. With the effort of seeking non toxic and non harmful materials, the flavonoids are getting much attention on the substitution of chemicals due to their high antioxidant property. With their specific properties, they are potentially useful for human health as protectors i.e. anti-carcinogens, anti-inflammatory agents, as well as inhibitors of platelet aggregation in '*in vivo*' and '*in vitro*' aspects.[20-21]

Some researches focused on the interactions between flavonoids and proteins.[22-23] These studies resulted in masking of antioxidant property on proteins and suggesting possible applications on the medical and food industries.[22,24-25] Through antioxidant masking on proteins, the oxidative damages involved in free radicals, which have noxious effect on cells provoking etiology of several diseases, can be protected.[25]

In this work, the catechin and quercetin, broadly researched flavonoids for bio applications were selected and crafted into proteins by laccase catalysed oxidation of α-casein. Flavonoids react with proteins of redox and oxidative coupling which resulted from one or two electron oxidation of the flavonoids.[26] This electron oxidation is eventually processed by the protein itself, and can be accelerated by enzymatic oxidation. The flavonoids oxidized products by enzymatic oxidation, i.e. aryloxylradical, quinines, and quinonoid compounds may react with nucleophilic or oxidizable amino acid residues. By this means, an irreversible modification of the proteins is produced by covalent coupling or oxidation.[24-25]

The oxidation of proteins by radicals' reaction can generally lead to various types of oxidative modification in addition to fragmentation and cross linking. Particularly, the detection of those modifications i.e. fragmentation and cross linking was generally performed by SDS-PAGE analysis [27-31].

The S1 and S2 components (23 kDa and 25 kDa) of α-casein are not usually detected by SDS-PAGE analysis appearing as one band around 35 kDa. When α-casein from bovine milk was oxidized by laccase in the presence of flavonoids, the remarkable band was detected above 200 kDa (line 6 and line 8 in Figure 4). They formed huge molecular weight too high to enter the gel resulting in staying at the entrance of the gel.[31] Interestingly, the casein protein incubation via laccase presented the fragmentation of proteins showing smaller molecular weight appearance than mother compounds MW around 35 kD (line 4 in Figure 4).

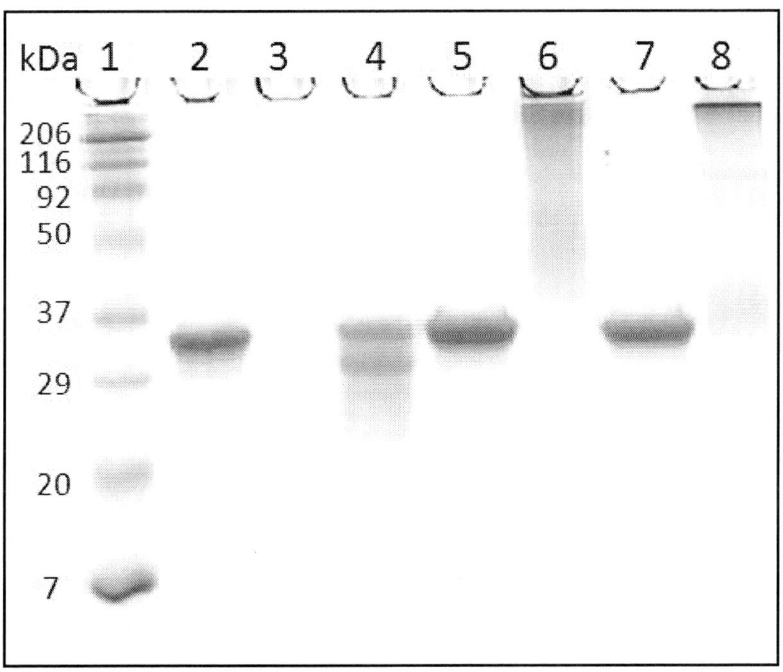

Figure 4. SDS-PAGE analysis of incubated α-casein with flavonoids. Lane 1: Mw marker, Lane 2: α-casein, Lane 3: Laccase, Lane 4: α-casein incubated with 5 units/mL of laccase, Lane 5: α-casein incubated with catechin, Lane 6: α-casein incubated with catechin and laccase, Lane 7: α-casein incubated with quercetin, Lane 8: α-casein incubated with quercetin and laccase.

The modification of α-casein by laccase catalysis performed in the presence of flavonoids results in rearrangements of molecules that were monitored by ultraviolet and visible spectrum.

When the enzymatic reaction was processed in casein solutions without flavonoids, the band shape was merely changed showing the absorbance increase at two positions (Figure 5). The increase of absorbance at two positions (absorbance around at 245 nm and 318 nm) was expressed by a linear relation between the dosage of laccase and absorbance intensity in α-casein. However, in the presence of catechin, the increase of absorbance intensity of bands after enzymatic incubation was detected and new absorption bands were detected as well. The band at 430 nm is such a band, which has appeared after enzymatic incubation. Its intensity has linearly increased depending on the laccase dosage. In casein and catechin incubated via laccase another new band emerged when more than 3 units of laccase were introduced in the reaction medium.

In the absence of laccase, the shape of bands were quite similar with that of protein itself showing no changes of the rearrangement of molecules involved by protein-flavonoids binding interaction. The polymerization of flavonoid compounds by enzymatic catalysis can be responsible for these results. The flavonoid polymers formed larger size of molecules which are increased of aromatic property and the density of hydroxyl groups comparing with catehcin/quercetin monomers.[32] Those polymers could have more affinity with proteins enhancing the potential of crosslink proteins. However, the most important factors of protein-flavonoids complex formation caused by crosslinking of protein molecules via phenols are broadly related to molecular size and complexity of flavonoids (trimer×dimer×monomer) which is primarily dependent on the number of reactive sites in the molecules being able to associate with proteins.[32-33] When a quercetin was applied as a binding bridge in protein modification, the band change by laccase dosage was less remarkable than catechin. The chemical structure of flavonoids can be responsible for binding protein. The substitution pattern and position of functional groups (hydroxyl group –OH) in flavonoids can determine the reaction pathway.[33] Moreover, the quercetin solution prepared in 10 % ethanol formed small unsolved solid particles indicating a decrease of solubility by time that could disturb the interaction with protein in liquor condition as well as instrumental analysis. Due to their physical changes, amplified at higher concentration, the sample solutions were further diluted using buffer solution as same ratio (20 times) then UV-visible spectrum was monitored and that can be also reason for shape of bands.

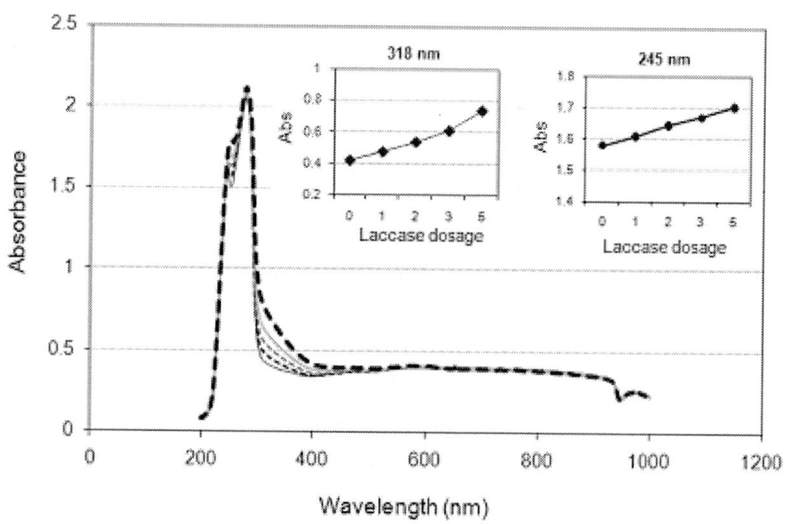

(A) Casein

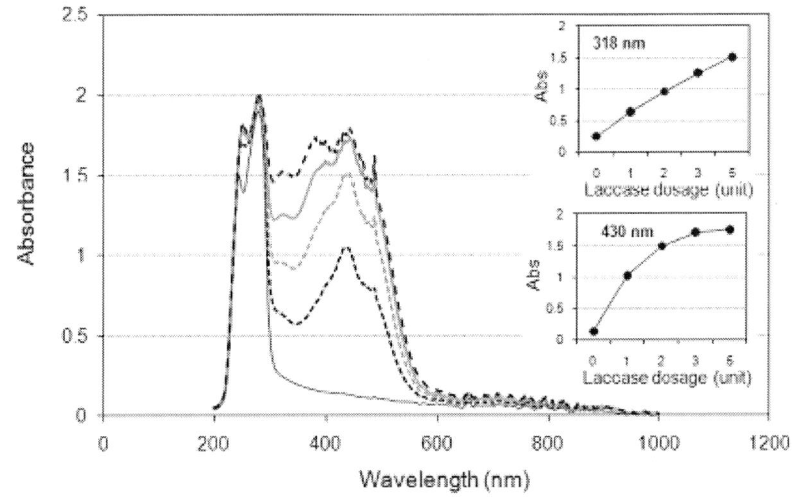

(B) Casein and Catechin

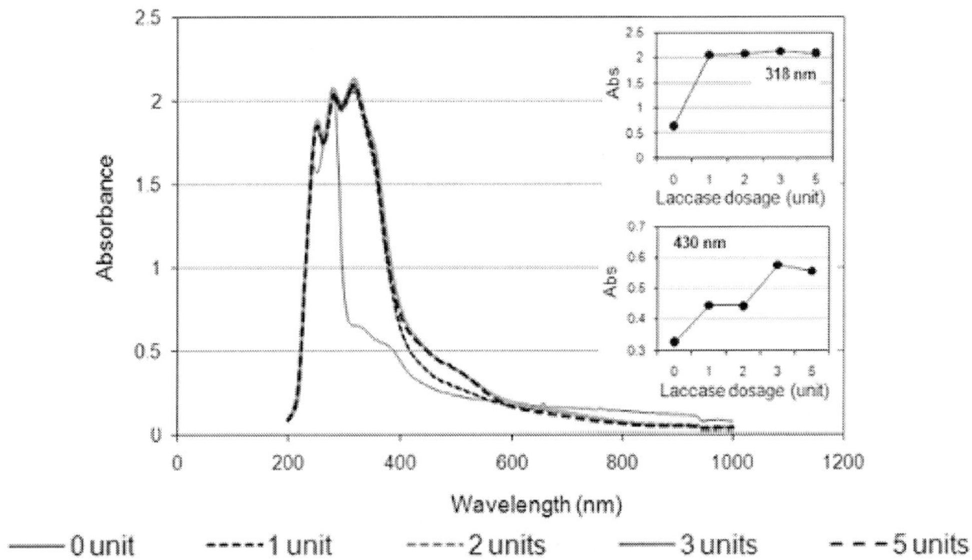

(C) Casein and Quercetin

Figure 5. The UV-Visible spectrum of α-casein protein incubated via laccase with and without flavonoids. (A) α-casein proteins incubated with different dosage of laccase such as 0, 1, 2, 3, and 5 unit/mL-1. Intensity values of at 318 and 245 nm were taken and presented linear relation. (B) α-casein proteins incubated with different dosage of laccase such as 0, 1, 2, 3, and 5 unit/mL-1 in the presence of catechin. New band was detected at 430 nm and intensity value was taken and presented as graph. (C) α-casein proteins incubated with different dosage of laccase such as 0, 1, 2, 3, and 5 unit/mL-1 in the presence of quercetin.

The surface electrical properties of proteins were analysed in terms of zeta potential measurement to verify the protein stability in reaction aqueous medium and presented in Table 2. The charge generation is determined depending on the amino acids locating on the surface of protein structures. The pH of reaction condition plays a major role in the magnitude of the zeta potential, particularly the isoelectric point of proteins.[34] However, other ions from the solution environment can modify the surface charge when they bind the proteins. By this reason, the modified casein by laccase oxidation was characterized conditional??? on the reaction factors in the presence of flavonoids. The measurements of zeta-potential by dynamic light scattering are generally sufficient to detect the oxidation degree of proteins in the medium. The excess of protein oxidation could be resulted in increasing of the zeta-potential values closing to the isoelectric point (ζ=0). The casein presented zeta potential (ζ) value around at -31 mV in the buffer solution at pH 7.4. After an enzymatic incubation at 50 °C for 24 hours in the presence of phenolic molecules an increase of ζ absolute values was detected in α-casein. That might be caused by the covalent binding of the reactive o-quinones into molecules of α-casein.

Table 2. Zeta potential (ζ) values of α-casein obtained before and after enzymatic oxidation. IC: incubation with laccase (2 U/mL) for 24 hours at 50 °C

	Casein	**Casein + Catechin**	**Casein + Quercetin**
Before **IC**	-30.9 ± 0.2	-27.8 ± 3.5	-29.1 ± 0.7
After **IC**	-29.3 ± 0.3	-37.5 ± 3.8	-35.2 ± 3.2

In this study, the anchoring of new polymers resulting from laccase catalyzed oxidation was carried out onto flax fabrics. The anchoring process was performed onto flax fabrics previously cationized with PDDA and non-cationized.

Firstly, the cationization was carried out in order to positively increase the fabrics surface charge, improving the affinity of the protein-flavonoid conjugates. The cationic layer deposited onto flax fabrics was evaluated by coomassie blue dye colouration. This dye is commonly used for staining and detection of proteins due to its affinity with amino groups. The coomassie dye is negatively charged and stable in anionic form in solution, therefore it can have broad affinity with any type of protonated amino groups.[20] Hence, flax fabrics positively charged can highly react with coomassie blue dyes through ionic interaction. The colour strength of flax fabrics after coloration with coomassie dye was measured in terms of k/s. The k/s value are related with the amount of dye absorbed onto the surface of the fabrics.

Flax is a natural lignocellulosic fabric consisting of mainly cellulose compounds. Generally, phenolic polymers generated by laccase do not have affinity towards cellulose resulting in non-covalent fixation.[35] Raw flax contains the aromatic lignin compounds at the surface responsible for the natural colouration of the fabrics and can be removed by bleaching processes. In this research, the natural lignin compounds, obtained after scouring, were used to provide grafting points for protein-flavonoid conjugates attachment. Laccase oxidizes not only flavonoid phenolic compounds (catechin, quercetin) but also the phenolic lignin compounds of flax fabrics to reactive o-quinones.[36] Therefore, the grafting points (reactive quinones) can further chemically react with the functional groups of amino acids in proteins or flavonoid o-quinones via enzymatic or non enzymatic pathway.

In this research, the charge of proteins might be responsible for the established interactions with the cationized flax fabrics. α-casein has the isoelectric point at pH 5. In accordance with is charge behaviour depending on pH, it gets negative charge at pH 7.4, where the reactions were carried out. The proteins negatively charged can definitely react with the cationized fabric surface through ionic interactions.

The anchoring efficiency of the conjugated protein-flavonoid onto flax fabrics was performed by colour strength measurement analysis. After incubation of flax fabrics with protein-flavonoid solutions, the colour generation, resulting from catechin or quercetin oxidation by laccase, was evaluated by means of k/s determination. The values are presented as checksum of k/s, obtained in the visible region from 400 to 700 nm (Figure 6). The colour generation is proportional to the amount of conjugate attached to the surface of the fabrics and can be presented as measurement of the anchoring efficiency. The k/s results also present the effect of reaction factors i.e. sort of proteins and flavonoids, enzymatic reaction, and the charge property of flax fabric.

Comparatively to polyquercetin, polycatechin presented higher k/s values indicating higher affinity to flax fabrics. The polyphenolic compounds also presented higher affinity onto fabrics in the absence of protein at the conjugate. In the case of protein-flavonoid conjugate, some grafting points are occupied by the functional groups of the protein, therefore the amount of possible sites for polymer attachment decreased. As previously stated, a higher level of oxidation was achieved when casein is present on the conjugate, influencing therefore the amount of product that will be attached onto flax fabrics.

The cationic layer previously deposited onto fabrics seems to be the most determinant factor for an improvement of the affinity of conjugates onto fabrics surface, presenting a remarkably increase of k/s values (2~3 times higher) comparing with the non-cationized samples. The results also show low levels of attachment when the grafting points available, obtained by laccase catalysis of lignin, were considered. In this case the increase of k/s values was only about 105%, in comparison with the 800% of increase when cationized fabrics were used.

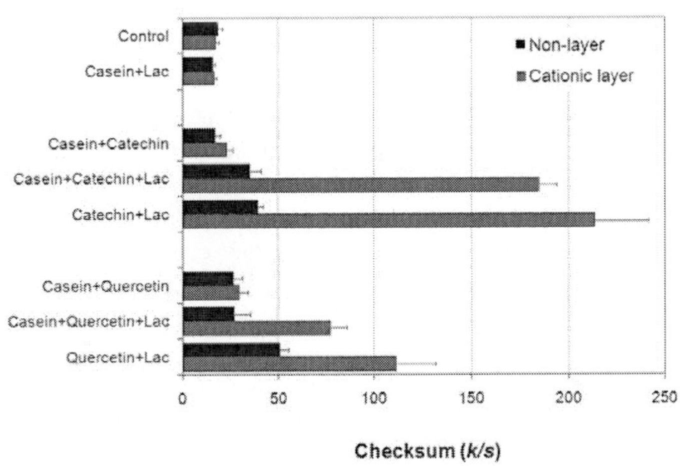

Figure 6. Estimation of casein-flavonoid conjugate binding onto flax fabrics measured in terms of colour strength (k/s obtained in the visible range from 400 to 700 nm).

The flavonoids are well known for their capacity as antioxidants by free radical scavenging and the OH group of B-ring, possessing electron donating properties and being a radical target, formed phenolic radicals (FIO').[37] By this reason, the antioxidant ability was expected on casein-flavonoid conjugates attached onto flax fabrics and evaluated by following the $ABTS^{\bullet+}$ cation decolourisation assay method and presented as the scavenging capability of $ABTS^{\bullet+}$ cation (Table 3).[38-39] Scoured or cationized samples, with non-attachment of conjugates presenting 0 % of antioxidant activity, were considered as controls.

Table 3. Antioxidant activity (%) of casein-flavonoid conjugates anchored onto flax fabrics; control: scoured or cationized samples, without conjugates attachment (0% antioxidant activity)

	Non-cationized	Cationized
Casein	-18	6.04
Casein+Catechin+Laccase	3.47	48.58
Casein+Quercetin+Laccase	15.85	96.23

Superior capacity level of the $ABTS^{\bullet+}$ cation scavenging was detected on previously cationized fabrics comparing with non cationized ones. The poly-quercetin developed much higher antioxidant activity on flax fabrics than poly-catechin. It can be explained by the structure differences between catechin and quercetin flavonoids used in these experiments. Generally, it is considered that a higher number of hydroxyl group substituent's in flavonoid structure results in a superior antioxidant activity [40]. Although, both catechin and quercetin possess the same number of substituted hydroxyl groups, quercetins possess a 4-oxo functional group and insaturation in the C-ring unlike catechin, which can play a role for an enhancement of the radical scavenging activity.

CONCLUSION

The above described studies highlight two different methods for the fabrication of proteinaceous casein microspheres and film materials. Using sonochemical radiation, we have succeeded in attaching a drug loaded proteinaceous casein microsphere to cotton and polyester fabrics by a one-step process that lasts a few minutes. We have found that if during the sonication time a drug and a piece of fabric are added to the precursor mixture containing an aqueous solution of a protein plus an over layering organic solvent, the drug will be encapsulated inside the microspheres and the spheres will be bonded to the fabric. This could be explained as due to the difference in the chemical structure of the fabrics. The coated fabrics are therefore suggested to be used as antibacterial fabrics. The coated bandages revealed an antibacterial activity stronger than that of a commercial TTCL tablet. The second method highlights the ability of laccase to catalyze the oxidation of proteins (α-casein) and flavonoids promoting the formation of conjugates with antioxidant properties. The level of enzymatic oxidation and further conjugation was measured and α-casein seems to be a promising protein for this purpose. Casein-flavonoid conjugates were successfully attached onto previously cationized flax fabrics. The protein-flavonoid conjugates can improve proteins natural properties being promising products to be used in medical, food and polymer industries where antioxidant ability is an essential feature.

REFERENCES

[1] K. S. Suslick, M. W. Grinstaff, *J. Am. Chem. Soc.* 1990, 112, 7807, and references therein.

[2] K. S. Suslick, M. W. Grinstaff, *Abstr. Pap. Am. Chem. Soc.* 1994, 207, 57-INOR.

[3] M. W. Grinstaff, K.S. Suslick, *Proc. Natl. Acad. Sci. USA.* 1991, 88, 7708 –7710.

[4] K. S. Suslick, M. W. Grinstaff, K. A. Kolbeck, M. Wong, *Ultrason. Sonochem.* 1994, 1, S65–S68.

[5] K. S. Suslick, M. W. Grinstaff, *Abstr. Pap. Am. Chem. Soc.* 1991, 201, 197-POLY.

[6] M. Wong, K. S. Suslick, *Abstr. Pap. Am. Chem. Soc.* 1995, 210, 587-INOR.

[7] U. Angel (Shimanovich), D. Matas, S. Michaeli, A. Cavaco- Paulo, A. Gedanken, *Chem. Eur. J.* 2010, 16, 2108 –2114.

[8] K. S. Suslick, Science 1990, 247, 1439.

[9] M. Del Duca, E. Jeager, M. O. Davis, F. Hovarka, *J. Acoust. Soc. Am.* 1958, 30, 301.

[10] B. Lippitt, J. M. McCord, I. Fridovich, *J. Biol. Chem.* 1972, 247, 4688.

[11] N. M. Green, *Biochem. J.* 1965, 94, C23.

[12] G. Sun, X. Xu, J. R. Bickett, J. F. Williams. *Ind. Eng. Chem. Res.* 2001,40,1016-1021.

[13] W. Whyte, R. Hodgson, B. V. Bailey, J. Graham. *Br. J. Surg.* 1978, 65, 469-474.

[14] A. N. Neely, M. P. Maley. *J. Clin. Microbiol.* 2000, 38(2), 724-726.

[15] S. F. Bloomfield, E. Scott. *J. Appl. Microbiol.* 1997, 83, 1-9.

[16] A. Hambraeus. *J. Hyg.* 1973, 71, 787-797.

[17] T. L. Vigo, M. A. Benjaminson. *Text. Res.* J. 1981, 51(7), 454-465.

[18] U. Angel (Shimanovich), C. M. Silva, A. Cavaco-Paulo, A. Gedanken. *Isr. J. Chem.* 2010, 50, 524 – 529.

[19] S. Petrulyte. *Dan. Med. Bull.* 2008, 55, 72-7.

[20] J.E. Chung, M. Kurisawa, H. Uyama, S. Kobayashi *Biotechnol. Lett.* 2003, 25, 1993–1997.

[21] P.J. Pietta. *Nat. Prod.* 2000, 63, 1035-1042.

[22] M.J.T.J.Arts, G.M.M. Haenen, L.C. Wilms, S.J.N. Beetstra, C.G.M.Heijnen, H. Voss, A. Bsat. *J. Agric. Food Chem.* 2002, 50, 1184-1187.

[23] C.D. Kanakis, P.A. Tarantilis, M.G.Polissio, S.Diamantoglo, H.A.Tajmir-Riahi. *J. Mol. Struct.* 2006, 798, 69–74.

[24] O.M. Andersen, K.R. Markham. 2006. Flavonoids: Chemistry, Biochemistry and applications. Chapter 9. Flavonoi-Protein Interactions. CRC Pubication, New York.

[25] M.J.T.J.Arts, G.R.M.M. Haenen, H.P. Voss, A. Bast. *Food Chem. Toxicol.* 2001, 39, 787–791.

[26] R. George, W. Witt, C. Kreutzfeldt. *J. Biochem. Biophys. Methods.* 1986, 4, 221-9.

[27] M.L. Mattinen, M. Hellman, P. Permi, K. Autio, N. Kalkkinen, J. Buchert. J. Agric. Food Chem. 2006, 53, 8883–8890.

[28] T. Michon, M. Chenu, N. Kellershon, M. Desmadril, *J. Gue'guen. Biochem.*1997, 36, 8504-8513.

[29] G. Oudgenoeg, R. Hilhorst, S.R. Piersma, C.G. Boeriu, H. Gruppen, M. Hessing, A.G.J.Voragen, C. Laane. *J. Agric. Food Chem.* 2001, 49, 2503- 2510.
[30] E. Selinheimo, P. Lampila, M.L. Mattinen, J. Buchert. *J. Agric. Food Chem.* 2008, 56, 3118-3128.
[31] C.L. Steffensen, M.L. Andersen, P.E. Degn, J.A. Nielsen. *J. Agric. Food Chem.* 2008, 56, 10202–12010.
[32] K.J. Siebert, *LWT*. 2006, 39, 987–994.
[33] D. Makris, J. Rossiter. *Food Chem.*2002. 77, 177-185.
[34] Y. Mukai, E. Iritani, T. Muras. *J. Membrane Sci.* 1997, 137, 271-275.
[35] S.G. Burton. *Current Organic Chemistry*. 2003, 7, 1317-1331.
[36] K. Lundquist, P. Kristersson. Biochem. J. 1985, 229(1), 277–279.
[37] D.Amić, D.Davidović-Amić, D. Bešlo, N. Trinajstić. *Croat. Chem. Acta*. 2003, 76, 51-61.
[38] D. Iveković, S. Milardović, M. Roboz, B. S. *Grabarić. Analyst*. 2005, 130, 708-714
[39] G. Miliauskas, P. R. Venskutonis, T. A. Beek. *Food Chemistry*. 2004, 85(2), 231-237.
[40] G. Cao, E. Sofic, R. L. Prior. *Free Radical Biology and Medicine*. 1997, 22(5), 749-760.

In: Casein: Production, Uses and Health Effects
Editors: A. M. Ventimiglia, et. al, pp. 159-176
ISBN: 978-1-62100-129-4

Chapter 8

GOAT CASEINS POLYMORPHISM: IMPLICATION IN CHEESEMAKING AND HUMAN HEALTH

Antonella Santillo [*] ***and Marzia Albenzio***
Department of Production and Innovation in Mediterranean Agriculture and Food Systems (PrIME). University of Foggia, Italy.

ABSTRACT

The genetic polymorphisms of milk proteins are of significance as they are associated to quantitative and qualitative parameters in milk. Extensive investigation of goat casein polymorphism revealed the presence of high numbers of alleles on the four casein loci. Goat casein polymorphism is associated with different casein synthesis levels and influence also fat content, micellar size and mineralization, and sensory attributes in milk therefore playing an important role both in milk coagulation performance and cheese quality. Genetic polymorphisms of milk proteins is also associated to different degrees of allergic reaction and to the generation of a wide spectrum of casein derived peptides able to exert multifunctional biological activities.

Selection and breeding of particular goat genetic lines could be performed for the production of milk destined to cheesemaking performance and of milk with hypoallergenic and functional properties.

Keywords: goat, casein polymorphism, milk processing, bio-functionality.

INTRODUCTION

Goat husbandry is widespread due to high adaptability of goat to different and extreme environments. Specialized dairy systems are mainly practiced in Europe and United States whereas extensive systems are located in developing countries. Numerous varieties of goat milk cheeses are produced worldwide, depending on diversity in locality, milk composition,

[*] Corresponding author. Tel. +39-0881-589326 Fax +39-0881-589331
E-mail address: a.santillo@unifg (A. Santillo)

and manufacturing techniques used. The genetic polymorphisms of milk proteins are of importance as they are associated to quantitative and qualitative parameters in milk. Casein loci in goat are highly polymorphic and numerous studies have been carried out to characterize the caseins both at a genetic and phenotypic level. Genetic polymorphisms of milk proteins play an important role both in eliciting different degrees of allergic reaction and for derived peptides with functional properties.

The purpose of the present review is to report information on the relationship of goat casein polymorphisms and milk coagulating performance and bio-functionality of goat milk.

Goat Husbandry and Milk Composition

Goat (*Capra hircus*) was domesticated as long ago as 7000 BC and has been reared for milk, meat and fibre production. The world total number of goats was 867.9 million and increased by 16% from 1999 to 2009 (FAOSTAT, 2009). The majority of the worlds' goats live in developing countries as Africa and Asia, but they thrive in all environments whether temperate, tropical, arid, humid, semi-humid, hot or cold desert. Goats nibble selectively and swallow a small bolus obtaining more energy per bite and therefore they need a proportionately smaller volume of food than cattle. Furthermore, they also consume a variety of plants and plant parts, some of them containing high levels of tannins and toxins; this is possible because they have proportionately larger livers equipped to more effectively detoxify compounds (Huston & Hart, 2003).

The management systems in goat husbandry are ascribed to i) intensive ii) semi-intensive and iii) extensive system. Intensive and semi-intensive system requires equipment for feeding, watering, milking and is practised especially for milk production in southern Germany, France, Italy, Spain, Greece, and USA. Extensive system is practiced in regions where pasture and land is available and goats are reared mostly for meat and fibre. Furthermore, in pastoral societies goats are used by pastoralists as part of their subsistence living as the Barbari breed in India (Wani, 2003) or the Beduin goat reared in the extreme desert in the Middle East (Degen, 2007).

Goats have principally cisternal milk (about 70%) and only small volume of milk comes from the alveoli during milking by the action of oxytocin. The larger cisterns play an important role in the storage of milk between milkings and can greatly affect the removal of milk at the time of milking. Machine milking parameters and settings need to be well integrated into goat dairy operations for they can have large impacts on stripping volumes, teat and thickness changes, oxytocin release and efficiency of milk removal (Marnet & McKusik, 2001).

The composition of goat's milk (Table 1) varies mainly with breed, nutritional and environmental factors, stage of lactation, parity and season (Tziboula-Clark, 2003). Casein micelles in goat milk differ from cow milk in less complete sedimentation rate, greater β-CN solubilisation, smaller size of micelles, more calcium and phosphorous, less salvation and low heat stability (Jenness, 1980).

Table 1. Gross composition of goat's milk

Component	Average (%, w/w)	Range (%, w/w)
Total solids	13.02	9.95-21.5
Fat	4.2	2.46-7.76
Crude protein	3.52	2.49-5.06
Casein	2.9	2.33-4.63
Lactose	4.52	3.62-6.30
Ash	0.8	0.69-0.89

Tziboula-Clarke, 2003.

Small ruminant dairying is usually based on local breeds that are well adapted to their production areas. In the European countries autochthonous goat breeds of the Mediterranean basin are characterized by high adaptability to climatic extremes and poor pasture. Milk protein characteristics of the autochthonous Garganica goat breed, reared on the Gargano promontory (Southern Italy), were studied by Albenzio et al. (2006; 2009). The study of the electrophoretic profile of Garganica goat milk (Figure 1) evidenced a band with the lowest electrophoretic mobility ascribed to para-κ casein. In the electrophoretic zone of the γ-CNs as detected in bovine milk (Bastian & Brown, 1996) and in ewe milk (Trujillo et al., 1997; Albenzio et al., 2004), appeared a group with at least six bands that was identified as β-CNs degradation products named γ-like components. The detection of a higher number of γ-like components in goat milk compared to bovine and ewe milk could be ascribed to the genetic polymorphism of caprine β-CN. It has been reported that in goat milk κ-CN shows a higher mobility than bovine κ-CN to such an extent that its bands overlap β-CN in starch gel (Russo et al., 1977) and in SDS-PAGE (Trujillo et al., 1997) under alkaline conditions. In Garganica goat milk the presence of two bands in the zone included between β-CN and α-CN was observed, which are ascribed to κ-CN polymorphisms in agreement with Caroli et al. (2001) and Angiolillo et al. (2002) in German and Italian goat breeds, respectively. The α_s-CN complex displayed three bands of almost equal intensity ascribed to α_{s1}- and α_{s2}-CNs.

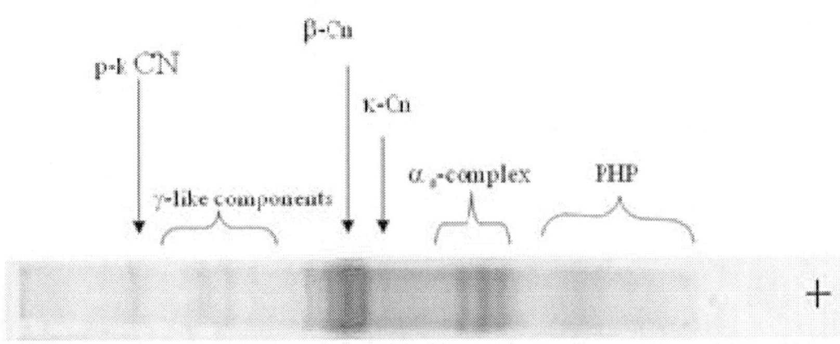

Adapted from Albenzio et al., 2006.

Figure 1. Urea-PAGE electrophoretogram of Garganica goat milk.

Lipids are present in goat milk in the form of globules, which are characteristic abundant in size less than 3.5 μm; this is advantageous for digestibility and a more efficient lipid metabolism compared with cow milk fat (Park 1994; Park et al., 2007). In milk of small ruminant higher contents of C6:0, C8:0, C10:0, and C12:0 than cow's milk were found (Alonso et al., 1999; Goudjil et al., 2004) due to differences in polymerization of the acetate produced by the rumen bacteria in goats. The main characteristic of goat milk fat is the high content in short- and medium- chain fatty acids which has at least twice as many C6:0-C10:0 fatty acids as cow milk fat (Raynal-Ljutovac et al., 2008). Medium-chain fatty acids, especially caprylic (C8:0) and capric (C10:0), have become accepted treatment for patients with malabsorption symptoms, a variety of metabolic disorders, cholesterol problems and infant malnutrition, because of their unique role of providing energy to the human metabolism instead of lipids to adipose tissue, and because of their ability to limit and dissolve serum cholesterol (Kalser, 1971; Tantibhedhyangkul & Hashim, 1978; Delacroix-Buchet & Lamberet, 2000; Parodi, 1999; Sébédio et al., 1999; Williams, 2000). Conjugated linoleic acids (CLA) content in milk is affected by many factors: animal feeding strategies i.e. pasture feeding and diets with plants and marine oils have been effective in enriching milk of small ruminants (Chilliard et al., 2005; Luna et al., 2005; Addis et al., 2005; Nudda et al., 2005; Cabiddu et al., 2006). Evidence from animal and/or human trials supports an influence of CLA on body composition, i.e. lowering of body weight and fat mass, on inhibition of carcinogenesis, on modulation of the immune system, and antidiabetic and antiatherogenic effect (Collomb et al., 2006).

Polymorphisms of Goat Casein and Cheesemaking Properties

The goat casein genes are linked on chromosome 6 as a cluster spanning about 250 kb (Hayes et al., 1993; Popescu et al., 1996) including CSN1S1 (α_{s1}-CN), CSN1S2 (α_{s2}-CN), CSN2 (β-CN), and CSN3 (κ-CN). Extensive investigation in goat milk revealed the presence of high numbers of alleles at the four casein loci: eighteen alleles at CSN1S1 locus, seven allels at CSN1S2 locus, eight alleles at CSN2 locus, sixteen alleles at CSN3 locus have been recently reported (Kupper et al., 2010).

The genetic polymorphisms of milk proteins are of importance as they are associated to synthesis rates and phosphorylation of the peptide chain of the corresponding casein in milk (Martin 1993, Grosclaude et al., 1994, Park et al., 2007, Caroli et al., 2007, Albenzio et al., 2009). Indeed, alleles ascribed to CSN1S1 are grouped into four classes according to α_{s1}-CN content in milk: strong alleles (A, B1, B2, B3, B4, B', C, H, L, and M) with a milk α_{s1}-CN content of 3.5 g/L; intermediate alleles (E and I) with 1.1 g/L; weak alleles (F and G) with 0.45 g/L; and null alleles (O1, O2, and N) without α_{s1}-CN. Alleles referred to CSN1S2 locus are characterized by three different levels of synthesis: A, B, C, E, and F are associate with α_{s2}-CN milk content of 2.5 g/L, D with lower levels and O allele with the lack of the specific casein. At least eight alleles have been found in goat CSN2 (Chessa et al., 2008): six of these alleles (A, A1, C, E, O and O') were characterized at the DNA level whereas variants B and D were described only at the protein level. The variants A, B, and C were associated with a normal β-CN content whereas O and O' with null content. Among ruminants goat shows the highest number of alleles at CSN3: thirteen are characterized by variations at the protein level whereas three showed silent mutations (Prinzenberg et al., 2005).

Albenzio et al. (2009) carried out a study on the genetic polymorphism of caseins in the milk of the Garganica goat: within the tested population, the distribution of CSN1S1, CSN1S2, and CSN2 genotype is shown in Table 2. A total of 6 haplotypes were observed: the haplotype *FAA* (CSN1S1 CSN1S2 CSN2) was the predominant one, followed by the haplotypes *AAB* and *AAA*. According to Sacchi et al. (2005) Garganica differs from other Southern breeds as Maltese and Jonica but is similar to the Northern breeds with a low frequency of the AAD haplotype. Although a complex distribution of genotype on casein loci was observed, the exact test of sample differentiation based on haplotype frequencies grouped the Garganica goat milk samples into two clusters characterized by the highest frequency of strong (S-CSN1S1) or weak (W-CSN1S1) alleles at CSN1S1. Protein and casein contents were higher in the group characterized by strong allele than in the group with weak allele at CSN1S1 (Table 3). Moreover, Albenzio et al. (2009) set a multiple covariance analysis including casein genotype, SCC, goat milk composition as factors able to account for milk coagulation properties: casein genotype was the factor that accounted for a significant percentage of the total variability for goat milk renneting parameters. Several studies showed that milk components (Grosclaude 1988, Ambrosoli et al., 1988; Barbieri et al., 1995; Pierre et al., 1996; Clarke & Sherbon, 2000) and α_{s1}-CN specifically (Ryniewicz et al., 1996) improved milk coagulation properties. In fact, Clarke & Sherbon (2000) found that goat milk that lacked α_{s1}-CN contained significantly lower percentage of milk components and tended to have poor coagulation properties. Also, micellar size and mineralization, cheese yield, and sensory attributes were influenced by CSN1S1 polymorphisms (Pirisi et al., 1994; Rameuf et al., 1995; Lamberet et al., 1996; Martin et al., 1999; Albenzio et al., 2009). In addition, κ-CN polymorphisms i.e. levels of glycosylation and phosphorylation affected the susceptibility of goat milk to clotting enzymes (Amigo et al. 2000) with important technological implication by influencing the coagulation stages of renneting.

Table 2. Distribution of Garganica goats by CSN1S1, CSN1S2, and CSN2 genotypes

Genotype		
	Total	
	N	**%**
CSN1S1		
AA	13	18.3
AB	14	19.8
AF	17	23.94
BF	8	11.28
BE	1	1.41
CF	6	8.45
BN	2	2.82
F0	1	1.41
AE	1	1.41
FF	1	1.41
AN	1	1.41
AC	2	2.82
BC	2	2.82

Table 2. (Continued).

Genotype		
	Total	
	N	%
C0	1	1.41
NN	1	1.41
Total	71	100
CSN1S2		
AA	17	23.94
A0	26	36.7
AF	24	33.8
F0	1	1.41
FF	3	4.22
Total	71	100
CSN2		
AA	20	28.17
BB	9	12.68
A0	17	23.94
AB	17	23.94
B0	6	8.45
00	2	2.82
Total	71	100

Albenzio et al., 2009.

Table 3. Gross composition of Garganica goat's milk

Item	Groups§		SEM	Effect, P
	S-CSN1S1	W-CSN1S1		
pH	6.73	6.7	0.02	NS[†]
SCC, $\log_{10}$, cells/ml	2.75	2.64	0.06	NS
Fat, %	5.79	3.63	0.11	***
Protein, %	4.27	3.45	0.06	***
Casein, %	3.3	2.66	0.04	***
Lactose, %	4.65	4.7	0.03	NS

§ S-CSN1S1= Group with high frequency of strong allele at CSN1S1; W-CSN1S1: group with high frequency of weak allele at CSN1S1

[†]NS, not significant,*** P<0.001

Adapted from Albenzio et al., 2009.

Results on calcium amount in milk from goats with different α_{s1}-CN genotype are conflicting: Grosclaude et al. (1994) observed higher amounts of calcium in milk from α_{s1}-CN-AA goats' than for α_{s1}-CN-FF goats whereas an opposite results was obtained from Schmidely et al. (2002).

Casein genotype influenced also milk fat content in milk (Zullo et al., 2005): higher synthesis of fat was found in the mammary gland of the α_{s1}-CN-AA goats than in the α_{s1}-CN-FF goats (Grosclaude et al., 1987; Mahé et al., 1993; Schmidely et al., 2002). However, the effect of α_{s1}-CN polymorphism on milk fat concentration is not yet completely explained (Chilliard et al., 2006). The relationship between casein polymorphisms and concentration and yield of fat in milk is not univocally reported in cows' (Ng-Kway-Hang, 1998) and ewes'milk (Pirisi et al., 1999). The fat concentration of cream from goats with the CSN1S1-FF genotype was 49% lower than that of AA genotype goats (Pitel & Delacroix-Buchet, 1994), whereas post milking lipolysis was higher (Lamberet et al., 1996). Neveu et al. (2002) reported that these two characteristics could be related to differences in the structure and/or size of milk fat globules or to other factors linked to differences in secretory mechanisms between FF and AA goats.

Polymorphisms of casein lead to amino acid substitution in the protein chain which in turn is responsible besides cheesemaking properties for the differences in digestibility and flavour of goat milk products (Rysstad et al., 1990). Taurine, glycine, and glutamic acid are the major free amino acids in goat milk (Rutherfurd et al., 2008). It was suggested (Rysstad, et al., 1990) that the relatively low amount of acetaldehyde observed in goats' milk yogurt is caused by feedback inhibition of threonine aldolase produced by the high amount of free glycine present in goats milk. This result is interesting for yogurt production from goat milk being acetaldehyde the main volatile aroma compound in yogurt. Peptides formed from goat milk casein by proteases taste much less bitter than those from cow milk casein (Pelissier & Manchon, 1976).

In the Mediterranean and many Eastern European countries, goat milk has gained economic importance mainly via its utilization in cheesemaking. Numerous varieties of goat milk cheeses are produced worldwide, depending on diversity in locality, milk composition, and manufacturing techniques used. There are large differences in the manufacture of goats' cheeses, but there are certain common steps that include the filtration of the milk, pasteurization or heating, and draining, moulding and ripening of the curds. Differences among goat milk cheeses are mainly attributed to the nature of physical and chemical changes during ripening, which are also influenced by chemicals, cultures or flavour ingredients added to curd during processing (Park, 2001). Most French goat milk cheeses are soft ripened cheese or soft lactic, either fresh or ripened (Raynal-Ljutovac et al., 2008); the well known Kefir is a lactic-alcoholic fermented milk beverage produced using goat milk (Quirós et al., 2005).

Cacioricotta goat cheese is the most common dairy product manufactured in Southern Italy (Pasqualone et al., 2003): its processing technology usually requires a starting step consisting in milk heating at temperature of about 90-95°C (Figure 2). During the high temperature step whey proteins are entrapped in the curd and could contribute to increased essential amino acids supplies such as cysteine, isoleucine, leucine, lysine, threonine and tryptophane (Raynald-Ljutovac et al., 2008).

Raw goat milk

↓

Heating of milk to 90°C

↓

Cooling of milk at 40°C

↓

Addition of artisanal kid rennet paste (50g/100)

↓

Coagulation of milk at 35°C after ca 30 min

↓

Cutting of coagulum

↓

Removal of the whey

↓

Curd held at ca. 20°C for 12 h

↓

Dry-salting of curd (1-4 days)

↓

Ripening at ca. 15°C and 95% relative humidity for 1 week

Adapted from Albenzio et al., 2006.

Figure 2. Production protocol of artisanal Cacioricotta cheese.

Polymorphisms of Goat Casein and Functional Properties of Goat Milk

Consumption of milk and dairy products, especially from goat is associated with beneficial health effects beyond its pure nutritional value (Silanikove et al., 2010). Park et al. (2007) reported that goat milk is more readily digested in the stomach because it forms a finer curd than cow milk following acidification. Moreover, 96% of goat casein was found to be completely hydrolysed *in vitro* by trypsin compared with cows casein with only 76-90% (Jasińska, 1995). As concerns whey protein only a small amount of goat's β-lg remains undigested in comparison to cow's after treatment with human gastric and duodenal juice (Almaas et al., 2006). Goat milk also contains a range of free amino acids that may be utilized directly by intestine (Duggan et al., 2002); taurine is particularly high being 20-40 fold higher than cow milk (Mahaia & Al-Kanhal, 1992) and it is involved in salt formation, osmoregulation, antioxidation, calcium transport and in the central nervous system. Thus, goat milk is a valuable source of taurin both for human neonate and adult.

Casein and whey proteins are the two main protein groups in milk that can be degraded into numerous peptides fragments by enzymatic proteolysis and serve as source of bioactive peptides. Bioactive peptides have been defined as specific protein fragments that have a positive impact on body functions or conditions and may ultimately influence health (Park, 2009). The size of active sequences may vary from two to twenty amino acid residues, and many peptides are known to reveal multifunctional properties (Meisel & FitzGerald, 2003) as opiate, antithrombotic, antihypertensive, immunomodulatory, metal-binding and antimicrobial activities.

Physiologically and functionally active peptides are produced *in situ* during gastrointestinal digestion (Korhonen & Pihlanto, 2007). Furthermore, in milk the liberation of peptide sequences is ruled by the action of main enzymatic system naturally occurring in milk, enzymes from microorganism such as lactic acid bacteria, and enzymes associated to milk processing.

Genetic polymorphism in goat milk protein associated with different milk protein composition may be responsible for the generation of a wide spectrum of casein-derived peptides (Meisel, 1997; Korhonen & Pihlanto, 2006) therefore the study of potential bioactivity of peptide sequences liberated upon hydrolysis is of particular interest. Santillo et al. (2009) investigated the effects of indigenous proteolytic enzymes on the release of bioactive peptides from Garganica goat milk. The authors evidenced the principal role of serine proteinase, especially plasmin, in the liberation of several peptides with potential bioactivity deriving from β-CN and α_{s2}-CN. Casein-derived peptides identified in Garganica goat milk were reported in Table 4. Almost 90% of the peptides identified shared structural homology with previously described bioactive peptides in caprine and bovine milk and dairy products. Some of the peptides have been shown to exert ACE-inhibitory activity (Quirós et al., 2005; Yamamoto et al. 1994), or displayed structural homology with peptides with antihypertensive (Yamamoto et al., 1994; Quirós et al., 2007) or antioxidant activity (Rival et al., 2001). As an example the sequence PYVRYL showed an ACE-inhibitory activity with an IC_{50} value of 2.4 μM in caprine Kefir (Quirós et al., 2005).

In fermented milks, the type of starter used is one of the main factor which influences the synthesis of bioactive peptides during food processing. Indeed, proteolysis, either from LAB or ripening strains, induces increased amounts of peptides and free amino acids (Raynald-Ljutovac et al., 2008). Several studies (Nakamura et a., 1995; Yamamoto et a., 1994; Meisel et al., 1997; Matar & Goulet, 1996; Rokka et al., 1997; Ganjam et al., 1997) reported that starter inoculated for fermented milk production enrich the product with antihypertensive, ACE inhibitory, opioid, immunostimulating, modulation of cancer cell proliferation peptides. Cell wall associated proteinase from *Lb. helveticus* CP790 liberated several casokinins (Yamamoto et al., 1994); in milk prefermented with *Lactobacillus casei* ssp. *rhamnosus*, and subsequently digested with pepsin and trypsin, was demonstrated the presence of ACE inhibitory peptides (Rokka et al., 1997).

Table 4. Casein-derived peptides identified in the fraction collected by RP-HPLC from the soluble nitrogen fraction of caprine milk incubated up to 7d

Fraction[a]	Elution time (min)	Observed mass[b]	Calculated mass[c]	Protein fragment	Sequence[d]	Score[e]
GP[1]						
F1	20.2	642.77	1925.0267	β-CN f(113-128)	**K.VKETMVPKH**KEMPFPK.Y	53
F2	22.3	667.45	1332.8129	β-CN f(180-191)	S.LSQP**KVLPVPQK.**V	26
F3	23.6	627.43	1252.6598	β-CN f(192-202)	K.VVPQ**RDMPIQA.F**	13
F4	27.8	724.37	1446.8	β-CN f(148-159)	N.**LHLPLPL**LQSWM.H	22
F5	28.8	712.6	1423.7823	β-CN f(205-217)	**L.LYQEPVLGPVRGP.F**	38
F6	29.2	700.92	1399.7282	β-CN f(152-203)	K.VVPQ**RDMPIQAF.**L	13
F7	29.9	905.88	2714.53	β-CN f(180-203)	S.LSQPK**VLPVPQK**VVPQ**RDMPIQAF.**L	19
F8	30.6	667.36	1332.6424	β-CN f(123-133)	K.EMPFPKYPVEP.F	36
F9	30.7	687.97	1373.7918	β-CN f(142-153)	T.LTDVEK**LHLPLP.L**	23
F10	31.8	857.13	2567.4682	β-CN f(180-202)	S.LSQPK**VLPVPQK**VVPQ**RDMPIQA.F**	19
F11	32.4	602.35	1202.6006	β-CN f(166-176)	P.LSPTVMFPPQS.V	18
F12	34.8	552.86	1103.5685	β-CN f(170-179)	T.VMFPPQSVLS.L	45
F13	37.4	795.02	1587.9236	β-CN f(140-153)	S.LTLTDVEK**LHLPLP.L**	13
F14	40.4	997.58	1993.1401	β-CN f(205-222)	**L.LYQEPVLGPVRGPFPILV.**V	53
F15	46	530.45	1587.9236	β-CN f(140-153)	S.LTLTDVEK**LHLPLP.L**	25
GP-P[1]						
F1	20.7	642.99	1926.3496	β-CN f(113-128)	K.**VKETMVPKH**KEMPFPK.Y	25
F2	22.3	445.33	1333.6186	β-CN f(180-151)	S.LSQPK**VLPVPQK.**V	13
F3	24.9	667.95	2001.3546	α_{s2}-CN f(205-221)	K.AMKPWTQPKTNAI**PYVR.Y**	19
F4	28.3	760.08	2277.6854	α_{s2}-CN f(205-223)	K.AMKPWTQPKTNAI**PYVRY.L**	21
F5	30	687.89	1374.624	β-CN f(142-153)	T.LTDVEK**LHLPLP.L**	95
F6	30.1	605.35	1209.3934	α_{s2}-CN f(214-223)	K.TNAI**PYVRY.L**	22
F7	35	530.36	1588.8855	β-CN f(140-153)	S.LTLTDVEK**LHLPLP.L**	21

[a]GP: Garganica pasteurized caprine milk; GP-P: Garganica pasteurized caprine milk incubated with pepstatin.

[b]The observed mass is the experimental m/z value.

[c] The calculated mass is the relative molecular mass calculated from the matched peptide sequence.

[d] Boldface characters represent the sequence of previously described bioactive peptides.

[e] The ion score for individual peptide is calculatyed by Mowse scoring algorithm form Mascot search (http://www.matrixscience.com).

Santillo et al., 2009.

Moreover, protolysis induced by fermentation and ripening process contributes primarily to cheese flavor but could also play a role in nutrition. Several bioactive peptides have been found in cheese and their activity depends on the intensity of proteolysis occurring during the ripening stage. However, it seems that bioactive peptides liberated during cheese ripening can be degraded into inactive fragments as a result of further proteolysis, especially in long ripened cheese (Addeo et al., 1992; Meisel et al., 1997) or in dependence of pH and salt concentration (Muehlenkamp & Warthensen, 1996).

Goat milk has gained importance in human nutrition as significant segment in many populations of developed countries is afflicted with cow milk allergies and gastro-intestinal disorders (Haenlein, 2004). In patients with cow milk allergy IgE-binding epitopes on α_{s1}-, α_{s2}- , κ-CNs, α-lactoalbumin, and β-lactoglobulin were recognized (Jarvinen et al., 2001; 2002; Natale et al., 2004) Genetic polymorphisms of goat milk protein influence the presence and level of synthesis of each protein fraction in milk playing an important role in eliciting different degrees of allergic reaction (Park, 1994; Saini & Gill, 1991) therefore goat milk can be considered as a proper alternative to human milk in infant nutrition (Slačanac et al., 2010). Low levels of α_{s1}-CN in goat milk means that its casein profile is closer to human milk than that of cow milk (Clark & Sherbon, 2000) as human milk lacks α_{s1}-CN. Indeed, milk from animals possessing mild alleles can be employed to produce milk for allergic subjects (Roncada et al. 2002). Due to high frequency of the weak allele F for CSN1S1, the presence of null allele for the same locus and the high frequency of A0 genotype at CSN1S2 locus in Garganica breed population, such milk could be exploited for cow milk allergy (CMA) subjects feeding (Albenzio et al., 2009). Campanozzi et al. (2008) studied the inflammatory response to Garganica goat milk in infant with CMA; these authors evaluated cytokines production by peripheral blood mononuclear cells (PBMCs) stimulated with whole milk, casein and β-lactoglobulin from cow and Garganica goat milk. In this study goat milk proteins lowered the production of proinflammatory cytokines and enhanced the release of antinflammatory ones from PBMC (Table 5).

The genetic information on goat CSN1S1 is not sufficient to select the animals suitable for the creation of hypoallergenic goat milk lines. The allergenicity of milk should be assessed by appropriate clinical studies before selecting the animals within a certain group of CSN1S1 genotype. Moreover, all casein genes should be considered to fix the most hypoallergenic casein combinations in these hypoallergenic lines (Ballabio et al., 2011).

Table 5. TNF-α, IL-10 and IL-12_{p70} (pg/mL) produced by Peripheral B lood Mononuclear Cells in infants affected by cow milk allergy

Item	TNF-α				IL-10		IL12p70		
	Infants with Cow Milk Allergy	Controls	*p*	Infants with Cow Milk Allergy	Controls	*p*	Infants with Cow Milk Allergy	Controls	*p*
Cow's caseins	1697 ± 749	435±118	<0.05	408±117	142±69	<0.05	132±19	49±3	<0.05
Goat's caseins	622±319	347±63	NS	610±65	202±9	<0.05	94±15	40±5	NS
Cow's β-lg	3090±1273	455±154	<0.05	244±43	63±7	<0.05	261±91	120±10	<0.05
Goat's β-lg	589±372	548±87	NS	109±23	262±12	NS	217±40	182±11	NS

Adapted from Campanozzi et al., 2008.

Conclusion

In conclusion goat milk is valuable from a nutritional point of view in terms of both protein and lipid fraction. The complex polymorphism of its casein fractions makes goat milk interesting for its high potential in cheese making and in human health exploitation.

It could be useful to select animals with strong alleles at CSN1S1 and CSN2 to increase casein content in milk and also because of the strong influence of casein genotype on fat content, milk coagulation properties, micellar size and mineralization, cheese yield and sensory attributes in milk. Moreover, animals with weak or null alleles for CSN1S1 and CSN1S2 should be used in breeding programs aimed at obtaining genetic lines for the production of milk for people afflicted with cow milk allergies and gastro-intestinal disorders.

Extensive investigation is still needed to completely define casein genotype and protein profile in goat milk for optimizing breeding programs for specific production, and could also represents a suitable strategy for preserving genotype of autochthonous goat breeds.

References

Addeo, F., Chianese, L., Salzano, A., Sacchi, R., Cappuccio, U., Ferranti, P., Malorni, A. (1992). Characterization of the 12% trichloroacetic acid-insoluble oligopeptides of Parmigiano-Reggiano cheese. *J. Dairy Res.* 59, 401-411.

Addis, M., Cabiddu, A., Pinna, G., Decandia, M., Piredda, G., Pirisi A., Molle, G. (2005). Milk and cheese fatty acid composition in sheep fed mediterranean forages with reference to conjugated linoleic acid *cis*-9, *trans*-11. *J. Dairy Sci.* 88, 3443–3454.

Albenzio, M., Caroprese, M., Marino, R., Santillo, A., Taibi, L., Sevi, A. (2004). Effects of somatic cell count and stage of lactation on the plasmin activity and cheese-making properties of ewe milk. *J. Dairy Sci.* 87, 533-542.

Albenzio, M., Caroprese, M., Marino, R., Muscio, A., Santillo, A., Sevi, A. (2006). Characteristics of Garganica goat milk and Cacioricotta cheese. *Small Rum. Res.* 64, 35-44.

Albenzio, M., Santillo, A., d'Angelo, F., Sevi, A. (2009). Focusing on casein gene cluster and protein profile in Garganica goat milk. *J. Dairy Res.* 76, 83-89.

Almaas, H., Cases, A.L., Devold, T.G., Holm, H., Langsrud, T., Aabakken, L., Aadnoey, T., Vegarud, G.E. (2006). In vitro digestion of bovine and caprine milk by human gastric and duodenal enzymes. *Int. Dairy J.* 16, 961-968.

Alonso, L., Fontecha, J., Lozada, L., Fraga, M.J., Juárez, M. (1999). Fatty acid composition of caprine milk: major, branched-chain, and trans fatty acids. *J. Dairy Sci.* 82: 878-884.

Ambrosoli, R., DiStasio, L., Mazzocco, P. (1988). Content of alpha$_{s1}$-casein and coagulation properties in goat milk. *J. Dairy Sci.* 71, 24-28.

Amigo, L., Recio, I., Ramos, M. (2000). Genetic polymorphism of ovine milk proteins: its influence on technological properties of milk- a review. *Int. Dairy J.* 10, 135-149.

Angiolillo, A., Yahyaoui, M.H., Sanchez, A., Pilla, F., and Folch, J.M. (2002). Characterization of a new genetic variant in the caprine κ-casein gene. *J. Dairy Sci.* 85, 2679-2680.

Ballabio, C., Chessa, S., Rignanese, D., Gigliotti, C., Pagnacco, G., Terracciano, L., Fiocchi, A., Restani, P., and Caroli, A.M. (2011). Goat milk allergenicity as a function of αs1-casein genetic polymorphism. *J. Dairy Sci.* 94, 998-1004.

Barbieri, M.E., Manfredi, E., Elsen, J.M., Ricordeau, G., Bouillon, J., Grosclaude, F., Mahe, M.F., Bibe, B. (1995). Infuence du locus de la caseine αs1 sur les performances laitieres at les parametres genetiques des chevres de race Alpine. *Genet. Sel. Evol.* 27, 437-450.

Bastian, E.D. & Brown, R.J. (1996). Plasmin in milk and dairy products: an update. *Int. Dairy J.* 6, 435-457.

Cabiddu, A., Addis, M., Pinna, G., Spada, S., Fiori, M., Sitzia, M., Pirisi, A., Piredda, G., Molle, G. (2006). The inclusion of a daisy plant (*Chrysanthemum coronarium)* in dairy sheep diet. 1: Effect on milk and cheese fatty acid composition with particular reference to c18:2 cis-9, trans-11. *Liv. Sci.* 101, 57-77.

Campanozzi, A., Foglia, M., D'Apolito, M., Ciavarella, M.E., Albenzio, M., Marinari, A., Lorusso, C., Catucci, A., Mastrangelo, L., Popolo, P., Di Giulio, L., Cavaliere, P., Santangelo, B., Giardino, I., Sevi A., Pettoello Mantovani, M. (2008). Levels of inflammatory cytokines from peripheral lymphocytes of children with cow's milk allergy. *Digestive and Liver Disease.* 40: A94-A95.

Caroli, A., Jann, O., Budelli, E., Bolla, P., Jäger, S., Erhardt, G. (2001). Genetic polymorphism of goat κ-casein in different breed and characterization at DNA level. *Anim. Genetics.* 32, 226-230.

Caroli, A., Chiatti, F., Chessa, S., Rignanese, D., Ibeagha-Awemu, E.M, Erhardt, G. (2007). Characterization of the casein gene complex in West Africa goats and description of a new *as1*-casein polymorphism. *J. Dairy Sci.* 90, 2989-2996.

Chessa, S., Rignanese, D. Kupper, J., Pagnacco, G., Erhardt, G., Caroli, A. (2008). Short communication: The β-casein (CSN2) silent allele c1 is highly spread in goat breeds. *J. Dairy Sci.* 91, 4433-4436.

Chilliard, Y., Rouel, J., Ferlay, A., Bernard, L., Gaborit, P., Raynal-Ljutovac, K., Lauret, A. (2005) Effects of type of forage and lipid supplementation on goat milk fatty acids and sensorial properties of cheeses. In: Future of the Sheep and Goat Dairy Sector, Special issue of the International Dairy Federation, No. 0501, Part 5, pp. 297–304.

Chilliard, Y., Rouel, J., Leroux, C. (2006). Goats alpha-s1 casein genotype influences its milk fatty acid composition and delta-9 desaturation ratios. *Anim. Feed Sci. Tech.* 131, 474-487.

Clark, S., & Sherbon, J.W. (2000). Genetic variants of αs1-CN in goat milk: breed distribution and associations with milk composition and coagulation properties. *Small Rum. Res.* 38, 135-143.

Collomb, M., Schmid, A., Sieber, R., Wechsler, D., Ryhanen, E.L. (2006). Conjugated linoleic acids in milk fat: Variation and physiological effects. *Int. Dairy J.* 16, 1347–1361.

Degen, A.A. (2007). Sheep and goat milk in pastoral societies. *Small Rum. Res.* 68, 7-19.

Delacroix-Buchet, A., & Lamberet, G. (2000). Sensorial properties and tipicity of goat dairy products. 7th Int. Conf. on goats, Tours, France. 15-21 May 2000. *Tome* 2: 559-563.

Duggan, C., Gannon, J., Walker, W.A. (2002). Protective nutrients and functional foods for the gastrointestinal tract. *Am. J. Clin. Nutr.* 75, 789-808.

El-Agamy, E.I. (2007). The challenge of cow milk protein allergy. Small Rum. Res. 68 64-72

FAOSTAT. (2009). http://faostat.fao.org/site/573/DesktopDefault.aspx?PageID=573 Accessed March 9, 2011.

Ganjam, L.S., Thornton, W.H., Marshall, R.T., MacDonald, R.S. (1997). Antiproliferative effect of yogurt fractions obtained by membrane dialysis on cultured mammalian intestinal cells. *J. Dairy Sci.* 80, 2325-2329.

Goudjil, H., Fontecha, J., Luna, P., de Lafuente, M.A., Alonso, L., Juárez, M. (2004). Quantitative characterization of unsaturated and trans fatty acids in ewe's milk fat. *Lait*, 84, 473-482.

Grosclaude, F., Mahe, M.F. Brignon, G., DiStasio, L., Jeunet, R. (1987). A Mendelian polymorphism underlaing quantitative variations of goat α_{s1}-casein. *Gen. Sel. Evol.* 19, 399-412.

Grosclaude, F. (1988). Le polymorphisme génétique des principales lactoprotéines bovines. Relation avec la qualitè, la composition et les aptitudes fromagéres du lait. INRA, *Prod. Anim.* 1, 5-17.

Grosclaude, F., Ricordeau, G., Martin, P., Remeuf, F., Vassal, L., Bouillon, J. (1994). Du gène au fromage: le polymorphisme de la caséine alphas1 caprine, ses effets, son évolution. *INRA Productions Animales* 7, 3-19.

Hanlein, G.F.W. (2004). Goat milk in human nutrition. *Small Rum Res.* 51, 155-163.

Hayes, B., Petit, E., Bouniol, C., Popescu P. (1993). Localisation of th alpha-s2-casein gene (CASA2) to the homologues cattle, sheep and goat chromosomes 4 by in situ hybridization. *Cyt. Cell Gen.* 64, 282-285.

Huston, J.E., & Hart, S.P. (2003). Goat husbandry: Feeding management. In H. Roginski, J.W. Fuquay, & P.F. Fox Ed Encyclopedia of Dairy Sciences, Cornwall Academic Press pp 1235-1243.

Jarvinen, K.M., Chatchatee, P., Bardina, L., Beyer, K., Sampson, H.A. (2001). IgE and IgG binding epitopes on alpha-lactalbumin and betalactoglobulin in cow's milk allergy. *Int. Arch. Allergy Immunol.* 126 (2), 111–118.

Jarvinen, K.M., Beyer, K., Vila, L., Chatchatee, P., Busse, P.J., Sampson, H.A. (2002). B-cell epitopes as a screening instrument for persistent cow's milk allergy. *J. Allergy Clin. Immunol.* 110 (2), 293–297.

Jasińska, B. (1995). The campàrison of pepsin and trypsin action on goat, cow, mare and human caseins. *Rocz,. Akad. Med. Bialymst.* 40, 486-493.

Jenness, R. (1980). Composition and characteristcs of goat milk: review 1968-1979. *J. Dairy Sci.* 63, 1605-1630.

Kalser, M.H. (1971). Medium chain triglycerides. *Adv. Internal Med.* 17, 301-322.

Korhonen, H., & Pihlanto, A. (2006). Bioactive peptides: production and functionality. *Int. Dairy J.* 16, 945-960.

Korhonen, H., & Pihlanto, A. (2007). Technological options for the production of health-promoting proteins and peptides derived from milk and colostrum. *Current pharmaceutical design* 13, 829-43.

Küpper, J., Chessa, S., Rignanese, D., Caroli A., Erhardt, G. (2010). Divergence at the casein haplotypes in dairy and meat goat breeds. *J. Dairy Res.* 77, 56-62.

Lamberet, G., Degas, C., Delacroix-Buchet, A., Vassal, L. (1996). Effect of characters linked to A and F caprine α-s1-casein alleles on goat flavours: cheesemaking with protein-fat exchange. *Lait* 76, 319-361.

Luna, P., Fontecha, J., Juárez, M., de la Fuente, M.A. (2005). Changes in the milk and cheese fat composition of ewes fed commercial supplements containing linseed with special reference to the CLA content and isomer composition. *Lipids*. 40, 445-454.

Mahé, M.F., Manfredi, E., Ricordeau, G., Piacère, A., Grosclaude, F. (1993). Effects of the αs1-casein polymorphism on goat dairy performance: a within-sire analysis of alpine bucks. *Gen. Sel. Evol.* 26, 151-157.

Marnet, P.G., & McKusik, B.C. (2001). Regulation of milk ejection and milkability in small ruminants. *Liv. Prod.* Sci. 70, 125-133.

Martin, P. (1993). Polymorphisme génétique des lactoprotéines caprines. Lait 73, 511-32.

Martin, P., Ollivier-Bousquet, M., Grosclaude, F. (1999). Genetic polymorphism of caseins: a tool to investigate casein micelle organization. *Int. Dairy J.* 9, 163-171.

Matar, C., & Goulet, J. (1996). β-casomorphin in milk fermented by a mutant of *Lactobacillus helveticus. Int. Dairy J.* 6, 383-397.

Mehaia, M.A., & Al-Kanhal, M.A. (1992). Taurin and other free amino acids in milk of camel, goat, cow, and man. *Milchwissenschaft*, 47, 351-353.

Meisel, H. (1997). Biochemical properties of bioactive peptides derived from milk proteins: potential nutraceuticals for food and pharmaceutical applications. *Liv. Prod. Sci.* 50, 125-138.

Meisel, H. & FitzGerald, R. J. (2003). Biofunctional peptides from milk proteins: Mineral binding and cytomodulatory effects. *Current Pharmaceutical Design*, 9, 1289–1295.

Muehlenkamp, M.R. & Warthensen, J.J. (1996). β-casomorphins: analysis in cheese and susceptibility to proteolytic enzymes from *Lactobacillus lactis* spp *cremoris. J. Dairy Sci.* 79, 20-26.

Natale, M., Bisson, C., Monti, G., Peltran, A., Perono Garoffo, L., Valentini, S., Fabris, C., Bertino, E., Coscia, A., and Conti, A. (2004). Cow's milk allergens identification by two-dimensional immunoblotting and mass spectrometry. *Mol. Nutr. Food Res.* 48, 363-369.

Nakamura, Y., Yamamoto, N., Sakai, K., Okubo, A., Yamazaki, S., Takano, T. (1995). Purification and characterization of angiotensin-I converting enzyme inhibitors from sour milk. *J. Dairy Sci.* 78, 777-783.

Neveu, C., Riaublanc, A., Miranda, G., Chich, J.F., Martin, P. (2002). Is the apocrine milk secretion process observed in the goat species rooted in the perturbation of the intracellular tran sport mechanism induced by defective alleles at the alpha(s1)-Cn locus? *Reprod. Nutr. Dev.* 42, 163-172.

Ng-Kway-Hang, K.F. (1998). Genetic polymorphism of milk proteins: relationship with production traits, milk composition and technological properties. *Can. J. Anim. Sci.* 78, 132-147.

Nudda, A., McGuire, M.A., Battacone, G., Pulina G. (2005). Seasonal variation in conjugated linoleic acid and vaccenic acid in milk fat of sheep and its transfer to cheese and ricotta. *J. Dairy Sci.* 88, 1311-1319.

Park, Y.W. (1994). Hypo-allergenic and therapeutic significance of goat milk. *Small Rum. Res.* 14, 151-159.

Park, Y. W. (2001). Proteolysis and lipolysis of goat milk cheese. *J. Dairy Sci.* 84, E84–E92.

Park, Y.W., Juárez, M., Ramos, M., Haenlein, G.F.W. (2007). Physico-chemical characteristics of goat and sheep milk. *Small Rum. Res.* 68, 88-113.

Park, Y.W. (2009). Bioactive components in goat milk. In Bioactive components in milk and dairy products. Ed. Park Y.W. Chapter 3, 43-82. Wiley Blackwell. 2121 State Ave, Ames, Ioua, 50014-8300, USA.

Parodi, P.W. (1999). Conjugated linoleic acid and other anticarcinogenic agents of bovine milk fat. *J. Dairy Sci.* 82, 1339-1349.

Pasqualone, A., Caponio, F., Bilancia, M.T. (2003). Effects of milk heating on taurine content and compositional characteristics of Apulian Cacioricotta goat's cheese. *Milchwissenschaft*, 58, 39-41.

Pelissier, J.P. & Manchon, P. (1976). Comparative study of the bitter taste of enzymic hydrolysates from cow, ewe and goat caseins. *J. Food Sci.* 41, 231.

Pierre, A., le Quere, J.L., Famelart, M.H., Rousseau, F. (1996). Cheese from goat milks with or without αs1-CN. In: Production and Utilization of Ewe and Goat Milk Proc. of the IDF/Greek National Committee of IDF/CIRVAL, Crete, Greece pp. 322.

Pirisi, A., Collin, P., Laurent, F., Scher, J., Parmentier, M. (1994). Comparison of milk composittion, cheesemaking properties and textural characteristics of the cheese from 2 groups of goats with a high or low rate of α-s1-casein synthesis. *Int. Dairy J.* 4, 329-345.

Pitel, G. & Delacroix-Buchet, A. (1994). Compte-rendu d'activité, Station de Recherches Laitières. INRA, Jouy-en-Josas, France.

Popescu, P., Long, S., Riggs, PWJ, Schumutz, S., Fries, R., Galagher, D.S. (1996). Standardization of cattle karyotype nomenclature: report of the committee for the standardization of the cattle karyotype. *Cyt. Cell Gen.* 74, 259-261.

Prinzenberg, E.M., Gutscher, K., Chessa, S., Caroli, A., Erhardt, G. (2005). Caprine k casein (CSN3) polymorphism: new developments in molecular knowledge. *J Dairy Sci.* 88, 1490-1498.

Quirós, A., Hernández-Ledesma, B., Ramos, M., Amigo, L., Recio, I. (2005). Angiotensin-converting enzyme ihnibitory activity of peptides derived from caprine Kefir. *J. Dairy Sci.* 88, 3480-3487.

Quirós A., Ramos, M., Muguerza, B., Delgado, M. A., Miguel, M., Aleixandre, M., Recio, I. (2007). Identification of novel antihipertensive peptides in milk fermented with *Enterococcus faecalis*. *Int. Dairy J.* 17, 33-41.

Rameuf, F., Verdalet-Guzman, I., Lenoir L. (1995). Technological adjustment of goat milk containing low synthesis α-s1-casein variants *Int. dairy j.* 5, 381 392.

Raynal-Ljutovac, K. Lagriffoul, G., Paccars, P., Guillet, I., Chilliard, Y. (2008). Composition of goat and sheep milk products: an update. *Small Rum. Res.* 79, 57-72.

Rival, S. G., Fornaroli, S., Boeriu, C. G., Wichers, H. J. (2001). Caseins and casein hydrolysates. 1. Lipoxygenase inhibitory properties. *J. Agr. Food Chem.* 48, 27-32.

Rokka, T, Syväoja, E-L, Tuominen, J., Korhonen, H. (1997). Release of bioactive peptides by enzymatic proteolysis of *Lactobacillus* GG fermented UHT milk. *Milchwissenschaft* 52, 675–678.

Roncada, P., Gaviraghi, A., Liberatori, S., Canas, B., Bini, L., Greppi, G.F. (2002). Identification of caseins in goat milk. *Proteomics* 2, 723–726.

Russo, V., Chiofalo, L., Micari, P. (1977). Esame elettroforetico delle proteine del latte in popolazioni caprine allevate in Sicilia. *Zootec. Nutriz. Anim.* 3, 247-253.

Rutherfurd, S.M., Moughan, P.J., Lowry, D., Prosser, C.G. (2008). Amino acid composition determined using multiple hydrolysis times for three goat milk formulations. *Int. J. Food Sci. Nutr.* 59, 679-690.

Ryniewicz, Z., Krzyzewski, J., Gradziel, N., Galka, E. (1996). Relationship between the genetic variants of αs1-CN, chemical composition and technological properties of the milk of Polish goats (initial obervations). In: Production and Utilization of Ewe and Goat Milk Proc. of the IDF/Greek National Committee of IDF/CIRVAL, Crete, Greece pp. 272.

Rysstad, G., Knutsen, W.J., Abrahamsen, R.K. (1990). Effect of threonine and glycine on the acetaldehyde formation in goat's milk yogurt. *J. Dairy res*. 57, 401-411.

Sacchi, P., Chessa, S., Budelli, E., Bolla, P., Ceriotti, G., Soglia, D., Rasero, R., Cauvin, E., Caroli, A. (2005). Casein haplotype structure in five Italian goat breeds. *J. Dairy Sci*. 88, 1561-1568.

Saini, A.L. & Gill, L.S. (1991). Goat milk: an attractive alternate. *Indian Dairyman*. 42, 562-564.

Santillo, A., Kelly, A.L., Palermo, C., Sevi, A., Albenzio, M. (2009). Role of indigenous enzymes in proteolysis of casein in caprine milk. *Int. Dairy J*. 19, 655-660.

Schmidely, PH., Meschy, F., Tessier, J., Sauvant, D. (2002). Lactation response and nitrogen, calcium, and phosphorus utilization of dairy goats differing by the genotype for αsa-casein in milk, and fed diets varying in crude protein concentration. *J. Dairy Sci*. 85, 2299-2307.

Sébédio, J.L., Gnaedig, S., Chardigny, J.M. (1999). Recent advances in conjugated linoleic acid research. *Curr. Opin. Clin. Nutr. Metabolic Care*, 2: 499-506.

Silanikove, N., Leitner, G., Merin, U., Prosser, C.G. (2010). Recent advances in exploiting goat's milk: quality, safety and production aspects. *Small Rum. Res*. 89, 110-124.

Slačanac, V., Božanić, R., Hardi, J., Szabó, J.R., Lučan, M., Krstanović, V. (2010). Nutritional and therapeutic value of fermented caprine milk. *Int. J. Dairy Tech*. 63, 171-189.

Tantibhedhyangkul, P. & Hashim, S.A. (1978). Medium-chain triglyceride feeding in premature infants: effect on calcium and magnesium absorption. *Paediatrics*, 61, 537-545.

Trujillo, A.J., Guamis, B., Carretero, C. (1997). Hydrolysis of caprine ß-casein by plasmin. *J. Dairy Sci*. 80, 2258-2263.

Tziboula-Clarke, A. (2003). Goat milk. In H. Roginski, J.W. Fuquay, & P.F. Fox Ed Encyclopedia of Dairy Sciences, Cornwall Academic Press, pp. 1270-1279.

Wani, G.M. (2003). Goat Husbandry: Multipurpose management. In H. Roginski, J.W. Fuquay, & P.F. Fox Ed Encyclopedia of Dairy Sciences, Cornwall Academic Press, pp 1259-1270.

Williams, C.M. (2000). Dietary fatty acids and human health. Ann. Zootech. 49: 165-180.

Yamamoto, N., Akino, A., Takano, T. (1994). Antihypertenisive effect of the peptides derived from casein by an extracellular proteinasi from *Lactobacillus helveticus* CP790. *J. Dairy Sci*. 77, 917-922.

Zullo, A.C., Barone, A.M., Chianese, L., Colatruglio, P., Occidente, M., Matassino. D. (2005). Protein polymorphisms and coagulation properties of Cilentana goat milk. *Small Rum. Res*. 58, 223-230.

ental evidence, processing conditions can be manipulated by the food scientist ing on the nature of the product, in order to be able to control the mechanisms ed in emulsion destabilization. Nevertheless, the detailed molecular mechanisms ved in emulsion formation and stabilization, which presupposes understanding of ehavior of individual protein molecules, alone or in combination, at interfaces, ains in many cases under investigation.

ords: caseins, stability, oil droplet, interface, flocculation, emulsion, proteins.

1. General Aspects of Food Emulsions

An emulsion is typically defined as a colloidal dispersion or suspension of a liquid phase n the form of droplets in a liquid continuous medium [Dickinson, 1992]. The substance that makes up the droplets in an emulsion is called the dispersed or internal phase and the substance that makes up the surrounding liquid is known as the continuous or external phase. More commonly, in the food industry the term emulsion is used to describe a two-phase system consisting of two immiscible liquids, e.g. oil and water. Depending on the distribution of the oil and aqueous phases, emulsions can be classified in different types. For instance, a system which consists of oil droplets dispersed throughout an aqueous medium is referred to as an oil-in-water emulsion (o/w). Typical examples of o/w emulsions exist in the food industry such as mayonnaise, salad dressings, milk etc. Those systems in which water droplets are dispersed in an oil phase are called water-in-oil emulsions (w/o) and they include butter, margarine and others.

The process of converting two immiscible liquids such as oil and water into an emulsion is known as homogenization and the mechanical device designed to carry out this process is called a homogenizer. Homogenization is a crucial step in emulsion formation, which aims to extend the shelf-life of products and improve the mouthfeel characteristics. To form a fine emulsion, large deformable droplets must be broken down by the vigorous application of mechanical energy [Walstra, 1983]. During this process the liquids are subjected to intense mechanical shear so that one liquid disperses into the other in the form of fine droplets. In food processing this can be traditionally achieved by various types of devices such as high-speed blenders, colloid mills, high pressure valve homogenizers and ultrasonic homogenizers. In the case of dairy products, a homogenizer operating under high pressure is commonly used in which the oil and water are driven under pressure through a valve that causes a sudden sharp constriction in the flow [Santana et al., 2010]. The pressure energy applied at the valve is converted into kinetic energy and the reduction in particle size is achieved by a combination of turbulence and laminar shear [Dickinson, 1992]. Nevertheless, thermodynamically speaking, the process of homogenization is extremely inefficient, with most of the electrical power being wasted in the form of heat [Dickinson, 2009].

2. Fundamentals on Emulsion Stability

A food emulsion, in most cases, is a complex system which consists of three distinct parts: the fat or oil that is in the interior of the droplets; the interfacial material between the

In: Casein: Production, Uses and Health Effects
Editors: A. M. Ventimiglia, et. al, pp. 177-194

Chapter 9

RECENT ADVANCES IN UNDERSTANDIN OF CASEINS IN THE FORMATION AND STA OF EMULSIONS: A MOLECULAR APPR

Vassilios Raikos*
Department of Chemistry, University of Patras, 26504, Patras, Greece.

ABSTRACT

Milk proteins and particularly caseins (αs1-, αs2-, β- and ϰ-) are widely used in the food industry as emulsifiers. Caseins, thanks to their substantial surface activity, adsorb rapidly at the oil-water interface during emulsion formation, and facilitate the formation of a strong, protective film around the oil droplet. The structure and composition of the protein interfacial layer determine to a large extent the physic-chemical properties of the emulsion, which in turn play significant role in the shelf-life and acceptability of the food product. This paper describes the role of the molecular structure of caseins either as monomers or as proteinaceous mixtures in emulsion formation and stabilization. Evidence exists which indicates the conformational changes in the three-dimensional structure of proteins following adsorption at the interface. Furthermore, competitive adsorption of the milk proteins occurs during emulsion formation, which affects the protein composition of the interfacial layer. Additionally, this article reviews recent advances in understanding the effect of processing factors on the interfacial protein coverage, composition and molecular configuration of caseins in the oil-in-water emulsion. These factors include heat treatment, pH, and ionic strength, which seem to affect the protein functionality and thus, emulsion stability. This paper describes the key findings of research undertaken over the last 25 years, which aim to provide some quantitative and qualitative insight on how extrinsic parameters, such as thermal processing, determine the physicochemical behavior of food emulsions. According to

* Corresponding author: Vassilios Raikos
Department of Chemistry, Laboratory of Physical Chemistry, University of Patras, PC 26504, Patras, Greece
Tel.: +30 2610 997109,
Fax: +30 2610 997144,
E-mail: vas.raikos@gmail.com

lipid material and the aqueous phase (e.g. proteins); and the aqueous phase itself [Dalgleish, 2006]. The molecules that constitute the materials of the emulsion distribute themselves among these three regions according to their concentration and polarity [Wedzicha, 1988]. However, following their initial distribution, they don't remain static but there is a continuous exchange of molecules between the different regions. In other words, emulsion stability is a kinetic concept. This dynamic process is very often triggered by the various interactions that occur between the individual molecules (intermolecular) or even between different parts of the same molecule (intramolecular) and affect its three dimensional conformation and flexibility. Droplet aggregation, for example, may be triggered by interactions between protein layers adsorbed on different droplets. This of course depends on a number of factors, such as bulk protein concentration and protein surface coverage, the layer thickness, the chemical nature and the structure of the adsorbed material. This continual movement could also be due to some alteration in the environmental conditions of an emulsion, with pH, heating, ionic strength and calcium ion content being the most important. Furthermore, emulsions are usually complex food systems which contain numerous ingredients. For instance, polysaccharides are sometimes added as thickening agents to increase the viscosity of the continuous phase. Other ingredients of food emulsions, which are included to serve different purposes, may include minerals, gums, preservatives and vitamins. It is evident that most food emulsions are composed of diverse and structurally complex ingredients that are capable of being in constant motion due to inter- and intra-molecular interactions or when subjected to environmental changes. By definition, a stable emulsion exhibits no discernible change in the size distribution of the oil droplets, in their state of aggregation, or in their spatial arrangement within the sample vessel, over the time scale of observation [Dickinson & Stainsby, 1982].

Alterations in the spatial distribution and conformational rearrangements of the ingredients affect the physical stability of the emulsion, which is the term used to describe its ability to resist changes in its properties over time [McClements, 1999]. Creaming, sedimentation, flocculation, coalescence, phase inversion and Ostwald ripening are the major causes of physical instability exhibited by oil-in-water emulsions [Cerimedo et al., 2010]. The major manifestations of emulsion instability are represented schematically in Figure 1. Creaming, which is usually reversible, is the movement of oil droplets, under gravitational or centrifugal forces, to form a concentrated layer at the top of an oil-in-water emulsion. On the other hand, sedimentation describes the downward movement of droplets due to their higher density than the surrounding aqueous phase. Flocculation and coalescence are both types of droplet aggregation. Flocculation, which may or may not be advantageous for the quality of the product, is a process that takes place when two or more droplets come together thanks to attractive interactions between them and form an aggregate in which the droplets retain their individual integrity and thus, there is no disruption of the protective stabilizing layer at the interface. Flocculation may induce creaming instability as the increase in effective size of the particles results in upward movement of oil droplets and as a result instability due to flocculation is related to creaming instability and vice versa. Furthermore, depending on the causes of instability, flocculation may be referred to as bridging, when sharing of the adsorbed layer between two droplets occurs and is associated with low protein surface coverage or depletion, when a considerable excess of unadsorbed protein is concentrated in the bulk phase. Coalescence is a term used to describe the merging of two or more droplets to form a single larger droplet. This mechanism of emulsion instability is thermodynamically favorable because it is associated with a decrease in the contact area between oil and water and can result in creaming or sedimentation. This mechanism of instability may be due to thinning or breakage of the film between colliding droplets [Dickinson, 2009]. Coalescence, which is usually irreversible, is a much more severe form of instability compared to creaming but could have a positive role in ensuring desirable release of flavor components during eating [Dickinson, 1992]. Extensive coalescence in an oil-in-water emulsion results in the visible separation of the dispersed phase which is seen as free oil at the surface (oiling-off). Ostwald ripening is the process whereby large droplets grow at the expense of smaller ones because of mass transport of dispersed phase from one droplet to another through the intervening continuous phase [Kabalnov & Shchukin, 1992]. This type of physical instability is negligible in food emulsion systems due to the very low solubility of triglyceride in water. Finally, phase inversion is a term used to describe the conversion from an oil-in-water emulsion to a water-in-oil emulsion and vice-versa and is usually triggered by some alteration in the composition or environmental conditions of the system [Shinoda & Friberg, 1986].

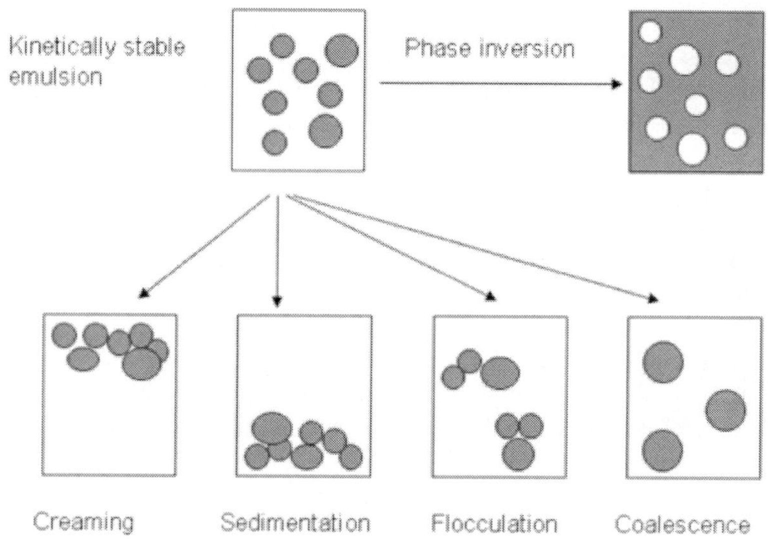

Figure 1. Major sources of physical instability of food emulsions.

3. Proteins at the Oil-Water Interface

Homogenization alone is not sufficient to produce stable emulsions on a long-term basis. The homogenized system (oil and water) separates rapidly into two distinct layers following the course of homogenization, the top layer consisting of oil (lower density) and the bottom layer consisting of water (higher density). The explanation of this process lies within the fact that emulsions are thermodynamically unstable systems. The contact between oil and water molecules is energetically unfavourable and therefore oil droplets tend to merge with their neighbouring ones, a process which eventually leads to phase separation [Israelachvili, 1992]. As a result, in order to prepare a stable emulsion, there must be some kind of surface-active material present to prevent the newly formed fine droplets from coalescing with their neighbours to form larger droplets. This surface active material, termed emulsifier, adsorbs to the surface of the freshly formed droplets and forms a protective coating around the droplets which stabilizes the emulsion. In other words, the role of the emulsifier is to contribute to emulsion stability by interfacial action [Dickinson, 1992]. The most commonly used emulsifiers in the food industry are amphiphilic molecules [they have both polar and non-polar regions] and they include proteins, small molecule surfactants and phospholipids (Figure 2).

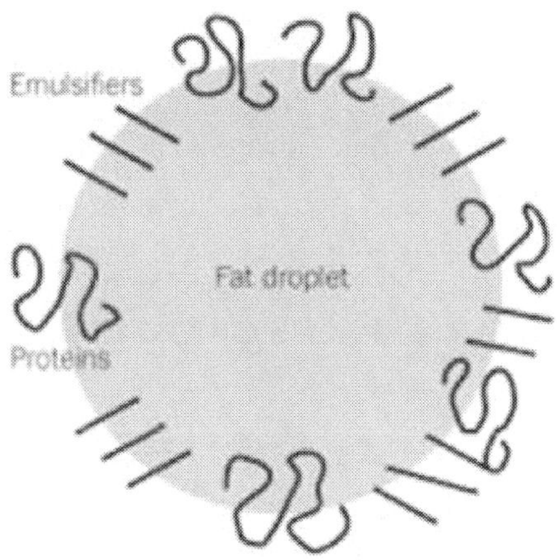

Figure 2. Surface-coated oil droplet.

Proteins represent an important class of molecules in emulsion technology and are very often characterized as functional food ingredients, thanks to their role in emulsion formation and stabilization. The adsorption of the emulsifier at an interface and the subsequent formation of a protective membrane around the droplet results from two energetically favorable interactions [McClements, 1999]. Firstly, adsorption reduces the contact area between the oil and water, which is thermodynamically unfavourable, and as a result the interfacial tension is lowered. Second, when the emulsifier is dispersed in an aqueous phase, some of the non-polar groups are in contact with water, which is energetically unfavourable due to the hydrophobic effect. However, when the amphiphilic molecule adsorbs to an interface, it can adopt a conformation (surface denaturation) where the non-polar groups are oriented towards the oil phase and the hydrophilic groups are located in the aqueous phase maximizing the contact with water. Protein adsorption at the oil-water interface is thermodynamically favorable because hydrophobic residues of the protein backbone are removed from the bulk aqueous phase and are oriented towards the oil phase, following protein structural rearrangement at the interface [Dickinson et al., 1988]. As a result, the adsorbed globular protein structure lies somewhere intermediate between the native state and the fully-denatured state, sometimes referred to as the molten globule state [Dickinson, 1998]. Thus, the ability of an emulsifier to "shield" the oil molecules from the water molecules, which results in decrease of the surface tension, and the hydrophobic effect are the driving forces favoring the adsorption of these amphiphilic molecules at the interface.

Furthermore, proteins can form films at the interface, providing electrostatic and steric repulsion between droplets, thus conferring long-term stability to the emulsion [van Aken &

van Vliet, 2002]. Milk proteins, for example such as caseins and whey proteins, are well known emulsifiers and hence are used as functional tools in a wide range of food emulsions [Ries et al., 2010]. Caseins are heat-stable proteins which adsorb strongly at fluid interfaces and form a steric protective layer which prevents droplets from immediate recoalescence and confers long-term stability during subsequent processing and storage [Dickinson & Stainsby, 1982]. Thus, for a protein to function as an effective emulsifier it must be able to i) effectively reduce the interfacial tension at the freshly-formed oil-water interface, ii) bind strongly to the interface once adsorbed via the formation of intra- or inter- molecular chemical bonds, and iii) protect the newly formed droplets against any destabilizing mechanism [Dickinson, 2009]. This protection against immediate recoalescence occurs initially via dynamic surface tension effects (the Gibbs-Marangoni mechanism) and at a later stage via repulsive colloidal interactions such as electrostatic and steric stabilization mechanisms [Dickinson, 1992].

4. Caseins: Molecular Aspects of their Role as Emulsifiers

4.1. General

The dominant protein fraction of milk is comprised of a group of disordered, flexible molecules known as caseins. Milk caseins represent approximately 80% of the total milk protein content and in their natural environment, the individual protein molecules tend to self-associate to form strong aggregates, named "casein micelles" [Wong et al., 1996]. These supramolecular structures are very often described as polydisperse protein particles sizing approximately 200nm and held together by calcium salts [Dickinson 1998]. Four main casein proteins exist, which show similarities in terms of their structure, but are not identical [Swaisgood, 1982]. These are α_{s1}-, α_{s2}-, β- and κ- casein, with α_{s1}-casein and β-casein being the major representatives as they comprise about 75% of the total milk casein found in bovines. Because of their abundance in milk, α_{s1}-casein and β-casein are the major determinants of the physic-chemical properties of casein-stabilized emulsions.

4.2. α_{s1}-casein

α_{s1}-casein is a calcium-sensitive phosphoprotein which is made up of about 200 amino acid residues, among which proline is highly abundant but completely lacks any cysteine residues [Swaisgood, 1982]. According to previous research, the α_{s1}-casein primary structure exhibits three distinct hydrophobic patches, which are located in the middle (~90-110) and at the ends (1-40 and 130-199) [Shimizu et al., 1986]. The molecule of α_{s1}-casein is a flexible disordered polyelectrolyte of substantial amphiphilic nature, which carries a high charge [-22e] at neutral pH [Dickinson, 2001]. Furthermore, experimental evidence exists according to which the hydrophobic and hydrophilic residues are randomly distributed along the polypeptide backbone [Dickinson & Matsamura, 1994]. α_{s1}-casein, when present in aqueous environment, can self-associate by means of hydrophobic linkages and form worm-like micelles [Horne, 1998].

Extensive research has been carried out in the past and numerous reports have been published in an effort to adequately describe the molecular configuration of α_{s1}-casein at the oil-water interface [Dalgleish, 1993; Dickinson, 1994; Dalgleish, 1996]. Today, the model which best describes the adsorbed protein conformation, based on both experimental and theoretical evidence, is the loop-like conformation [Dickinson, 1999, Dickinson, 2001]. According to this model, the protein forms a loop between residues 40 – 80, which extends somewhat away from the oil surface, as illustrated in Figure 3. This middle hydrophilic region is a highly charged area clustering eight phosphoserine residues. At both ends of this loop, strong hydrophobic areas on the protein backbone ensure that the molecule is firmly attached to the oil surface, hence forming a loop instead of a tail. Furthermore, α_{s1}-casein exhibits limited surface coverage and the monolayer formed is substantially thinner compared to that of β-casein [Brooksbank et al., 1993; Dalgleish, 1993].

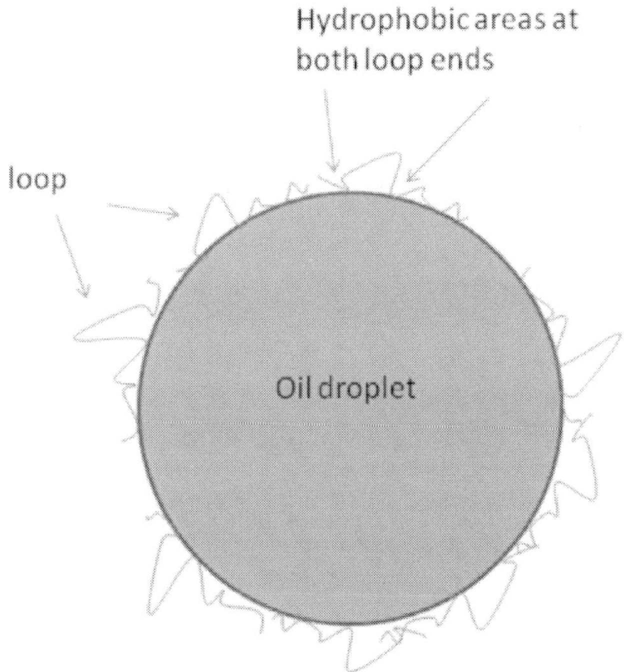

Figure 3. Schematic representation of the “loop-like” conformation of the α_{s1}-casein molecule adsorbed at the oil-in-water interface.

4.3. β-casein

β-casein, which is possibly the most extensively documented monomer of the casein family, is a flexible linear polyelectrolyte comprising of 209 amino acid residues [Swaisgood, 1982]. β-casein is a slightly bigger molecule compared to α_{s1}-casein with a molecular weight of 24kDa, carries a net charge of -15e at neutral pH and shows no intramolecular covalent bonding [Dickinson 2001]. Based on its non-random amino acid sequence distribution, the protein molecule is highly amphipathic with a distinct hydrophilic region (N-terminus domain) and a distinct hydrophobic region (C-terminus domain) [Rollema, 1992].

β-casein, which predominantly shows monolayer surface coverage, is more surface active compared to α_{s1}-casein and it exhibits a higher initial rate of lowering the interfacial tension [Dickinson, 1999]. Furthermore, evidence exists which suggests the formation of β-casein aggregates at the interface through hydrophobic interactions [Dauphas et al., 2005]. At the interface the protein adopts a somewhat different conformation compared to α_{s1}-casein, designated as the "train-loop-tail" model", illustrated in Figure 4 [Graham & Philips, 1979].

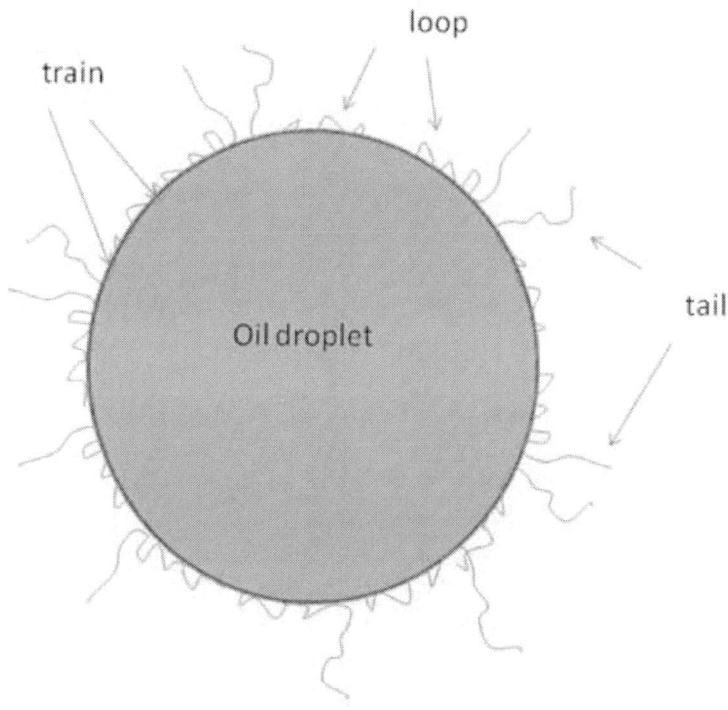

Figure 4. Schematic representation of the "train-loop-tail" conformation of the β-casein molecule adsorbed at the oil-in-water interface.

According to this model, the extensive hydrophobic regions of the molecule (~70% of the residues) are distributed very close to the surface and form a dense inner layer (1-2nm thick), which is very often described as "small loops" and "trains" [Dickinson, 2001]. The remaining hydrophilic part of the molecule of about 40-50 residues at the N-terminus, which includes many charged residues such as 5 phosphoserines, forms a "tail" extending approximately 10nm from the surface and confers steric stability to the emulsion droplet [Dalgleish, 1997]. This model is supported by experimental evidence based on different techniques, including the increased susceptibility of the N-terminus tail to trypsin compared to the rest of the adsorbed protein backbone, indicating that the hydrophilic region protrudes into the aqueous phase and therefore is easily accessible by the proteolytic enzyme [Leaver & Dalgleish, 1992].

4.4. Sodium Caseinate

Pure caseins or pure casein mixtures are commonly used to stabilize in oil-in-water emulsions. Nevertheless, in many food formulations, a casein derivative known as sodium caseinate is preferred due to its excellent emulsifying properties. At a first stage the production of commercially available sodium caseinate involves acidification of whole casein to pH 4.6. Then pH is readjusted to 6.7 with the addition of NaOH, and the process is completed with pasteurization followed by spray-drying [Dickinson 2001]. Sodium caseinate is aqueous solution of pH around 7 exists as a mixture of free casein monomers (α_{s1}-casein, β-casein, α_{s2}-casein and ϰ-casein) together with aggregated forms of ϰ-casein [Hadjsadok et al., 2008; Semenova et al., 2009]. The self-assembled protein particles also known as "sub-micelles", have a diameter of 10-20 nm [Creamer & Berry, 1975]. The more aggregated state of sodium caseinate compared to the simple mixtures of casein monomers, affects the interfacial properties of the proteins and is responsible for the enhanced emulsion stability exhibited by the sodium-caseinate formulations.

5. Competitive Adsorption of Casein Mixtures at the Interface

In several food formulations it is common practice to use mixtures of casein proteins for emulsion stabilization. As a result, a major concern of the food technologist is to understand and control the adsorption behavior of the different emulsifier molecules when different protein populations are present. This is particularly important for the organoleptic properties of the food product, because the protein surface composition on a long term basis determines the stability of the emulsion.

As a general rule, in binary mixtures of proteins, the emulsifier molecule that arrives first at the interface during the emulsification process is the dominant species with respect to surface composition and little free reversible exchange between adsorbed and unadsorbed protein occurs afterwards [Dickinson et al., 1989]. While this observation is particularly true for globular proteins, disordered flexible structures such as caseins, behave in a different manner. In the later case, extended competitive adsorption occurs, which seems to be dependent on various factors. First of all, thermodynamic factors determine the surface composition of the interfacial layer in systems containing mixtures of proteins [Dickinson, 2010]. Furthermore, molecular aspects of the protein structure play significant role to the adsorption behavior of the individual molecules. According to Gong & Szleifer [2004], the size and shape of the adsorbing protein molecules or their overall charge or charge distributions can determine the protein-protein interactions at the interface and/or in the bulk phase. β-casein is more effective in lowering the overall free energy following adsorption compared to α_{s1}-casein, and therefore being more surface active and slightly more hydrophobic the former predominates at the oil-water interface [Mitchell et al., 1970, Dickinson et al., 1988]. Furthermore, exchange experiments have demonstrated that β-casein, when added to the aqueous phase of a α_{s1}-casein stabilized emulsion, it can systematically displace α_{s1}-casein from the interface, as illustrated in Figure 5 [Dickinson et al, 1988, Anand & Damoradan, 1996].

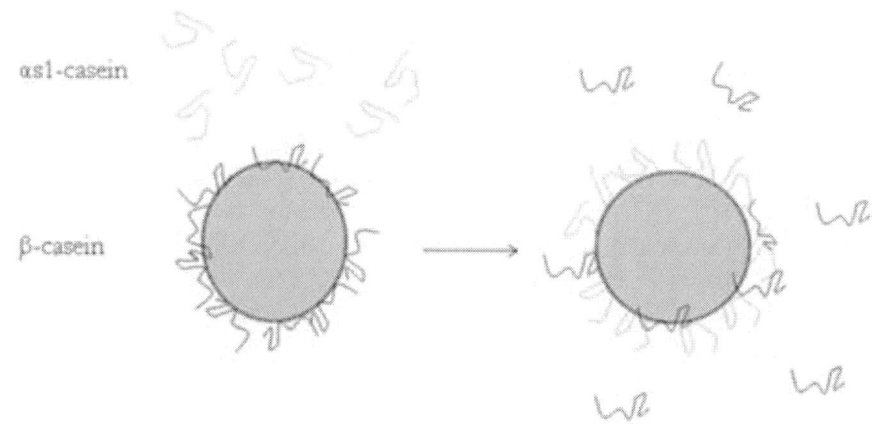

Figure 5. Competitive displacement of α_{s1}-casein at the interface of oil-in-water emulsion droplet by the addition of β-casein in the aqueous phase.

The same studies showed that α_{s1}-casein is also able to displace β-casein from the interface but to a lesser extent (Figure 6).

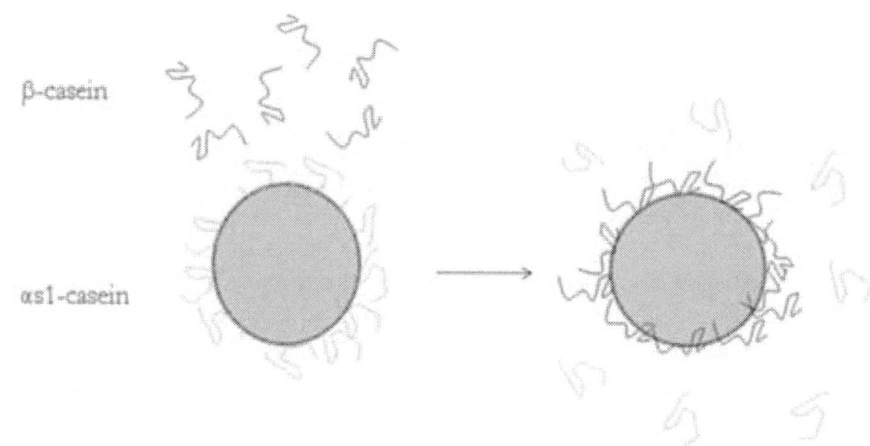

Figure 6. Competitive displacement of β-casein at the interface of oil-in-water emulsion droplet by the addition of α_{s1}-casein in the aqueous phase.

As a result, these studies indicate that these two disordered proteins are reversibly adsorbed at the interface and exchangeability depends mainly on molecular and thermodynamic factors. The above hypothesis is further supported by protein exchangeability studies in which no preference for α_{s1}-casein or β-casein to adsorb at the interface was observed, using sodium caseinate to stabilize the emulsions [Robson & Dalgleish, 1987]. These studies indicate that competitive adsorption between β-casein and α_{s1}-casein is limited and little exchange and displacement occurs between the two proteins at the interface. This is possibly due to the more aggregated form of sodium caseinate, which restricts the molecular mobility and flexibility of the individual proteins at the interface. Nevertheless, other factors such as protein concentration, NaCl and $CaCl_2$ addition were shown to determine the protein surface composition in mixed protein systems [Srinivasan et al., 1996; Srinivasan et al., 2000; Ye & Singh, 2001]. According to these studies, when the caseinate content is low, β-casein predominates at the interface, whereas a larger proportion of α_{s1}-casein was present at high protein levels. Furthermore, raising of the ionic strength of the environment by the addition of NaCl and $CaCl_2$ resulted in enhanced adsorption of a_s protein at the interface of the emulsion droplets.

6. Overview of the Factors Affecting the Emulsifying Ability of Caseins

The ability of caseins to form and stabilize emulsions is not dependent solely on the protein properties but on a number of extrinsic factors which, either individually or in synergy, determine the emulsion environment and thus the protein functionality. These factors affect the emulsion properties by manipulating the protein surface composition and structure of the stabilizing layer and mainly include the pH, ionic strength, calcium ion content and temperature. Thus, the effect of the above factors on the emulsion characteristics depends primarily on the protein content present during emulsion formation.

β-casein stabilized-emulsions remain unaffected in terms of stability when NaCl is added to concentrations up to 2M [Dickinson et al., 1998]. Furthermore, the addition of NaCl does not have an impact on emulsion stability, even when the pH is adjusted near the isoelectric point of β-casein (pH 5.5) and thus lowering the overall charge carried by the stabilizing layer covering the droplet surfaces. This indicates that β-casein stabilized-emulsions are protected against flocculation by a steric mechanism rather than by electrostatic repulsive forces [Dickinson, 2010]. Nevertheless, the addition of calcium ions appears to have a negative impact on the stability of β-casein emulsions [Dauphas et al., 2008]. In particular, emulsions exhibit flocculation when calcium chloride is added and the extent of flocculation is temperature-dependent. According to the authors, the formation of large protein aggregates is the main driving force of the destabilizing mechanism. The binding of calcium ions to β-casein molecules results in weakening the steric repulsion between emulsion droplets whereas the temperature rise increases hydrophobic interactions between casein molecules.

On the contrary, α_{s1}-casein stabilized-emulsions exhibit increased sensitivity to destabilizing mechanisms when the ionic strength increases [Dickinson et al., 1998]. This effect induced by the salt addition, which is detectable even at low concentrations (NaCl, 0.05M), is even more profound when combined with pH adjustment near the isoelectric point of α_{s1}-casein. As α_{s1}-casein stabilized-emulsions are prone to flocculation, which is brought about by the addition of electrolytes and the lowering of the pH, one could assume that the charge distribution of the surface coated droplets determines to a large extent the stability of the system. Furthermore, emulsions formed with α_{s1}-casein are highly sensitive to the addition of calcium ions, with flocculation occurring with the addition of just 5mM $CaCl_2$ [Dickinson et al., 1987].

The addition of salt to sodium caseinate-stabilized emulsions is more likely beneficial rather than detrimental like in the case of α_{s1}-casein stabilized-emulsions. Experimental evidence shows that emulsions made with sodium caseinate are not flocculated by the addition of NaCl, either added before or after emulsion formation [Srinivasan et al., 2000; Surh & McClements, 2008]. On the contrary, the addition of NaCl increases the protein surface coverage and determines the composition of the droplet layer, with α_{s1}-casein exhibiting enhanced adsorption in the form of aggregates, as stated by the authors [Srinivasan et al., 2000]. This results in decreased protein concentration in the bulk and consequently the critical concentration required for depletion flocculation is not reached. This could be the main reason accounting for the improved creaming stability of caseinate emulsions in the presence of NaCl. On the other hand, sodium-caseinate emulsions are highly sensitive to calcium ions and can be destabilized by the addition of $CaCl_2$, particularly when the pH is lowered towards the protein's isoelectric point [Dickinson & Eliot, 2003]. As illustrated in Figure7, increasing the $CaCl_2$ concentration at pH 5.6, the emulsion shows a bimodal distribution, with increased particle size reflecting the droplet flocculation.

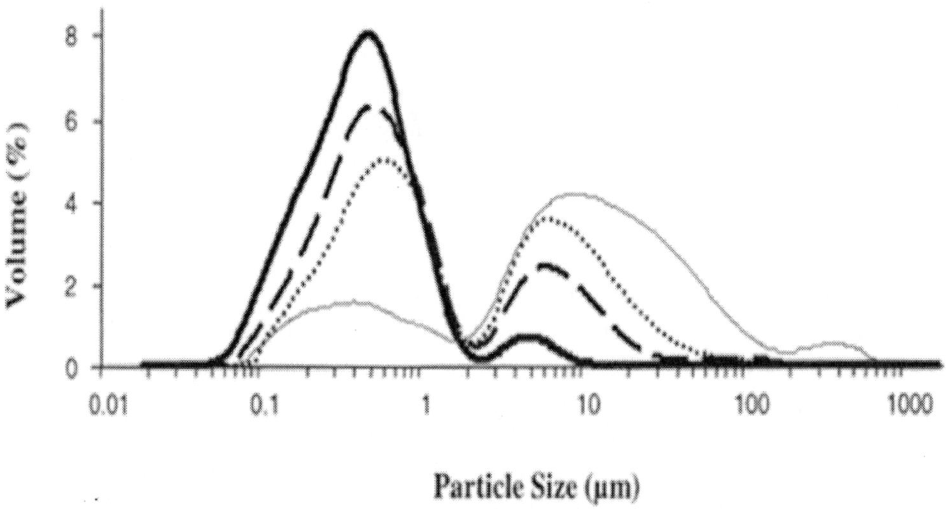

Reproduced from [Dickinson & Eliot, 2003] with permission.

Figure 7. Particle-size distribution of emulsions (45 vol.% oil) made with 8.1 wt.% sodium caseinate solution adjusted to pH 5.6 with 0.5 wt. % GDL with various concentrations of CaCl2 added prior to emulsification: ——(thick line), 0 mM; - - - , 10 mM; . . . . . ., 15 mM; : ——(thin line), 20 mM. Differences between samples reflect differences in the extent of flocculation of the droplets.

Moreover, Ye & Singh [2001] reported that the addition of calcium ions prior or after homogenization increases surface protein composition, mainly because of enhanced adsorption of a_s casein at the interface. Depending on the caseinate content the addition of $CaCl_2$ can impair emulsion stability by the formation of large flocs. Figure 8 shows the effect of increasing $CaCl_2$, added prior to homogenization, on 0.5% sodium-caseinate oil-in-water emulsions with emulsions containing 5mM $CaCl_2$ showing large number of small particles aggregated in the form of clusters.

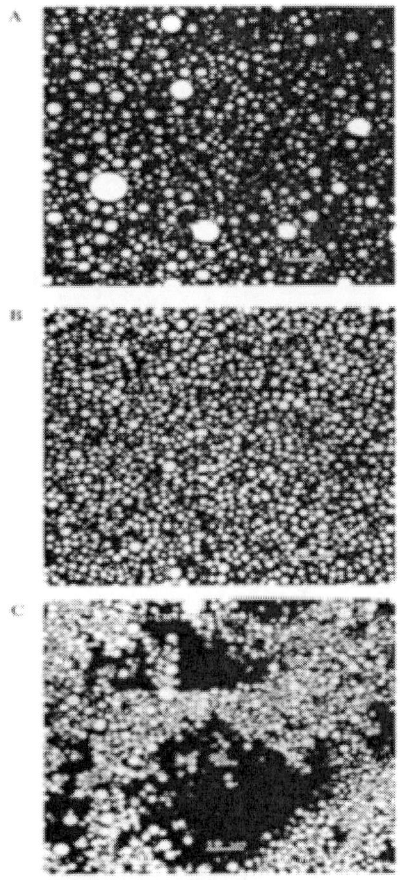

Reproduced from [Ye & Singh, 2001] with permission.

Figure 8. Confocal micrographs of 0.5% sodium caseinate-stabilized oil-in-water emulsions: a) 0 mM, b) 3 mM, or c) 5 mM CaCl2, was added to caseinate solution prior to emulsion formation.

In emulsions containing higher amounts of protein (3%), irreversible flocculation of oil droplets occurred when the concentration of added $CaCl_2$ exceeded 20mM. Finally, heat treatment affects the molecular structure of milk proteins at the interfaces of oil-in-water emulsions and in aqueous media [Raikos, 2010]. Caseins are heat-stable proteins and seem to have a protective effect against thermal denaturation. Nevertheless, heat-treatment of sodium caseinate solutions in controlled chemical environment resulted in significant improvement of emulsifying properties, which was attributed to exposure of previously hidden hydrophobic domains on the protein backbone [Jahaniaval, et al., 2000]. Dickinson & Eliot (2003) reported that hydrophobic interactions between protein molecules become stronger when sodium caseinate emulsions were thermally processed in the range 30-45 °C. These temperature-induced molecular changes resulted in the formation of a solid-like emulsion gel. Other studies indicate that temperatures above 15 °C promote cross-linking between protein molecules in sodium-caseinate emulsions, as illustrated in Figure 9 [Euston & Mayhill, 2001].

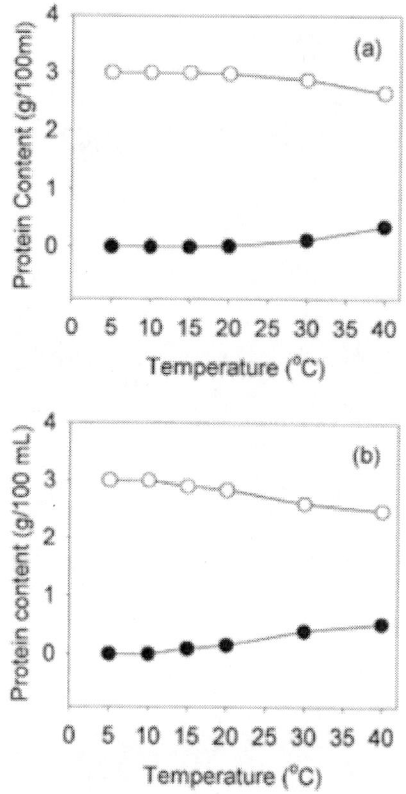

Reproduced from [Euston & Mayhill, 2001] with permission.

Figure 9. The proportion of casein and polymerized high-molecular-weight material in whole emulsions as a function of temperature after 28 days storage. (a) Milk-fat emulsions; (b) soybean oil emulsions; (○) casein; (●) high-molecular-weight material.

These interactions can be beneficial or not on the emulsion stability, depending on whether they are mediated between proteins adsorbed on the same oil droplet or between molecules found on neighboring droplets respectively. In the first case cross-linking increases coalescence stability by forming a more compact layer around the oil droplet, whereas in the second case bridging flocculation is likely to occur with subsequent reduced creaming stability.

7. Concluding Remarks

Caseins represent an important class of bioactive molecules naturally occurring in mammalian milk. The casein family consists of four major protein fractions, namely α_{s1}-casein, β-casein, α_{s2}-casein and κ-casein, which possess distinct molecular features. These heat stable, disordered, flexible molecules are well known in the food industry, thanks to their functionality as emulsifiers. During the last three decades extensive and detailed research has improved our understanding on the physico-chemical properties of casein-stabilized oil-in-water emulsions. The behavior, including the shelf-life and acceptability of the end product, depends on the molecular characteristics of the protein interfacial layer and the bulk properties. This in turn, is very commonly affected by extrinsic factors which are involved in food processing, such as thermal processing, the addition of electrolytes, pH etc. Research based upon experimental evidence using primarily spectroscopic and microscopic techniques in conjunction with computer simulations provided useful knowledge on how the above factors can manipulate the functionality of caseins and shed light on the molecular mechanisms involved in emulsion stability. At this moment, although our perspective on casein functionality has dramatically changed over the last 25 years, much remains unknown with respect to the structure-function relationship of the casein molecules. Furthermore, the need for further exploitation of the casein functionality in various food applications is always a challenge. Different approaches have been suggested in the past for manipulating the properties of molecules involved in the formation and stabilization of emulsions, including genetic engineering. Nevertheless, up-to-date the progress made towards the goal of using molecular biology and genetic engineering techniques for manipulating the molecular behavior of proteins in complex colloidal systems is not remarkable. Instead, the use of a different approach such as the chemical modification of protein molecules could be more realistic and promising for altering the functionality and predicting the behavior of caseins in the future.

REFERENCES

Anand, K., & Damodaran, S. (1996). Dynamics of exchange between α_{s1}-casein and β-casein during adsorption at air-water interface. *Journal of Agricultural and Food Chemistry*, 44, 1022-1028.

Brooksbank, D.V., Leaver, J., & Horne, D.S. (1993). Adsorption of milk proteins to phosphotidyl-glycerol and phosphotidylcholine liposomes. *Journal of Colloid and Interface Science*, 161, 38-42.

Cerimedo, M.S.A., Iriart, C.H., Candal, R.J., & Herrera, M.L. (2010). Stability of emulsions formulated with high concentrations of sodium caseinate and trehalose. *Food Research International*, *42*, 1482-1493.

Creamer, L.K, & Berry, G.P. (1975). A study of the properties of dissociated bovine casein micelles. *Journal of Dairy Research*, 42, 169-183.

Dalgleish, D.G. (1993). The sizes and conformations of the proteins in adsorbed layers of individual caseins on lattices and in oil-in-water emulsions. *Colloids and Surfaces B: Biointerfaces*, 1, 1-8.

Dalgleish, D.G. (1996). Conformations and structures of milk proteins adsorbed to oil-water interfaces. *Food Research International*, 29(5-6), 541-547.

Dalgleish, D.G. (1997). Adsorption of protein and the stability emulsions. *Trends in Food Science & Technology*, 8(1), 3013-3018.

Dalgleish, D.G. (2006). Food emulsions-their structures and structure-forming properties. *Food Hydrocolloids*, *20*, 415-422.

Dauphas, S., Amestoy, M., Liamas, G., Anton, M., & Riaublanc, A. (2008). Modification of the interactions between β-casein stabilized oil droplets with calcium addition and temperature changing. *Food Hydrocolloids*, 22 (2), 231-238.

Dauphas, S., Mouhous-Riou, N., Metro, B., Mackie, A.R., Wilde, P.J., Anton, M., & Riaublanc, A. (2005). The supramolecular organisation of β-casein: effect on interfacial properties. *Food Hydrocolloids*, 19, 387-393.

Dickinson, E, Semenova, M.G., & Antipova, A.S. (1998). Salt stability of casein emulsions. *Food Hydrocolloids*, 12 (2), 227-235.

Dickinson, E. (1992). An introduction to food colloids. Oxford Science Publishers, Oxford, UK.

Dickinson, E. (1994). Protein-stabilized emulsions. *Journal of Food Engineering*, 22, 59-74.

Dickinson, E. (1999). Caseins in emulsions: interfacial properties and interactions. *International Dairy Journal*, 9, 305-312.

Dickinson, E. (2009). Hydrocolloids as emulsifiers and emulsion stabilizers. *Food Hydrocolloids*, *23*, 1473-1482.

Dickinson, E. (2011). Mixed biopolymers at interfaces: Competitive adsorption and multilayer structures. *Food Hydrocolloids*, In press corrected proof, Available online 10 December 2010.

Dickinson, E., & Eliot, C. (2003). Defining the conditions for heat-induced gelation of a caseinate-stabilized emulsion. *Colloids and Surfaces B: Biointerfaces*, 29, 89-97.

Dickinson, E., & Matsamura, Y. (1994). Proteins at liquid interfaces: role of the molten globule state. *Colloids and Surfaces B: Biointerfaces*, 3, 1-17.

Dickinson, E., & Stainsby, G. (1982). *Colloids in Food*. London: Applied Science

Dickinson, E., (1998). Proteins at interfaces and in emulsions: stability, rheology and interactions. *Journal of the Chemical Society, Faraday Transactions*, 94 (12), 1657-1669.

Dickinson, E., (2001). Milk protein interfacial layers and the relationship to emulsion stability and rheology. *Colloids and Surfaces B: Biointerfaces*, 20, 197-210.

Dickinson, E., (2010). Flocculation of protein-stabilized oil-in-water emulsions. *Colloids and Surfaces B: Biointerfaces*, 81, 130-140.

Dickinson, E., Murray, B.S., & Stainsby, G., (1988). Protein adsorption at air-water and oil-water interfaces. In E. Dickinson & G. Stainsby (Eds.), *Advances in Food Emulsions and Foams* (pp.123). Elsevier Applied Science, London, United Kingdom.

Dickinson, E., Rolfe, S.E., & Dalgleish, D.G. (1988). Competitive adsorption of α_{s1}-casein and β-casein in oil-in-water emulsions. *Food Hydrocolloids*, 2, 397-405.

Dickinson, E., Rolfe, S.E., & Dalgleish, D.G. (1989). Competitive adsorption in oil-in-water emulsions containing α-lactalbumin and β-lactoglobulin. *Food Hydrocolloids*, 3, 193-203.

Dickinson, E., Whyman, R.H., & Dalgeish, D.G. (1987). Colloidal properties of model oil-in-water food emulsions stabilized separately by α_{s1}-casein, β-casein and κ-casein. In E. Dickinson (Ed.), *Food Emulsions and Foams*, (pp. 40-51). Royal Society of Chemistry, London.

Euston, S.R., & Mayhill, P.G. (2001). Time- and temperature-dependent changes in the adsorbed layer on caseinate-stabilized emulsion droplets. *Food Research International*, 34, 369-376.

Gong, P., & Szleifer, I. (2004). Competitive adsorption of model charged proteins: the effect of total charge and charge distribution. *Journal of Colloid and Interface Science*, 278, 81-90.

Graham, D.E., & Philips, M.C. (1979). Proteins at liquid interfaces. *Journal of Colloid and Interface Science*, 70, 403-439.

HadjSadok, A., Pitkowski, A., Nicolai, T., Benyahia, L., & Moulai-Mostefa, N. (2008). Characterization of sodium caseinate as a function of ionic strength, pH and temperature using static and dynamic light scattering. *Food Hydrocolloids*, 22(8), 1460-1466.

Horne, D.S. (1998). Casein interactions: Casting light on the black boxes, the structure in dairy products. *International Dairy Journal*, 8(3), 171-177.

Israelachvili, J., N. (1992). Intermolecular and surface forces, Academic Press, London.

Jahaniaval, F., Kakuda, Y., Abraham, V., & Marcone, M.F., 2000. Soluble protein fractions from pH and heat treated sodium caseinate: physicochemical and functional properties. *Food Research International*, 33, 637-647.

Kabalnov, A., S., & Shchukin, E., D. (1992). Ostwald ripening theory: Applications to fluorocarbon emulsion stability. *Advances in Colloid and Interface Science*, *38*, 69-97.

Leaver, J., & Dalgleish, D.G. (1992). Variations in the binding of β-casein to oil-water interfaces detected by trypsin-catalyzed hydrolysis. *Journal of Colloid and Interface Science*, 149, 49-55.

McClements, D., J. (1999). Food emulsions; principles, practice and techniques, CRC Press, New York.

Mitchell, J.R., Irons, L., & Palmer, G.J. (1970). A study of the spread and adsorbed films of milk proteins. *Biochimica et Biophysica Acta*, 200, 138-150.

Raikos, V. (2010). Effect of heat treatment on milk protein functionality at emulsion interfaces. A review. *Food Hydrocolloids*, 24 (4), 259-265.

Ries, D., Ye, A., Haisman, D., & Singh, H. (2010). Antioxidant properties of caseins and whey proteins in model oil-in-water emulsions. *International Dairy Journal*, *20*, 72-78.

Robson, E.W., & Dalgleish, D.G. (1987). Interfacial composition of sodium caseinate emulsions, *Journal of Food Science*, 52, 1694-1698 .

Rollema, H.S. (1992). Casein association and micelle formation. In P.F. Fox (Ed.), *Advanced Dairy Chemistry* (vol. 4, pp. 173-244). London, Elsevier.

Santana, R.C., Perrechil, F.A., Sato, A.C.K., & Cunha, R.L. (2011). Emulsifying properties of collagen fibers: Effect of pH, protein concentration and homogenization pressure. *Food Hydrocolloids*, 25 (4), 604-612.

Semenova, M.G., Belyakova, L.E., Polikarpov, Y.N., Antipova, A.S., & Dickinson, E. (2009). Light scattering study of sodium caseinate + dextran sulfate in aqueous solution: Relationship to emulsion stability. *Food Hydrocolloids*, 23(3), 629-639.

Shimizu, M., Ametani, A., Kaminogawa, S., & Yamauchi, K. (1986). The topography of α_{s1}-casein adsorbed to an oil/water interface: an analytical approach using proteolysis. *Biochimica et Biophysica Acta*, 869, 259-264.

Shinoda, K., & Friberg, S. (1986). Emulsions and solubilization, John Wiley & Sons, New York

Srinivisan, M., Singh, H., & Munro, P.A. (1996). Sodium caseinate-stabilized emulsions: factors affecting coverage and composition of surface proteins. *Journal of Agricultural and Food Chemistry*, 44, 3807-3811.

Srinivisan, M., Singh, H., & Munro, P.A. (2000). The effect of sodium chloride on the formation and stability of sodium caseinate emulsions. *Food Hydrocolloids*, 14, 497-507.

Surh, J., & McClements, D.J. (2008). Influence of salt concentrations on the stabilities and properties of sodium-caseinate stabilized oil-in-water emulsions. *Food Science and Biotechnology*, 17 (1), 8-14.

Swaisgood, H.E. (1982). Chemistry of milk protein. In P.F. Fox (Ed.), *Developments in Dairy Chemistry*, (pp.1). Applied Science London, United Kingdom.

Van Aken, G.A., & Van Vliet, T. (2002). Flow-induced coalescence in protein-stabilized highly concentrated emulsions: Role of shear-resisting connections between the droplets. *Langmuir*, 18(20), 7364-7370.

Walstra, P. (1983). Formation of emulsions. In P. Becher, *Encyclopedia of emulsion Technology*, vol.1 (pp. 57-127). New York: Marcel Dekker.

Wedzicha, B.L. (1988). Distribution of low-molecular weight food additives in dispersed systems. In E. Dickinson & G. Stainsby (Eds.), *Advances in Food Emulsions*. Elsevier, London.

Wong, D.W.S., Camirand, W.M., & Pavlath, A.E. (1996). Structures and functionalities of milk proteins. *Critical Reviews in Food Science and Nutrition*, 36(8), 807-844.

Ye, A., & Singh, H. (2001). Interfacial composition and stability of sodium caseinate emulsions as influenced by calcium ions. *Food Hydrocolloids*, 15, 195-207.

In: Casein: Production, Uses and Health Effects
Editors: A. M. Ventimiglia, et. al, pp. 195-207
ISBN: 978-1-62100-129-4

Chapter 10

ARE DIETARY CASEINS RELATED TO THE ONSET AND EVOLUTION OF TYPE 1 DIABETES AND CELIAC DISEASE?

A.M. Calderón de la Barca*[*1] and M.E. Mejía-León[1]

[1] Coordinación de Nutrición. Centro de Investigación en Alimentación y Desarrollo, A.C. Hermosillo, Sonora, México.

ABSTRACT

Type 1 Diabetes (T1D) and Celiac Disease (CD) are autoimmune disorders, which share a genetic component associated to HLA-DQ2 and DQ8 haplotypes. However, more than 25% of any given population presents these haplotypes and only a small proportion develops T1D or CD. Therefore, other cofactors of the onset and exacerbation of these diseases has been looked for. Intestinal microbiota changes induced by infections and antibiotic treatments as well as feeding regimes for the first months of life seem to be important. The causal relationship between intake of wheat gluten and related cereal proteins (gliadins) and CD onset in genetically susceptible subjects is well established, and a similar effect has been observed in T1D animal models. Furthermore, there is evidence that bovine milk caseins such as those present in infant formulas correlate to T1D development and to the intestinal inflammatory response typically associated with CD. The autoimmune response cascade in both diseases appears to be related to the chemotactic properties of caseins and gliadins towards human leucocytes. Furthermore, gliadins and caseins are also partially resistant to digestion because of their high proline content, aside of partial similarity in their amino acid sequences. In this chapter we discuss the role of bovine milk caseins as a risk factor for T1D and CD onset and progression on subjects with genetic predisposition. We focus on breastfeeding practices as protective factors and on the gradual and moderate gluten and bovine milk introduction in the infant´s diet while still breastfeeding.

* E-mail amc@ciad.mx

INTRODUCTION

Type 1diabetes (T1D) is characterized by an absolute insulin deficiency caused by an autoimmune-mediated destruction of pancreatic beta cells (Fowler, 2010). Meanwhile, celiac disease (CD) is an autoimmune enteropathy resulting by permanent intolerance to wheat gliadins and other cereal prolamins in geneticallysusceptible individuals (Volta and Villanaci, 2011).

T1D and CD are the two most common autoimmune childhood diseases. The prevalence of those pathologies has grown dramatically during the past few decades, increasing its levels 2.5 and 4 times, respectively (Hermann et al., 2003). Thus, the current T1D global prevalence is 1:300 (Kantarova and Buc, 2007) and the prevalence for CD is around 1:100 to 1:300 (Briani et al., 2008).

These changes in prevalence could be associated to their genetic component. However, although genetic susceptibility is necessary for the development of T1D and CD, their etiology is multifactorial (Karvonen et al., 2000) and the prevalence has increased very fast compared to genetic changes. This is the reason to consider some environmental cofactors with an important role in the pathogenesis of both diseases, such as infant´s diet and infections.

Feeding regimes for the first months of life involve breastfeeding, infant formulas based on cow's milk proteins and the introduction of other foods in infant´s diet. There is evidence that wheat and bovine milk proteins like caseins could lead to an immunogenic response and to the onset of T1D and CD in susceptible children. The variable feeding patterns at this age represents an opportunity to perform a preventive approach for those disorders at their pre-pathogenic state.

GENETIC BASIS OF DISEASE

Celiac disease and type 1 diabetes are mainly associated with major histocompatibility complex genes genotypes. The human leukocyte antigen ‹‹HLA›› gene complex is located on the short arm of chromosome 6. This region is comprised by a group of highly polymorphic genes. The class II HLA is the major determinant of genetic predisposition in both CD and T1D, mainly DQ and DR fractions (Kantarova and Buc, 2007). These genes account for over 50% of genetic factors for both disorders (Tollefsen et al., 2006; Mychaleckyj et al., 2010).

In Europe, around 90-95% of celiac patients have HLA-DQ2 haplotype, composed of alleles <<DQA1*0501/DQB1*0201>>, while the other 5-10% have the HLA-DQ8 haplotype <<DQA1*0301/DQB1*0302>> (Margaritte-Jeannin, 2004). In Amerindian celiac patients the haplotype HLA-DQ8 is predominant (Araya et al., 2000). These same haplotypes are present in about 90% of T1D patients (Fowler, 2010). However, a higher risk has been observed associated to HLA-DQ8, conversely to CD where HLA-DQ2 provides the greatest risk (Bratanic et al., 2010). Interestingly, more than 25% of any population presents these haplotypes and only a small proportion develops T1D or CD. This observation confirms the influence of other factors besides genetics in both pathologies, as shown in Figure 1. In this chapter we will focus mainly on diet.

There are also other non-HLA genes associated with T1D and CD, such as the MYO9B. This gene encodes for proteins involved in intestinal permeability regulation. Thus, polymorphisms at this point increase individual susceptibility for both diseases. Other examples are PARD3 and MAGI2 genes in CD, whereas in T1D, the insulin gene region or locus IDDM2, contributes for an additional risk of 10% (Visser et al., 2009).

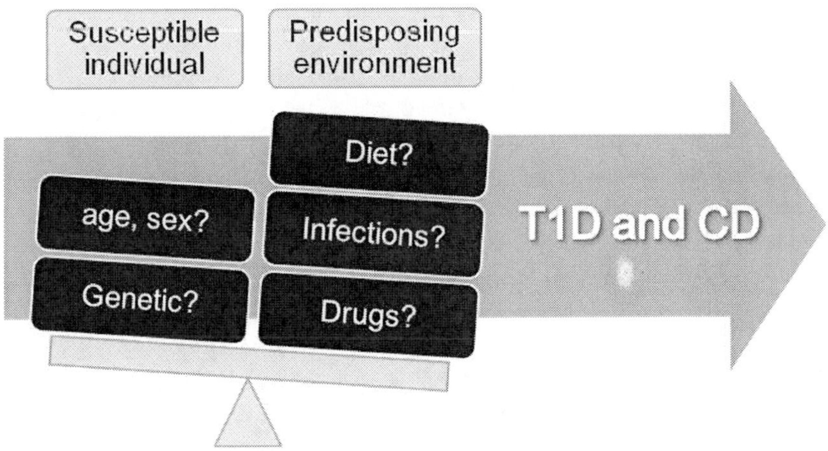

Figure 1. Risk factors interaction during pre-pathogenic period, and its possible relationship to disease development. T1D: Type 1 Diabetes, CD: Celiac Disease.

DIET AND IMMUNITY

Diet is an important source of antigens, which the gut immune system must process and must respond. It is also the source of nutrients that may have some positive or negative effects on T1D and CD disease onset or prevention. Diet is able to modify the immune response in a favorable or negative manner affecting the integrity of the intestinal mucosa and the gastrointestinal microflora balance.

Nowadays, the most important associated dietary factors in T1D and CD onset are breastfeeding, age of introduction of foods to infant´s diet and current intake of nutrients (Di Sabatino and Corazza, 2009; Fasano et al., 2009; Ivarsson et al., 2002). They all influence modulating the immune response and participating in the maintenance of the integrity of the intestinal barrier.

Since T1D and CD are the most common autoimmune disorders in preschool children age appearing with the classical symptoms, dietary factors in early life are very important. Cow´s milk is the first element incorporated in the human diet, either replacing or supplementing breastfeeding (Calder et al., 2006). That is the reason why cow´s milk has been initially considered as a possible risk factor involved in T1D and CD etio-pathogenesis, either by the properties of bovine milk itself or by the absence of maternal breast milk as a protective factor.

Non-breastfed infants are exposed to cow-milk proteins from infant formulas and other dietary proteins introduced earlier than to breastfed infants. Such regimes increase the risk of infections and therefore drug treatments, modifying the microbiota.

In CD the major food antigens are wheat gliadins and other similar prolamins from rye and barley. Some gliadin-derived peptides cross the intestinal barrier, and inside of the lamina propria, these peptides are recognized by antigen presenting cells and elicit a T cell-adaptive immune response, dominated by Th1-type cytokines. Wheat gliadin-peptides also activate epithelial and innate immune cells, contributing to CD pathogenesis (Stepniak and Koning, 2006).

In addition to antibodies against wheat proteins, in subjects with active disease there are increased serum antibodies against other food proteins such as those present in cow′s milk. In contrast, healthy people show low levels of antibodies against food antigens to which they are exposed. Thereby, in absence of food allergy in celiac patients, the explanation of this physiological phenomenon could be that a small fraction of food proteins or long peptides cross undigested the damaged intestinal wall (Kristjánsson et al., 2007). Then, they are presented to the T cells, resulting in antibody production and activation of Th1 cells which favor the local inflammatory infiltrate and villous atrophy (Volta and Villanacci, 2011).

In some studies conducted in diabetes-prone biobreeding rats ‹‹DP-BB›› and non-obese diabetic mice ‹‹NOD››, when animals were fed with a gluten-free hydrolyzed casein diet, the development of T1D was delayed and prevented from 90% to 50%. This is possible through proposed mechanisms such as avoiding the activation of the auto-reactive T cell pool or by skewing those cells to a less pathogenic phenotype. Another mechanism could be the reduction of dietary diabetogenic triggers and the induction of islet neogenesis (Visser et al., 2009). However, these hypotheses have not been extrapolated to human beings due to lack of clinical studies to prove it.

Thereby, dietary prevention of autoimmune diabetes in animal models suggests that changes in the intestinal immune system may be responsible of the induction of the autoimmune destruction of beta cells in pancreas (Westerholm-Ormio et al., 2003). This suggests a link between diet, the gut immune system and development of T1D, establishing their relationship over time.

CASEINS AND GLIADINS

Milk proteins may be divided into two groups: the caseins' fraction and the whey proteins fraction. The fraction of caseins represents around 80% of cow′s milk proteins while its proportion is lower in human milk reaching 40% at the mature period (Calderón de la Barca et al., 1996). They are a heterogeneous group of phosphorylated proteins, which have a strong tendency to associate with each other, leading to the formation of micelles, depending of pH and temperature (Thorn et al., 2009). These biomolecules are classified based on their electrophoretic mobility in 4 types: αS1-casein, s2-casein, β-casein and κ-casein. The β and αS1 types are the most abundant in milk micelles.

One of the most allergenic proteins in cow milk is β-casein. It is the second most abundant protein in cow's milk, after αS1-casein, constituting 28% of the total milk proteins. β-casein has at least 13 genetic variants differing in just one amino acid and each variant fits into one of few main categories known as A1, A2 and B. It is a 209 amino acids phosphoprotein that contains proline and lack of disulphide bonds. It has no secondary structure and presents diminished tertiary interactions, suggesting that the epitopes of this protein are generally linear (Cocco et al., 2007). It is an important issue in relation to intestinal immune response to dietary peptides.

During last 15 years, the consumption of milk with A1 and B variants of β-casein has been related with the development of some diseases such as T1D (Bell et al., 2006). Thus, increased titers of antibodies anti-β-casein has been found in children with newly diagnosed T1D and in patients with latent autoimmune diabetes of adult ‹‹LADA››. Also epidemiological studies showed a significant association between the intake of A1 milk and the incidence of T1D (Laugesen and Elliott, 2003).

Looking for reasons of the β-caseins effect on T1D, it was firstly found that *in vitro* the successive gastrointestinal proteolytic digestion of bovine β-casein variants A1 and B, but not variant A2, produced the bioactive peptide β-casomorphin 7 (BCM-7). In hydrolyzed milk with variant A1 of β-casein, BCM-7 level is 4-fold higher than in A2 milk (Kamiński et al., 2007). In subsequent studies, it was found that people who consumes milk with high levels of β-casein A1 produces bioactive peptides such as the opiod BCM-7. This peptide is absorbed through gut epithelium in newborns and young infants, but not in adults (Bell et al., 2006). This occurs because there is increased intestinal permeability in children compared with healthy adults. Therefore, young infants are more likely to experience the opioid effects of casein. On the opposite way, it has been recently suggest that BCM-7 can protect diabetic rats from hyperglycemia and free radical-mediated oxidative stress (Yin et al., 2010). Anyway, there are no known relationship between opiod peptides and autoimmunity.

Other interesting explanation of effects of β-casein on T1D patients has been found (Woodford, 2008). The sequence of variants A1 and B of β-casein (SLVYPF**PGPI**HN), and the larger fragments containing it, such as the sequences of β-casein variant A2 (**PGPIP**), are responsible for the induction of an immune response towards β-casein. It is because, by cross reactivity, the response would be directed towards the homologous sequence of the glucose transporter GLUT2 inside insulin-producing β-cells in the pancreas, causing damage to the cells that produce insulin. The GLUT2 sequences are the following: (409–420) SFFEIG**PGPIP**W, (412–423) EIG**PGPIP**WFMV, and (414–425) G**PGPIP**WFMVAE, where the **PGPI** sequence as in β-casein is three times repeated. In comparison to bovine β-casein, human β-casein has the following sequence (48–59): SLVYPF**VEPIPY**. Therefore, no cross reactivity between GLUT2 and human β-casein is found out.

On the other hand, cow's milk has long been associated as an etiologic factor for CD; some patients on a strict gluten-free diet still present symptoms when consuming milk and it cannot be attributed to lactose intolerance (Nieminen et al., 2001). Therefore, a possibility was that CD-related symptoms after milk consumption could be due to the presence of gluten proteins in cows' milk coming from wheat-containing feed. Recently, it was clearly demonstrated that there are not gluten peptides in milk from cows fed with wheat (Dekking et al., 2009).

We evaluated, by Western blot, sera from patients with CD for IgA reactivity to bovine and human caseins. CD IgA strongly recognized gliadin subunits and softly α- and β-caseins, but not κ-casein (Cabrera-Chávez and Calderón de la Barca, 2009) neither human β-casein (Cabrera-Chávez et al., 2009). It is a distinctive identification, which agrees with IgA reactions to a casein mixture found by Berti et al. (2003).

The serum IgA response of patients with CD to bovine milk could be related to gliadins and caseins sharing epitopes recognized by anti-gliadin IgA antibodies. The most abundant fragment derived from gliadins is a 33-mer of the alpha fraction (residues 57 to 89) with six T-cell epitopes that stimulate the acquired immunity of the intestinal lamina propria cells from celiac patients (Rodrigo, 2009). The 33-mer sequence LQLQPFPQPQL**PYPQ**PQL**PYPQ**PQL**PYPQ**PQPF (Darewicz et al., 2007) has the motif **PYPQ** repeated three times, which is found in a bovine β-casein peptide. The 7-residues peptide at positions 177-183 of bovine β-casein AVPYPQR was obtained by tryptic/chymotryptic digestion and remained undigested even after microbial peptidases digestion (Deutsch et al., 2000). In contrast, the human β-casein does not present the 7-residues peptide but PIPQ or PVPQ into the sequence VLPVPQP (residues 75-81) and VLPIPQE (residues 161-167), according to total sequence by Greenberg et al. (1984). The aliphatic character of Ile and Val contrasts with the aromatic of Tyr, inducing a very different conformation and antibodies´ recognition.

Additionally, we have found that CD patients IgA immunoreactivity to caseins was age dependent, while human milk caseins were not recognized by IgA from CD patients of any age (Cabrera-Chávez et al., 2009). Recognition by IgA could reflect a differential manifestation of the effects of bovine caseins on the intestinal barrier because the high intrinsic resistance to hydrolysis of bovine β-casein (Deutsch et al., 2000) and its homology to T-cell stimulatory α-gliadins.

Kristjánsson et al. (2007), found that the 50% of celiac patients developed an intestinal inflammatory reaction induced by cow´s milk proteins. In their study, they analyzed the local inflammatory reaction to gluten and cow´s milk proteins, performed rectal challenges in subjects with CD in remission. The response due to caseins was similar in magnitude to that of gluten producing elevated concentrations of nitric oxide and myeloperoxidase after challenge by pathways not identified yet. The release of myeloperoxidase correlates with gut mucosal granulocyte activation. It could suggest an innate as well as adaptive immune response to cow´s milk proteins, especially to casein.

Moreover, caseins, as well as gliadins, have some other common characteristics. Among those, emphasize that they both have chemotactic properties to human leucocytes (Kristjánsson et al., 2007). This means that both caseins and gliadins have the ability to produce a chemical stimulation to attract and accumulate leukocytes at the site of the reaction. In this situation, this phenomenon occurs as part of the initial inflammatory response. In consequence, this condition favors the lymphocyte and other white cells intraepithelial infiltrate which has been mostly related with intestinal villi shortening and atrophy (Di Sabatino and Croazza, 2009).

It is important to note that in some studies, casein milk suspensions were used trying to mimic the physiological stage, as occurs in genetic susceptible individuals who follow a casein-rich diet. However in other studies it has been used other protein concentrates from milk such as whey powder and calcium caseinate. This has caused controversy due to structural and behavioral differences between the calcium caseinate, whey and micellar casein. In this dilemma, calcium caseinate appears to have intermediate properties between whey and micellar caseins, but closer to the latter, in other words, to the milk in this natural form (Reitelseder et al., 2011). Thereby, when micellar casein is consumed, it clots in the stomach resulting in large protein globules, which leaves the stomach much more slowly than acid-soluble whey proteins (Phillips, 2011). As a matter of fact, caseins extracted from fresh milk show some difference in recognition by IgA from CD patients that these obtained from yogurt (Cabrera-Chávez et al., 2009).

Differences in immune reactions found among studies, could be also related to different digestion treatments of caseins. As it was previously commented, several large undigested β-casein peptides remained after tryptic/chymotryptic and peptidases digestion (Deutsch et al., 2000). However, according to Berti et al. (2003), the immunoreactivity of IgA from celiac patients to bovine caseins completely disappeared after hydrolysis. The casein hydrolysis (pepsin or pancreatin:substrate ratios of 1:30 for 24 h pepsin and posterior 3 h pancreatin) in their study is far away of the physiological conditions in intestinal lumen, but an extensive laboratory hydrolysis.

Intestinal Mucosa and Disease

The mucosal immune system, especially in the gastrointestinal tract, is covered by a specialized epithelium including both cells and lymphoid structures. It should be emphasized that most antigens enter the organism through the mouth. Based on this, diet is an essential factor to consider and manage as a trigger of many diseases, such as DM1 and CD.

The gut epithelium provides one of the body´s major interfaces between the organism and the external environment. The first lines of defense at the intestinal lumen are gastric, pancreatic and brush border proteases. Those enzymes protect against potentially toxic diet proteins, including gluten peptides and caseins (Shan et al., 2002). Additionally to them, the gut associated lymphoid tissue ‹‹GALT›› is responsible for processing and responding to a wide variety of antigens including those that come from food, bacteria, viruses and other microorganisms. This system is integrated by epithelial cells as a physical barrier, as well as the cellular component highlighting dendritic and T cells (Calder et al., 2006).

The immune system also ensures tolerance to self, to food and other environmental components. Nevertheless, there are anomalies in susceptible individuals that may lead to disease. Then, in the pathogenesis of CD and T1D it is considered that either the physical barrier or the immune response is disrupted. It allows wheat gluten peptides, and other proteins such as zeins and caseins, reach the lamina propria and trigger an inflammatory response (Di Sabatino and Corazza, 2009). In relation to those mechanisms, it has been proposed a zonulin-associated paracelular mode to explain the loss of tight cell junctions and consequently protein passage trough intestinal epithelium (Visser et al., 2009), as shown in Figure 2. This pathway appears to be physiologically involved in other functions, including the movement of fluids, macromolecules and leukocytes between the bloodstream and the intestinal lumen and vice versa.

In the NOD mice model has been detected that diabetogenic immune cells are activated in the mesenterial immune system already before pancreatic islets are infiltrated. That emphasizes the role of the gut immune response in T1D onset (Westerholm-Ormio et al., 2003). Transferring this to humans, this result suggest the importance of mucosal integrity, a proper functioning immune system and an adequate diet management in the first year of life, prior to the diagnosis of disease, typically around the age of 2 years.

Historically, it had been accepted the idea that CD is a disorder linked to gluten intake. However, recent studies have found a mucosal inflammatory response, similar to that elicited by gluten, produced by cow´s milk proteins in about half of the patients with CD. Considering this finding, casein in particular, seems to be involved in this reaction since it has been ruled α-lactalbumin as immunogenic inductor (Kristjánsson et al., 2007). It explains why celiac patients who are on gluten-free diet sometimes have persistent symptoms of disease.

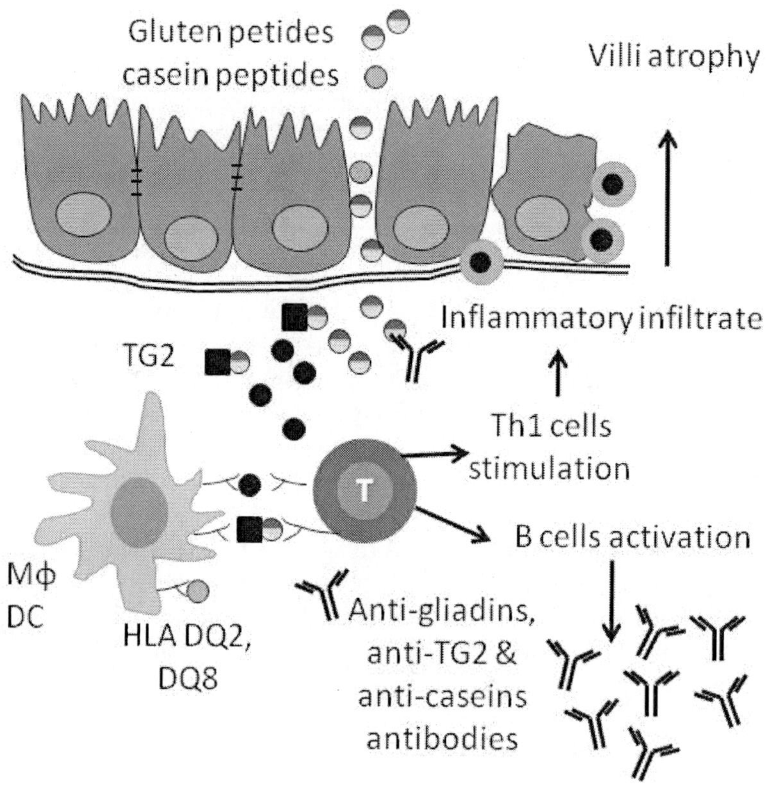

Modified from Sollid and Lundin, 2009.

Figure 2. Paracelular pathway. Protein peptides pass through disrupted tight junctions to the lamina propria and become deaminated by the enzyme transglutaminase 2 (TG2). Then, these peptides are presented to the CD4+ T cells, by HLA DQ2 and DQ8 molecules, promoting gut damage and antibodies release. M: Macrophage, DC: Dendritic Cell.

Moreover, in TID patients, there is also evidence of immunological activity in the small intestine. They show elevated concentrations of tumor necrosis factor alpha ‹‹TNF-α›› and interferon gamma ‹‹IFN-γ›› positive cells than those of healthy controls (Visser et al., 2009). In other studies, enhanced immune responses to casein and others cow´s milk proteins have been reported in serologic samples of newly diagnosed T1D children (Westerholm-Ormio et al., 2003). Both results suggest an inflammatory response similar to what occurs in CD patients, in which gluten is not the only trigger.

Breastfeeding as a Protective Factor against Infections and Autoimmunity

Generally, the children diet triggers an immune process resulting in a phenomenon of tolerance or intolerance to food components. However, there is an interval during which, in humans, the ability to develop oral tolerance against the introduction of a new antigen is decreased. This incident is also influenced by the amount of antigen and type of food, particularly in cases in which the cooking process altered protein immunogenicity. It is also necessary to consider that the immune response can be modified by other exposures. The most important of these is breastfeeding, for its multiple properties (Ivarsson et al., 2002; Fasano, 2009).

Recent studies suggest that the infant nutritional pattern plays a critical role in development of CD and other autoimmune disorders such as T1D. Breastfeeding appears to reduce the risk of both pathologies; in part, because it aids to maintaining favorable microorganisms in intestinal flora. For example, in children with breastfeeding scheme, *Bifidobacterium* is frequently prevalent; whereas in formula-fed infants predominate pathogenic bacteria such as *Bacteroids, Clostridium* and *Streptococcus* (Visser et al., 2009). Another breastfeeding advantage consists in its immunomodulator effect; since it performs a vital role in preventing potentially diabetogenic viral infections like those caused by coxsackie and rotavirus, at this stage of life (Couper et al., 1999).

This should be considered, as these viruses cause a transient increase in gut permeability. This facilitates the prolamins and cow's milk proteins passage through the intestinal mucosa and, in consequence, the overstimulation of tissue transglutaminase, encouraging immunogenicity (Di Sabatino and Corazza, 2009). In addition, there is evidence that viral infections as it by rotavirus promote the secretion of interferon α, leading to CD (Green and Jabri, 2003).

It has been found a significant association with coxsackie virus B-4 at the time of T1D diagnosis. Likewise, prospective studies in susceptible children have shown an increased incidence of enterovirus infections in children who later develop T1D, and a temporal relationship between infection and autoimmunity (Yeung, 2011). Thus, there is evidence that repeated rotavirus infections cause autoimmune diseases in predisposed subjects (Settyet al., 2008). Rotavirus produces the neutralizating protein VP7, which shows homology with tissue transglutaminase related to CD pathogenesis (Di Sabatino and Corazza, 2009). This protein also has similarity with T cell epitopes on tyrosin-phosphatase autoantigen associated with AI-2 molecule ‹‹anti-islet›› and with an isoform of glutamic-acid decarboxylase (Knip, 2003), partnering with development of both diseases. Moreover, in bacteria, *Vibrio cholerae* produces zonula occludens toxin ‹‹Zot›› (Visser et al., 2009). This enterotoxin reversibly opens the tight junctions between intestinal cells, favoring the paracelular pathway (Fasano, 2009).

In clinical practice, the presence of infection is usually associated with antibiotics administration. These drugs have among their effects, the ability to alter the gut microflora composition. Thus, its use can promote colonization by pathogenic microorganisms, such as those mentioned, which may induce the T1D and CD onset.

Besides the protective effects against infections and microflora maintenance, another breast milk benefit, is that while breastfeeding is practiced, the introduction of other food to the diet is delayed. In this food group may be mentioned wheat and cow´s milk, since they are the first to be introduced to the infant diet.

In this context, Fasano (2009), propose a preventive approach to EC, which applies also for DM1. Their idea is based on the fact that the immune system matures dramatically in the first 12 months of life. In their research, they studied 700 children with genetic susceptibility and family history of CD in a dietary regimen that encourages breastfeeding, avoiding entirely the ingestion of gluten in the first year of life. This strict approach aimed to delay the onset of CD, or prevent it completely. Interestingly, their results suggest a reduction by four fold in CD development. These findings can be explained in terms of that this strategy helps to the development of immune system tolerance to gluten, and probably other dietary antigens such as bovine milk caseins, as occurs in healthy people, avoiding an overreaction against them, and thus decreasing the propensity to CD.

Moreover, in multivariate analysis, it was found that complementary feeding with cow´s milk-based formulas is a risk factor even more important than a short breastfeeding history (Knip, 2003). In CD the current recommendation is aimed at maintaining breastfeeding at least the first 6 months of life, delaying the exposure to cow´s milk proteins. The latter should be introduced in a progressive manner in moderate amounts (Di Sabatino and Corazza, 2009; Ivarsson et al., 2002), while continuing breastfeeding. Thus, it can help to reduce the risk for CD (Setty et al., 2008). Such recommendation is the base of PREVENT CD, which is the European scientific research project with the goal to prevent CD. The project examines the influence of the dietary history in the prevention of CD and studies the possibilities of induction of tolerance for gluten in genetic predisposed children for CD (http://www.phadia.com/en-US/Laboratories/Autoimmunity/Partnerships/Prevent-Celiac-Disease/).

At present, the changes of environmental factors seem to be the most effective strategies for T1D and CD prevention (Knip, 2003). The trial to reduce insulin-dependent diabetes mellitus in the genetically at risk ‹‹TRIGR study›› was designed to determine whether the risk of developing T1D can be reduced by elimination of cow milk proteins in the diet of infants during the first 6 to 8 months of life. In this trial, the use of a hydrolyzed formula versus cow´s milk standard formula is being compared, as a likely way to reduce the risk of acquiring the disease. The preliminary results suggest the possible prevention of beta cell autoimmunity, as a result of changes in the infant´s diet (Aakerblom, 2011).

CONCLUSION

There is evidence of the role of diet in the development of T1D and CD in individuals with genetic susceptibility. The type of diet in childhood determines the classes of microorganisms of intestinal microflora, promoting or preventing disease development. Breastfeeding is recommended during the first 6 months of age because its immunomodulatory and bactericidal effects. Its practice also promotes the delay in introduction of foods, such as wheat gluten and casein´s cow's milk. These proteins have immunogenic potential and have been found associated with onset of both diseases.

The solution is not to avoid cow's milk, but its inclusion in infants' diet must be at the right time and in moderate amounts to prevent T1D and CD. This process must be performed along with other foods introduction, such as cereals. This would help to delay the onset of T1D and CD in the third of population with genetic vulnerability. Finally, in patients with CD, when symptoms persist despite a gluten-free diet, it is also recommend a casein-free diet to stop the progression of the disease and to reduce the clinical symptoms.

AKNOWLEDGMENT

This chapter was written as part of a project with financial support by the Mexican Council for Science and Technology (CONACyT) through the Grant SALUD-2009-C01-115212. The authors are grateful to MSci. Adriana Bolaños and Dr. Roger Sotelo for manuscript edition.

REFERENCES

Aakerblom, H. K. (2011). The trial to reduce IDDM in the genetically al risk (TRIGR) study: Recruitment, intervention and follow-up. *Diabetologia, 54,* 627-633.

Araya, M., Mondragón, A., Pérez-Bravo, F., Roessler, J., Alarcón, T., Ríos, G. & Bergenfreid, C. (2000). Celiac disease in a Chilean population carrying Amerindian traits. *Journal of Pediatric Gastroenterology & Nutrition,* 31, 381-386.

Briani, C., Samaroo, D. & Alaedini, A. (2008). Celiac disease: From gluten to autoimmunity. *Autoimmunity Reviews,* 7, 644-650.

Bratanic, N., Smigoc, D., Mendez, A., Bratina, N., Battelino, T. & Vidan-Jeras, B. (2010). An influence of HLA-A, B, DR, DQ, and MICA on the occurrence of celiac disease in patients with type 1 diabetes. *Tissue Antigens,* 76, 208-215.

Bell, S. J., Grochoski, G. T. & Clarke, A. J. (2006). Health implications of milk containing [beta]-casein with the A2 Genetic variant. *Critical Reviews in Food Science and Nutrition,* 46, 93-100.

Berti, C., Trovato, C., Bardella, M. T. & Forlani F. (2003). IgA anti-gliadin antibody immunoreactivity to food proteins. *Food and Agricultural Immunology,* 15, 217-23.

Cabrera-Chávez, F., Rouzaud-Sández, O., Sotelo-Cruz, N. & Calderón de la Barca A. M. (2009). Bovine milk caseins and transglutaminase-treated cereal prolamins are differentially recognized by IgA of celiac disease patients according to their age. *Journal of Agriculture and Food Chemistry,* 57, 3754-3759.

Cabrera-Chávez, F. & de la Barca, A. M. (2009). Bovine milk intolerance in celiac disease is related to IgA reactivity to α and β-caseins. *Nutrition,* 25, 715-716.

Calder, P. C., Krauss-Etschmann, S., de Jong, E. C. (2006). Early nutrition and immunity – progress and perspectives. *British Journal of Nutrition*, 96, 774-790.

Calderón de la Barca A. M., Bolaños-Villar, A., Román-Pérez R. (1996). Protein contents of milk surrogates and their sanitary regulation. *Salud Publica Mex,* 38, 268-275.

Cocco, R. R., Järvinen, K. M., Han, N., Beyer, K. & Sampson, H. A. (2007). Mutational analysis of immunoglobulin E-binding epitopes of b-casein and b-lactoglobulin showed a heterogeneous pattern of critical amino acids between individual patients and pooled sera. *Clinical and Experimental Allergy,* 37, 831-838.

Couper, J. J., Steele, C., Beresford, S., Powell, T., McCaul, K., Pollard, A., Gellert, S., Tait, B., Harrison, L. & Colman, P. (1999). Lack of association between duration of breast-feeding or introduction of cow's milk and development of islet autoimmunity. *Diabetes*, 48, 2145-2149.

Darewicz, M., Dziuba, J. & Minkiewicz, P. (2007). Computational characterization and identification of peptides for in silico detection of potentially celiac-toxic proteins. *Food Science and Technology International,* 13, 125-33.

Dekking, L., Koning, F., Hosek, D., Ondrak, T. D., Taylor, S. L., Schroeder, J. W., Bauer, M. (2009). Intolerance of celiac disease patients to bovine milk is not due to the presence of T-cell stimulatory epitopes of gluten. *Nutrition,* 25, 122-3.

Deutsch, S. M., Molle, D., Gagnaire, V., Piot, M., Atlan, D., & Lortal S. (2000). Hydrolysis of sequenced β-casein peptides provides new insight into peptidase activity from thermophilic lactic acid bacteria and highlights intrinsic resistance of phosphopeptides. *Applied and Environmental Microbiology,* 66, 5360-5367.

Di Sabatino, A. & Corazza G. R. (2009). Coeliac Disease. *The Lancet*, 373, 180-1493.

Fasano, A. Surprises from celiac disease. (2009). *Scientific American,* 301, 54-61.

Fowler, M. J. (2010). Diabetes: Magnitude and mechanisms. *Clinical Diabetes*, 28, 42-46.

Green, P. & Jabri, B. (2003). Coeliac disease. *The Lancet*, 362, 383-391.

Greenberg, R., Groves, M. I. & Dower, H. J. (1984). Human beta-casein amino acid sequence and identification of phosphorylation sites. *The Journal of Biological Chemistry,* 259, 5132-5138.

Hermann, R., Knip, M., Veijola, R., Simell, O., Laine A. P., Akerblom, H. K., Groop, P. H., Forsblom, C., Pettersson-Fernholm, K. & Ilonen, J. (2003). Temporal changes in the frequencies of HLA genotypes in patients with Type 1 Diabetes – indication of an increased environmental pressure? *Diabetologia,* 46, 420-425.

Ivarsson, A., Hernell, O., Stenlund, H. & Persson, A. (2002). Breast-feeding protects against celiac disease. *American Journal of Clinical Nutrition,* 75, 914-921.

Kamiński S., Cieslińska, A. & Kostyra, E. (2007). Polymorphism of bovine beta-casein and its potential effect on human health. *Journal of Applied Genetics*, 48, 189-98.

Kantarova, D. & Buc, M. (2007). Genetic susceptibility to Type 1 Diabetes Mellitus in humans. *Physiological Research,* 56, 255-266.

Karvonen, M., Viik-Kajander, M., Moltchanova, E. & Libman, I. (2000). Incidence of childhood type 1 diabetes worldwide. *Diabetes Care*, 23, 516-1526.

Knip, M. (2003). Environmental triggers and determinants of beta cell autoimmunity and type 1 diabetes. *Reviews in Endocrine & Metabolic Disorders,* 4, 213-223.

Kristjánsson G., Venge, P. & Hällgreen, R. (2007). Mucosal reactivity to cow's milk protein in coeliac disease. *Clinical and Experimental Immunology,* 147, 449-455.

Laugesen, M. & Elliott, R. (2003). Ischaemic heart disease, Type 1 diabetes, and cow milk A1 beta-casein. *The New Zealand Medical Journal*, 116, U295.

Margaritte-Jeannin, P., Babron, M. C., Bourgey, M., Louka, A. S., Clot, F., Percopo, S., Coto, I., Hugot, J. P., Ascher, H., Sollid, L. M., Greco, L. & Clerget-Darpoux, F. (2004). HLA-DQ relative risks for coeliac disease in European populations: A study of the European genetics cluster on coeliac disease. *Tissue Antigens,* 63, 562-567.

Mychaleckyj, J. C., Noble, J., Moonsamy, P., Carlson, J., Varney, M., Post, J., Helmberg, W., Pierce, J., Bonella, P., Fear, A., Lavant, E., Louey, A., Boyle, S., Lane, J., Sali, P., Kim, S., Rappner, R., Williams, D., Perdue, L., Reboussin, D., Tait, B., Akolkar, B., Hilner, J.

& Steffes, M., Erlich, H. (2010). HLA genotyping in the international Type 1 Diabetes genetics consortium. *Clinical Trials,* 7, S75-S87.

Nieminen, U., Kahri, A., Savilahti, E. & Farkkila, M. A. (2001). Duodenal disaccharidase activities in the follow-up of villous atrophy in coeliac disease. *Scandinavian Journal of Gastroenterology,* 36, 507-10.

Phadia A. B. Prevent Celiac Disease [online]. [2011-01- 15] Available from: http://www.phadia.com/en-US/Laboratories/Autoimmunity/Partnerships/Prevent-Celiac-Disease/

Phillips, S. M. (2010). A comparison of whey to caseinate. *American Journal of Physiology-Endocrinology and Metabolism*. 300, E610.

Reitelseder, S., Kjaer, M. & Holm, L. (2011). Reply to letter to the editor. *American Journal of Physiology-Endocrinology and Metabolism,* 300, E611-E612.

Rodrigo, L. (2009). Investigational therapies for celiac disease. *Expert Opinion on Investigational Drugs,* 18, 1865-1873.

Setty, M., Hormaza, L. & Guandalini, S. (2008). Celiac disease: Risk assessment, diagnosis and monitoring. *Molecular Diagnosis & Therapy,* 12, 289-298.

Shan, L., Molberg, O., Parrot, I., Hausch, F., Filiz, F., Gray, G., Sollid, L. & Khosla, C. (2002). Structural basis for gluten intolerance in celiac sprue. *Science*, 297, 2275-2279.

Sollid, L. M. & Lundin, K. E. A. (2009). Diagnosis and treatment of celiac disease. *Mucosal Immunology,* 2, 3-7.

Stepniak, D. & Koning, F. (2006). Celiac disease-sandwiched between innate and adaptive immunity. *Human Immunology,* 67, 460-468.

Thorn D. C., Ecroyd, H. & Carver, J. A. (2009). The two-faced nature of milk casein proteins: Amyloid fibril formation and chaperone-like activity. *The Australian Journal of Dairy Technology,* 64, 34-41.

Tollefsen, S., Arentz-Hansen, H., Fleckenstein, B., Molberg, O., Raki, M., Kwok, W., Jung, G., Lundin, K. & Sollid, L. (2006). HLA-DQ2 and DQ8 signatures of gluten cell epitopes in celiac disease. *Journal of Clinical Investigation,* 116, 226-2236.

Visser, J., Rozing, J., Sapone, A., Lammers, K. & Fasano, A. (2009). Tight junctions, intestinal permeability and autoimmunity: Celiac disease and type 1 diabetes paradigms. *Annals of New York Academy of Science,* 1165, 195-205.

Volta, U. & Villanacci, V. (2011). Celiac disease: diagnostic criteria in progress. *Cellular and Molecular Immunology,* 8, 96-102.

Westerholm-Ormio, M., Vaarala, O., Pihkala, P., Ilonen, J. & Savilahti, E. (2003). Immunologic activity in the small intestinal mucosa of pediatric patients with type 1 diabetes. *Diabetes,* 52, 2287-2295.

Woodford, K. A1 beta-casein, type 1Diabetes and links to other modern illnesses. International Diabetes Federation Western Pacific Congress [online]. (2008) [2011-01-15] Available from: http://www.guernsey.net/~wgcf/PageMill_Resources/A1_Beta-casein%20Woodford.pdf

Yeung, W. C., Rawlinson, W. D., & Craig, M. E. (2011). Enterovirus infection and type 1 diabetes mellitus: Systematic review and meta-analysis of observational molecular studies. *British Medical Journal*, 342, d35.

Yin, H. & Miao J. Y. (2010). Protective effect of beta-casomorphin-7 on type 1 diabetes rats induced with streptozotocin. *Peptides*, 31, 1725.

In: Casein: Production, Uses and Health Effects
Editors: A. M. Ventimiglia, et. al, pp. 209-221
ISBN: 978-1-62100-129-4

Chapter 11

TYPES OF CASEIN AND THEIR USE IN THE MANUFACTURE OF PROCESS CHEESE

Mamdouh El-Bakry
Department of Dairy Science and Technology, Faculty of Agriculture,
Cairo University, Giza, Egypt

ABSTRACT

Casein, the main milk protein that constitutes approximately 75% of that protein, is in the form of micelles with a radius of 0.1 μm. Most of the methods used for the preparation of casein from milk basically involve a removal of the fat of the milk followed by a destabilisation process of the desired protein by insolubilization. This process is achieved either by an isoelectric precipitation, i.e. by acidification in order to reach the isoelectric point of the casein, or an enzymatic destabilisation of the casein micelle by the addition of the rennin enzyme, which could be of animal, plant or microbial origin.

An important and essential use of casein is in the process cheese industry, particularly imitation cheese. Process cheese products are manufactured by blending different types of cheese, or dry casein in the case of imitation cheese, with water, fat, salts and other ingredients like flavour or colour enhancing compounds in the presence of the necessary thermal and mixing energy required.

This chapter reviews the different types of casein and their use in the manufacture of process imitation cheese products. These types of protein include mainly rennet casein, acid casein, caseinate, phosphocasein, enriched casein fractions, casein fragments and casein-whey protein aggregates. The most commonly used casein in the process cheese manufacture is the acid- and rennet casein and/or also the sodium caseinate. This is due to the fact that these protein types are the main cause of the physico-chemical stability of the cheese end-product obtained.

The current chapter illustrates recent research work done on the use of casein, as a principal source of protein, and its major impact on the physico-chemical (functional) properties, i.e. heat-induced flowability, viscoelastic properties and hardness, of imitation cheese. Traditionally, rennet casein is the casein type used in the manufacture of this type of cheese. Increased casein content resulted in increased viscoelastic properties and hardness, whereas a partial replacement of casein by whey proteins led to a reduced heat-induced flowability due to the interactions between casein and whey proteins. Interestingly, a partial substitution of the casein used by rice starches led to changes in the flowability and the hardness depending on the amylose content of the starch.

1. INTRODUCTION

The word milk comes from the old Germanic word *meluk,* which means the liquid secreted by the mammary glands of female mammals for suckling their young. Milk is considered as the real commodity in the food industry, because of milk proteins, which are the core of many industrial processes, due to their unique properties (De Kruif, 2003). The approximate composition of milk is in average as follows, 4.36% fat, 4.51% lactose and 3.87% protein. The total dry matter content is 13.2%. In general, milk can be viewed as a colloidal dispersion of protein and fat particles in a continuous phase (solvent) or watery solution of salts and lactose. Milk proteins constitute about 3.87% of milk. These proteins consist of casein, the main milk protein that constitutes about 3% of the milk, and whey proteins. Casein is in the form of micelles, with a radius of 0.1 μm, and whey proteins have a radius of 0.003 μm (Vasbinder, 2003).

Process cheese products, or cheese substitutes, are considered as non-standard cheese products that are used as an alternative for natural cheese due to the cost efficiency, feasibility in production, versatility in applications and usefulness and long shelf life (Meyer, 1973; Berger et al., 1989; Pellegrino et al., 1996). These cheese substitutes are, unlike natural cheeses, not made directly from milk or dehydrated whole milk, but rather from various combinations of other dairy and/or non-dairy ingredients (Zehren & Nusbaum, 2000; Guinee et al., 2004). These ingredients can include natural cheese or cheese curd, casein or caseinates, butter oil and vegetable ingredients such as vegetable oil, proteins and carbohydrate like starch. Imitation cheese, which is also called analogue cheese, is a form of process cheese products (Fox et al., 2000; Bachmann, 2001; Guinee et al., 2004).

Imitation cheese is a term usually applied to products manufactured by mixing water, various edible oil/fats, protein (usually, but not necessarily, rennet casein) and other ingredients in the presence of so-called emulsifying salts to obtain a homogeneous product having similar properties to natural cheese. This is generally achieved with the aid of a certain required level of heating and mechanical agitation (Guinee, 2003). Imitation cheese differs from other process cheese products in that the protein ingredients are dry ingredients like casein powder rather than natural cheese (Meyer, 1973).

The manufacture of imitation cheese has grown in recent years, due to a number of reasons. Firstly, the high consumer demand for prepared or convenience foods; in these foods imitation cheese is used as an alternative to natural cheese, e.g. Mozzarella-like imitation cheese in pizza, pies, lasagna and cordon bleu products, where cheese melting properties are crucial (Mounsey & O'Riordan, 1999). Secondly, imitation cheese is favoured by food manufacturers due to its cost-effectiveness, which arises from the ease of its manufacture and the replacement of more expensive milk ingredients, e.g. milk fat, by cheaper vegetable blend products (Mounsey & O'Riordan, 2008a). Thirdly, imitation cheese, unlike natural cheese, allows the manufacturer to tailor the product to specific requirements by altering formulation and processing parameters (Guinee et al., 2004).

This chapter is focusing on reviewing the different types of casein and their use in the manufacture of process imitation cheese products. These types of protein include rennet casein, acid casein, caseinate, phosphocasein, enriched casein fractions, casein fragments and casein-whey protein aggregates. In addition, recent research work done on the use of casein, as a principal source of protein, in imitation cheese is being illustrated. Most importantly, the major impact of the change of the protein ingredients used is being discussed, e.g. the impact on the physico-chemical (functional) properties, i.e. heat-induced flowability, viscoelastic properties and hardness.

2. Types of Casein

There are several types of casein depending on the manufacture process starting from the milk and also on the way of fractionation and preparation of the protein. The principal types of casein are rennet casein, acid casein and caseinate. The approximate chemical composition of these caseins, in the form of dried powder, is shown in Table 1. In addition there are other forms of casein that will be discussed in this section: phosphocasein, casein fragments, enriched casein fractions and casein-whey protein aggregates.

Table 1. Chemical composition of casein dried powder

Casein Type	Protein	Fat	Carbohydrates	Moisture	Ash
Rennet Casein	90.45	0.81	0.11	9.55	8.47
Acid Casein	97.21	0.81	0.11	9.56	1.78
Caseinate	92.89	1.23	0.21	4.24	4.38

Compiled from Fox, 2003; Walstra et al., 2006.

2.1. Rennet Casein

The manufacture of rennet casein depends on the action of an enzyme on κ-casein, which is located at the surface of casein micelles and plays a principal role in the stabilisation of the micelles. Rennet casein is obtained by proteolytic destabilisation of the casein micelles in milk by using chymosin, the main enzyme of the rennet, or chymosin-like proteinases, which cleave the phenylealanine$_{105}$–methionine$_{106}$ bond of κ-casein (Southward & Walker, 1980; Mulvihill, 1989). Chymosin cleaves κ-casein into the water soluble glycomacropeptide, responsible for the colloidal stability, and the precipitated colloidal para-κ-casein, which lost its stability due to the presence of calcium salts.

The manufacture steps can be summarized as follows. During the manufacture of rennet casein, the pH remains around 6.6-6.8, which is the pH of the milk used in the manufacture process. The milk fat is removed and the skimmed milk is pasteurized, warmed to approximately 30°C and renneted (~ 0.012- 0.014%). After this treatment at 30°C for about 40 minutes, the renneting step is complete and curd particles are formed. This curd is cooked at ~ 60°C to increase the firmness, inactivate the enzyme and encourage syneresis. The curd particles are washed and then finally dried. Rennet casein consists of para-casein micelles, which are native micelles lacking their stabilising layer of caseinomacropeptide (κ-casein-$_{(106-169)}$ peptide). In this respect, rennet casein curd is considered to be similar to cheese curd, and the only difference is the absence of starter cultures and rennet activity. Indeed in some instances, the former is marketed as dehydrated skimmed milk curd.

2.2. Acid Casein

Acid casein is defined as the product obtained after isoelectric precipitation at a pH of 4.6-4.7 of the caseins from milk, usually skimmed milk (Mulvihill, 1989). The formation of acid casein depends mainly on the fact that the solubility of the casein in milk reaches its minimum at the isoelectric point, whereas, in the contrary at this pH, the whey proteins remain soluble. Therefore, it is necessary to separate the formed casein curd from the whey, which is achieved by sedimentation or sieving in the presence of a centrifugal separation. This is followed by a washing step, to remove the residual whey components, and then a drying step of the precipitated casein. A distinction can be made between lactic acid casein and mineral acid casein. The former type of casein is obtained from acidification by lactic acid produced by a starter culture, e.g. Lactobacilli or Streptococci. In this case, the casein has a very low lactose content. Mineral acid casein is obtained from acidification by mineral acids, such as hydrochloric, sulphuric or nitric acid. The acidification dissolves the colloidal calcium phosphate in the casein micelles; where part of that colloidal calcium remains bound to the casein. Acid casein contains a low amount of calcium, 80 mg/ 100 g vs more than 300 mg/ 100g for the rennet casein, which remains bound to the proteins (Walstra et al., 2006). All types of acid caseins contain some residual plasmin and plasminogen, which are associated with the casein fraction. Accordingly, the proteolytic activity of plasmin can affect the stability of products containing this type of casein.

2.3. Caseinates

Caseinates are obtained by addition of alkali to acid casein. The increase in pH will dissolve the casein and, depending on the cation of the alkalising agent, different types of caseinate are obtained, i.e. sodium, potassium, calcium and ammonium caseinates (Dalgleish, 1997). Caseinate, in comparison to rennet casein and also acid casein, is highly soluble which allows the resulting solution, which is viscous in its nature, to form foams and emulsions exhibiting very good functional properties and numerous applications in the food industry. The functionality of calcium caseinate is quite different compared to that of sodium-, potassium- and ammonium caseinates, due to the particular interaction of calcium ions with the phosphoserine residues of the casein (Aimutis, 1995). Dissolved calcium caseinate is a milky colloidal dispersion due to calcium-induced formation of micelle-like casein aggregates. Sodium, potassium and ammonium caseinates form nearly clear solutions; in most cases slight opalescence is observed due to the association of the caseins (Dalgleish, 1997).

2.4. Phosphocasein

Phosphocasein, or casein micelles, is prepared by microfiltration and/or dialfilration processes of skim milk, during which whey proteins and serum caseins are permeated and leading to the retention of the casein micelles (Kelly et al., 2000). These casein micelles are to a high extent free of whey proteins and accordingly these micelles can be used in colloidal chemistry research where a model system for casein micelles is needed. Phosphocasein is a stable colloidal dispersion and is relatively inert compared to other milk proteins. As the casein is actually low in whey proteins, it has therefore high heat stability. In addition to this important property, phosphocasein shows high coaguability to rennin enzymes or their equivalents, which is less affected by heat treatments compared to fresh milk or skim milk.

2.5. Casein Fragments

In the past few decades, protein fragments derived from milk proteins attracted a lot of interest because of their biological activity (Korhonen & Pihlanto, 2006; Phelan et al., 2009). The bioactivity of casein-derived peptides ranges from nutritional aspects to effects on the cardiovascular, immune, nervous and digestive systems.

2.5.1. Caseinomacropeptide

Caseinomacropeptide (CMP), also known as glycomacropeptide, caseinoglycomacropeptide and casein macropeptide, is a κ-casein peptide that contains the amino acids (106–169) and is a product from the enzymatic cleavage of κ-casein by chymosin and/or chymosin-like enzymes. The CMP is a negatively charged partly glycosylated peptide. CMP can be isolated from natural cheese whey particularly rennet casein whey, as the latter provides a purer product due to the absence of possible contamination by starter cultures or salt used as essential ingredients in the cheese manufacture (Thoma-Worringer et al., 2006). CMP can be also derived from rennet-induced proteolysis of κ-casein containing casein mixtures or κ-casein rich fractions (Doultani et al., 2003). A variety of biological activities, such as immunosuppression, inhibition of pathogen invasion and induction of satiety, are assigned to CMP. Moreover, the CMP is entirely free from aromatic amino acids, which makes it a suitable protein source for patients suffering from phenylketonuria.

2.5.2. Casein Phosphopeptides

These are casein fragments that are keeping the serine phosphate residues, which constitute the metal and the calcium phosphate binding sites of α_S-, β- and κ-casein. Due to the presence of these binding sites, phosphopeptides can bind metal ions, especially the divalent ones, and may function as carriers for a variety of minerals. It has been demonstrated that casein phosphopeptides, particularly the peptides from β-casein which are available in pure form in commercial quantities, possess an anticariogenic effect (Sofia et al., 2005). Therefore, applications of these phosphopeptides has been proposed in the treatment of dental diseases. Some of the phosphopeptides have an immunomodulating effect (Phelan et al., 2009). In general, phosphopeptides are released in vivo, such as their formation in some cheeses, or they can be obtained by enzymatic hydrolysis using serine proteinases, e.g. trypsin.

2.5.3. Casein Peptides

These peptides are obtained through the hydrolyses of the different types of caseins or through the fermentation process of milk. The aim of the preparation of such peptides is to obtain biologically active peptides, i.e. peptides that have a biological beneficial activity in the human body. The first bioactive peptides discovered related to milk are the β-casomorphins, fragments of β-casein$_{(60–70)}$ segment (Korhonen & Pihlanto, 2006). According to Sofia et al. (2005) biologically active peptides from casein have been classified into 4 groups: biologically active peptides that affect the cardiovascular, nervous, immune and digestive system. Most common active peptides are those associated with the function of the cardiovascular system, the antihypertensive peptides (Phelan et al., 2009). These peptides have the ability to regulate the blood pressure and avoid hypertension that may lead to strokes and heart attacks, especially that it is known that various side effects are associated with the use of hypertension drugs. Furthermore, antihypertensive biologically active peptides were obtained from the fermentation of milk by starter cultures of specific strains of lactic acid bacteria (Gobbetti et al, 2004).

2.6. Enriched Casein Fractions

The preparation of the casein fractions depends mainly on the use of temperature in the dissociation of β-casein from the casein micelles, e.g. at 4°C, isolated β-casein is present in a monomeric form, where in milk, a substantial part of the micellar β-casein dissociates into the serum phase (Dalgleish & Law, 1988). Moreover, at this low temperature, β-casein becomes soluble at the isoelectric pH value. The sources being used in obtaining pure β-casein fractions are essentially rennet skim milk curd, sodium and/or calcium caseinate and phosphocasein (Ward & Bastian, 1996; Huppertz et al., 2006).

The concern in the preparations of certain enriched casein fractions is based on the feasibility of the fractionation methods used in industrial applications and also on the particular functional properties of the individual casein components obtained or purified (Maubois, 1998; Mulvihill & Ennis, 2003). Another benefit is that these methods allow for the investigation of specific properties of caseins. For instance, the study of the casein fraction which is responsible for the emulsifying potentiality, in order to develop novel products like process cheeses with certain desirable properties. This is due to the fact that emulsifying properties of the protein used are of crucial importance during the manufacture process (Ennis & Mulvihill, 1999).

2.7. Casein-Whey Protein Aggregates

The casein-whey protein aggregates, which is considered to be a co-precipitate, is a mixture of caseins and whey proteins in their denatured form, as a result of the heat treatment of milk to such a degree that most of the whey proteins are denatured (Maubois & Olivier,

1997). This is done in the presence of an acidification process to reduce the pH of milk close to the isoelectric point of casein to about 4.6 (usually achieved by mineral acids). This leads to the precipitation of casein, whereas the heat treatment leads to the denaturation of whey proteins and the formation of co-precipitates with the casein. In this process, the denaturation of the whey proteins leads to the formation of both, aggregates of whey proteins and also aggregates of casein-whey proteins. The interaction between casein and whey proteins leads normally to substantial changes in the functional properties of the casein. Casein-whey protein aggregates are indeed the total milk protein fraction but with properties considerably different from those of the native proteins (Mulvihill & Ennis, 2003).

3. The Use of Casein in the Manufacture of Process Imitation Cheese

3.1. Definition and Manufacture of Process Imitation Cheese

This chapter is focusing mainly on one type of process cheese, which is imitation cheese. Imitation cheese, a form of process cheeses, is defined as cheese-like products manufactured by blending water, fat, proteins and other ingredients into a homogeneous mass in the presence of the necessary heat, mechanical shear and emulsifying salts (Guinee, 2003). The manufacture of process imitation cheese is generally a simple and quick method which involves mixing the ingredients in a specified mixer or cooker in presence of the necessary heat to produce a stable molten o/w emulsion that sets to form a homogeneous cheese-like mass on cooling. The manufacturing protocol can be summarized as follows (Mounsey & O'Riordan, 1999). Initially, water and fat are blended together at ~ 50°C in the mixer used, in the presence of salt (NaCl), emulsifying salts (usually phosphates and/or citrates) and sorbic acid for 1-2 minutes, which is followed by the addition of casein (dry powder) and a further mixing of 1-2 minutes is performed. The temperature is increased to somewhere in the range of 70-85°C either by direct steam injection or by an external high temperature water bath, depending on the mixer used. Mixing is continued for ~ 5-7 minutes until a homogeneous, smooth plastic mass is formed and no free water or oil remains visible in the mixer. This is followed by the addition of citric acid leading to a final desired pH value of ~ 6 in the end product. After manufacture, the molten product is discharged immediately from the cooker, stored at ~4°C and vacuum packed. The mixer used for manufacture was mainly a Blentech twin-screw pilot scale cooker and it was additionally shown that it was possible to manufacture imitation cheese in a Stephan single-blade pilot scale cooker or in a Farinograph on a small scale (Noronha et al., 2008b; El-Bakry et al. 2011a,b).

3.2. Proteins Used in the Manufacture of Process Imitation Cheese

The protein ingredients are usually in the form of dry proteins, not natural cheese. Protein is arguably the most crucial ingredient in the manufacture of imitation cheese. This component forms the continuous hydrated matrix of the cheese (Noronha et al., 2008a). In addition, the protein stabilises the oil-in-water emulsion by reducing the interfacial tension at the aqueous phase-oil droplet interface and by increasing the viscosity of the aqueous phase therby decreasing the frequency of collisions between oil droplets (Shimp, 1985; Ennis et al., 1998; Ennis & Mulvihill, 1999). The sources of protein that have been used in the manufacture of imitation cheese include caseinate (Cavalier-Salou & Cheftel, 1991), whey protein (Mleko & Foegeding, 2000), vegetable proteins and also rennet casein, which has been investigated in numerous research work (Yang & Taranto, 1982; Cavalier-Salou & Cheftel, 1991; Mounsey & O'Riordan, 1999; Montesinos-Herrero et al., 2006; Noronha et al., 2008a), with the latter being the best, most commonly and widely used source of protein.

Rennet casein is the most commonly used protein source and regarded as the most desirable protein for imitation cheese manufacture, due to the significant contribution of the protein to the structure and the functionality of the cheese products (Aimutis, 1995). This type of casein is a protein which is insoluble in water due to the calcium-mediated protein-protein interactions, especially that rennet casein has a high mineral, and accordingly high calcium phosphate content, as the protein was precipitated at neutral pH. Therefore, in process imitation cheese manufacture, adequate casein hydration can only be achieved by the addition of the so called melting or emulsifying salts, which are sequestering compounds such as phosphates and citrates, in the presence of the necessary agitation and heating to temperatures up to ~ 70-90°C. The insoluble calcium para- casein is converted to the soluble para-caseinate, which binds water and subsequently fat, as the solubilised casein acts as emulsifier causing the stabilisation of the oil-in-water emulsion by reducing the tension at the aqueous phase-oil droplet interface (Ennis & Mulvihill, 1999; Guinee et al., 2004). Recently, the changes occurring during the manufacture of imitation cheese was deeply investigated (Noronha et al., 2008a; El-Bakry et al., 2011b). The initial stage of manufacture was mainly concerned with water uptake and the formation of a hydrated casein matrix, as ~75% of the added water was absorbed. This was followed by a fat emulsification phase, which was found to be a consequence of adequate protein hydration, and once sufficiently emulsified, the fat was fully incorporated into the matrix to form a homogeneous final cheese product.

The properties of imitation cheese products are tailored to meet specific applications (Southward & Walker, 1980; Mulvihill & Ennis, 2003). For instance, low fat high protein cheese or low sodium cheese can be easily manufactured. However, late-lactation rennet casein has to be avoided in the manufacture of imitation cheese, as it was reported that this protein contains high levels of plasmin, a trypsin-like enzyme (fibrinolysin E.C.3.4.21.7), (Farkye & Fox, 1990; O'Malley et al., 2000). Plasmin is a primary proteolytic agent found in late-lactation rennet casein-based imitation cheese, causing proteolysis during the storage of the cheese, where β-casein is the casein most susceptible to proteolysis.

Acid casein can also be used as a protein source in the manufacture of imitation cheeses, but because of its insolubility in aqueous media and solubility at an increased pH, this protein has to firstly be partially converted into sodium or calcium caseinate using alkali like sodium or calcium hydroxide in order to develop a functionality in the cheese obtained (Aimutis, 1995; Mulvihill & Ennis, 2003). However, acid casein is not the desired protein source used in process cheese manufacture due to the presence of plasmin which may adversely affect the functionality and also, due to its unsatisfactory organoleptic properties, e.g. soapy or musty flavour (Zehren & Nusbaum, 2000). The functional and organoleptic properties of cheeses manufactured with caseinates can be negatively affected for the same reasons as above-mentioned for acid casein.

3.3. Effect of Changing the Protein Ingredients

The type and concentration of protein used in the manufacture process have a major impact on the physico-chemical properties of process cheeses, in general. Joshi et al. (2004) and Dimitreli and Thomareis (2007) reported that increasing the casein content of process cheese resulted in an increase in the viscoelastic properties, especially the elastic modulus, and hardness. Increasing protein contents led to increased protein–protein interactions resulting in a denser network of the cheese product. There was an increase in the intra- and inter-protein strand reactions, the matrix elasticity, and hardness values. In addition, increasing the protein in process cheese resulted in increased apparent viscosity at high temperatures, whereas the change in the fat content did not show any effect (Damodaran, 1997; Dimitreli & Thomareis, 2004). An exponential relation was found between the protein content and the viscosity. This relation was attributed to the increased interactions between the hydrated protein molecules and the protein intermolecular interactions, which were affected by only the continuous phase -water and proteins- rather than the dispersed fat phase. It was found that at temperatures above 40°C, all the fat of process cheese is liquid, which indicated that the viscosity of the dispersed phase is very low compared to the one of the continuous phase. Furthermore, the presence of a thickening agent, the protein in the continuous phase of an oil-in-water emulsion (like casein in the case of process imitation cheese) largely determines the overall rheological properties of the system (McClements,1999).

Interestingly, it has generally been found that a partial replacement of the casein by whey proteins in process imitation cheeses resulted in reduced heat-induced flowability (Zuber et al., 1987; Savello et al., 1989). In addition, Mleko and Foegeding (2001) used whey protein isolate and whey protein polymers to investigate the functionality of whey proteins as partial casein replacers, at different levels of substitution, in imitation cheese. These studies showed that the fracture stress increased and meltability decreased with increasing levels of the whey proteins. It was suggested that whey proteins interacted with the casein network and that a strengthening of the structure occurred.

Recently, up to 15% of the rennet casein used in imitation cheese manufacture was replaced by rice starches. This replacement of casein with high amylose starches strengthened the cheese structure but this reinforcement was at the expense of melt properties, i.e. heat-induced flowability (Mounsey & O'Riordan, 2008a). On the other hand, when pre-gelatinised starches with reduced amylose levels were used, a decrease in hardness, cohesiveness and also flowability was observed (Mounsey & O'Riordan, 2008b,c). These changes in functionality, as a result of replacing the casein by starch, were related to the competitive dehydration the protein matrix by starch which might have caused weakening of the protein-protein interactions within the cheese matrix formed.

4. Conclusion

There are different types of casein, which include rennet casein, acid casein, caseinate, phosphocasein, enriched casein fractions, casein fragments and casein-whey protein aggregates. The preparation of these caseins is usually from milk, and depending mostly on the heat and/or acid treatments that leads to the formation of the different types of caseins.

The common caseins used in the manufacture of process imitation cheese is the acid casein, sodium caseinate and rennet casein, where the latter being the most common and best protein source for the manufacture of this type of cheese. This is due to the fact that this protein type is the main cause of the physico-chemical stability of the cheese end-product obtained. The current chapter illustrates recent research work done on the use of rennet casein and its major impact on the physico-chemical properties, such as heat-induced flowability, viscoelastic properties and hardness, of imitation cheese. Increased casein content resulted in increased viscoelastic properties and hardness, whereas a partial replacement of casein by whey proteins led to a reduced heat-induced flowability due to the interactions between casein and whey proteins. Interestingly, a partial replacement of the casein used by rice starches led to changes in the flowability and the hardness depending on the amylose content of the starch.

References

Aimutis, A.W. (1995). Dairy protein usage in processed and imitation cheeses. *Food Technology Europe*, 2, 30-34.

Bachmann, H. (2001). Cheese analogues: a review. *International Dairy Journal*, 11, 505-515.

Berger, W., Klostermeyer, H., Merkenich, K. & Uhlmann, G. (1989). *Processed Cheese Manufacture, A JOHA Guide*. BK Ladenburg GmbH., Wurzburg, Germany.

Cavalier-Salou, C. & Cheftel, J.C. (1991). Emulsifying salts influence on characteristics of cheese analogs from calcium caseinate. *Journal of Food Science,* 56, 1542-1547, 1551.

Dalgleish, D.G. (1997). Structure-function relationships of caseins. In Damodaran, S. & Paraf, A. (Eds), *Food Proteins and their Application* (pp. 199–223). New York, USA: Marcel Dekker.

Dalgleish, D.G. & Law, A.J.R. (1988). pH-Induced dissociation of bovine casein micelles. I. Analysis of liberated caseins. *Journal of Dairy Research*, 55, 529–538.

De Kruif, C.G. (2003). Caseins, the main protein fraction of milk. IOP Industrial Proteins, 11(2), 9-12.

Dimitreli, G. & Thomareis, A.S. (2004). Effect of temperature and chemical composition on processed cheese apparent viscosity. *Journal of Food Engineering*, 64, 265–271.

Dimitreli, G. & Thomareis, A.S. (2007). Texture evaluation of block-type processed cheese as a function of chemical composition and in relation to its apparent viscosity. *Journal of Food Engineering*, 79, 1364-1373.

Doultani, S., Turhan, K.N. & Etzel, M.R. (2003). Whey protein isolate and glycomacropeptide recovery from whey using ion exchange chromatography. *Journal of Food Science*, 68, 1389–1395.

El-Bakry, M., Beninati, F., Duggan, E., O'Riordan, E.D. & O'Sullivan, M. (2011a). Reducing salt in imitation cheese: effects on manufacture and functional properties. *Food Research International*, 44, 589-596.

El-Bakry, M., Duggan, E., O'Riordan, E.D. & O'Sullivan, M. (2011b). Casein Hydration and Fat Emulsification during Manufacture of Imitation Cheese, and Effects of Emulsifying Salts Reduction. *Journal of Food Engineering*, Vol. 103 (2), 179-187

Ennis, M.P. & Mulvihill, D.M. (1999). Compositional characteristics of rennet caseins and hydration characteristics of the caseins in a model system as indicators of performance in Mozzarella cheese analogue manufacture. *Food Hydrocolloids*, 13, 325–337

Ennis, M.P., O'Sullivan, M.M. & Mulvihill, D.M. (1998). The hydration behaviour of rennet caseins in calcium chelating salt solution as determined using a rheological approach. *Food Hydrocolloids*, 12, 451-457.

Farkye, N. Y. & Fox, P. F. (1990). Observation of plasmin activity in cheese. *Journal of Dairy Research*, 57, 413-418.

Fox, P.F. (2003). Milk proteins: general and historical aspects. In Fox, P.F. & McSweeney, P.L.H. (Eds), *Advanced Dairy Chemistry 1: Proteins*, 3rd edition (pp. 1-48). London: Kluwer Academic.

Fox, P.F., Guinee, T.P., Cogan T.M. & McSweeney, P.L.H. (2000). *Fundamentals of Cheese Science* (pp. 587-590). Gaithersburg Aspen Publishers, MD.

Gobbetti, M., Minervini, F. & Rizzello, C.G. (2004). Angiotensin I-converting-enzyme-inhibitory and antimicrobial bioactive peptides. *International Journal of Dairy Technology*, 57, 173-188.

Guinee, T.P. (2003). Cheese Analogues. In Roginski, H. et al. (Eds.), *Encyclopedia of Dairy Science Vol. 1* (pp. 428-434). Amsterdam, Netherlands: Academic Press.

Guinee T.P, Carić M. & Kaláb M. (2004). Pasteurized processed cheese and substitute/imitation cheese products. In Fox, P.F. et al. (Eds.), *Cheese: Chemistry, Physics and Microbiology, Vol. 2: Major cheese groups*. (pp. 349-394). London, New York: Elsevier Applied Science.

Huppertz, T., Hennebel, J.B., Considine, T., Rehman, S.U., Kelly, A.L. & Fox, P.F. (2006). A method for the large-scale isolation of β-casein. *Food Chemistry*, 99, 45–50.

Joshi, N.S., Jhala, R.P., Muthukumarappan, K., Acharya, M.R. & Mistry, V.V. (2004). Textural and rheological properties of processed cheese. *International Journal of Food Properties*, 7, 519 – 530.

Kelly, P., Kelly, J., Mehra, R., Oldfield, D., Raggett, E. & O'Kennedy, B. (2000). Implementation of integrated membrane processes for pilot scale development of fractionated milk components. *Lait* , 80, 139–153.

Korhonen, H. & Pihlanto, A. (2006). Bioactive peptides: production and functionality. *InternationalDairy Journal*, 16, 945–960.

Maubois, J.L. (1998). Fractionation of milk proteins. In Proceedings of the 25th International Dairy Congress, Vol. II. Dairy Science and Technology (pp. 74–86). International dairy Federation, Brussels, Belgium.

Maubois, J.L. & Ollivier, G. (1997). Extraction of milk proteins. In Damodaran, S. & Paraf, A., *Food Proteins and their Application* (pp. 579–595). New York: Marcel Dekker,.

Meyer, A. (1973). *Processed Cheese Manufacture*. London, UK: Food Trade Press Ltd.

Mleko, S & Foegeding, E. A. (2001). Incorporation of polymerized whey proteins into processed cheese analogs. *Milchwissenschaft*, 56, 612–615.

Mleko, S. & Foegeding, E. A. (2000). Physical properties of rennet casein gels and processed cheese analogs containing whey proteins. *Milchwissenschaft*, 55, 513-516.

Montesinos-Herrero, C., Cottell, D.C., O'Riordan, E.D. & O'Sullivan, M. (2006). Partial replacement of fat by functional fibre in imitation cheese: Effects on rheology and microstructure. *International Dairy Journal*, 16, 910–919.

Mounsey J.S. & O'Riordan, E.D. (2008a). Characteristics of imitation cheese containing native or modified rice starches. *Food Hydrocolloids*, 22, 1160-1169.

Mounsey J.S. & O'Riordan, E.D. (1999). Empirical and dynamic rheological data correlation to characterise melt characteristics of imitation cheese. *Journal of Food Science*, 64 (4), 701–703.

Mounsey J.S. & O'Riordan, E.D. (2008b). Influence of pre-gelatinised maize starch on the rheology, microstructure and processing of imitation cheese. *Journal of Food Engineering*, 84, 57-64.

Mounsey J.S. & O'Riordan, E.D. (2008c). Modification of imitation cheese structure and rheology using pre-gelatinised starches. *European Food Research and Technology*, 226, 1039–1046.

Mulvihill, D.M. (1989). Caseins and caseinates: manufacture. In Fox, P.F. (Ed), *Developments in Dairy Chemistry, Vol. 4* (pp. 97–130). New York, USA: .Elsevier Applied Science.

Mulvihill, D.M. & Ennis, M.P. (2003). Functional milk proteins: production and utilization. In Fox, P.F. & McSweeney, P.L.H. (Eds), *Advanced Dairy Chemistry, Vol. 1, Part B* (pp. 1175–1228). New York, USA: Kluwer Academic/Plenum Publishers.

Noronha, N., Duggan, E., Ziegler, G.R., O'Riordan E.D. & O'Sullivan M. (2008a). Investigation of imitation cheese matrix development using light microscopy and NMR relaxometry. *International Dairy Journal*, 18, 641-648.

Noronha, N., O'Riordan, E.D & O'Sullivan M. (2008b). Influence of processing parameters on the texture and microstructure of imitation cheese. *European Food Research and Technology*, 226, 385-393.

O'Malley, A.M., Mulvihill, D.M. & Singh, T.K. (2000). Proteolysis in rennet casein-based cheese analogues. *International Dairy Journal*, 10, 743-753.

Pellegrino, L., Resmini, P., De Noni, I. & Masotti, F. (1996). Sensitive determination of lysinoalanine for distinguishing natural from imitation Mozzarella cheese. *Journal of Dairy Science*, 79: 725-734.

Phelan, M., Aherne, A., FitzGerald, R. & O'Brien, N. (2009). Casein-derived bioactive peptides: biological effects, industrial uses, safety aspects and regulatory status. *International Dairy Journal*, 19, 643-654.

Savello, P.A., Ernstrom, C.A. & Kalab, M. (1989). Microstructure and meltability of model process cheese made with rennet and acid casein. *Journal of Dairy Science*, 72, 1-11.

Shimp, L.A. (1985). Process cheese principles. *Food Technology*, 39, 63-69.

Sofia V., Silva, F. & Malcata, X. (2005). Caseins as source of bioactive peptides. *International Dairy Journal*, 15, 1–15.

Southward, C.R. & Walker, N.J. (1980) The manufacture and industrial use of casein. *New Zealand Journal of Dairy Science and Technology*, 15, 201–217.

Thoma-Worringer, C., Sorensen, J. & Lopez-Fandino, R. (2006). Health effects and technological features of caseinomacropeptide. *International Dairy Journal*, 16, 1324–1333.

Vasbinder, A.J. & de Kruif, C.G. (2003). Casein–whey protein interactions in heated milk: the influence of pH. *International Dairy Journal*, 13, 669–677.

Ward, L.S. & Bastian, E.D. (1996). A method for isolating β-casein. *Journal of Dairy Science*, 79, 1332–1339.

Walstra, P., Wouters, J.T.M. & Guerts, T.J. (2006). *Dairy Technology*, 2nd edition. Boca Raton: CRC Press (Taylor & Francis Group).

Yang, C.S.T. & Taranto, M.V. (1982). Textural properties of Mozzarella cheese analogs manufactured from soybeans. *Journal of Food Science*, 47, 906-910.

Zehren, V.L. & Nusbaum, D.D. (2000). *Process Cheese*. Madison, Wisconsin, USA: Cheese Reporter Publishing, Co, Inc.

Zuber, F., Megard, D. & Cheftel, J.C. (1987). Contineuous emulsification and gelation of dairy ingredients by HTST extrusion cooker – production of processed cheese. *International Journal of Food Science and Technology*, 22, 607-626.

In: Casein: Production, Uses and Health Effects
Editors: A. M. Ventimiglia, et. al, pp. 223-232
ISBN: 978-1-62100-129-4

Chapter 12

HEALTH EFFECTS OF CASEIN

T.B. Ng[*]
School of Biomedical Sciences
The Chinese University of Hong Kong, Sha Tin,
Hong Kong China.

ABSTRACT

Casein is an important milk protein. β-casomorphins (β-CM)are opioid peptides produced by the digestion of β-casein. Two major(A^1 and A^2)alleles of β-casein exist in cattle populationsβ-casein A^1 and related minor variants bearing histidine at residue 67 yield bioactive β-CM-7whereas the A^2 variant with proline in that position does not generateβ-CM-7 by gastric and gut enzymes during digestion. A significant relationship was observed between consumption of β-casein A^1 and the incidence of type 1 diabetes. LDL oxidation byβ-CM-7 constitutes an important link between the risk of heart disease and milk protein consumption.It has been postulated that neurological conditions such as autism, schizophrenia, post-partum depression, and and sudden infant death syndrome may be related to β-casein intake since β-CM-7 reaches the brain cells and cause symptoms associated with the diseases Caseinophosphopeptides are produced on an industrial-scale and have been considered for application as components in 'functional foods' and pharmaceutical products randomized clinical trials are needed to reveal if dairy products could be a complementary treatment for oral health.

INTRODUCTION

Bovine milk and dairy products have an abundance of macronutrients such as protein, and micronutrients likecalcium, vitamin D, and magnesium. Dairy products have health benefits such as possibly preventing osteoporosis and colon cancer. An association between intake of bovine milk and the risk of type 1 diabetes, cardiovascular disease and neurological disorders such as autism and schizophrenia has been demonstrated. In the following currently

[*] Tel: 852-26098031

available evidence is presented. A significant relationship was observed between consumption of β-casein A^1 and the incidence of type 1 diabetes.

CASEIN AND BETA-CASOMORPHINS

Bovine milk protein is composed of about 80% casein and 20% whey (Shah, 2000). There are four components in casein: α_{s1}, α_{s2}, β, and κ. Human milk casein consists mainlyof βand κ. For exampleβ-casomorphins (β-CM), the peptides produced by the digestion of β-casein, have opioid activity. The β-CMs possess an exposed N-terminal tyrosine moiety and another aromatic residue (phenylalanine or tyrosine) in the third or fourth position.These structural features account for a high binding affinityto opioid receptors (Meisel and FitzGerald, 2000).

The first example of a bioactive β-CM was identified as β-CM-7 (Jinsmas et al., 1999). Two major(A^1 and A^2)alleles of β-casein exist in cattle populations (Jinsmas et al., 1999; Tailford et al., 2003). β-casein A^1 and related minor variants bearing histidine at residue 67 yield bioactive β-CM-7whereas the A^2 variant with proline in that position does not generateβ-CM-7 by gastric and gut enzymes during digestion.

Holstein cows produce milk with the A^1 variant comprising about 66% of the β-casein fraction (Tailford et al., 2003; USDA, 2002; USDA, 2001).On the other hand, the Guernsey breed with high frequency of the A^2 allele produces milk rich inA^2 protein (90%).Milk is pooled in the United States and most Western countries and thus the population consumes both A^1 and A^2 variants. in the United States, the vast majorityof the dairy cows are of the Holstein breed .Hence, most of the population drink milk with high levels of β-casein $A^1$and thus capable of producing β-CM-7.

β-CM-7 slows gastrointestinal motility in animals (Shah, 2000).). In infants, chronic constipation and anal fistula development were significantly correlated with consumption of larger amounts of bovine milk and shorter period of breast-feeding . Bottle-fed infants exhibited decelerated gut motilitydue to the morphine-like effect of β-CM-7. Breastfeeding maybe protective against while bovine milk mayenhance fistula development(Andiran et al., 2003).

β-CMs appear to be absorbed through the gastrointestinal tract of newborns and young infants (Sun et al., 1999; Sun and Cade, 1999; Sun et al., 2003). Adults, despite consuming bovine casein and showing β-CM-7 production in the gastrointestinal tract, fail to have circulating β-CMs (Svedberg et al., 1985). If β-CMs do get absorbed through the gastrointestinal tract, they must then localize in the central nervous system to produce an effect on the respiratory or endocrine system. Newborn infants have more permeable gastrointestinal tracts and immature nervous systems, making them more likely candidates than adults to experience the opioid effects of β-CMs from bovine casein, which has been the case in animal studies (Umbach et al., 1985; Singh et al., 1989). Besides, most infants consume nearly all of their nutrition from formula based on bovine milk, whereas adults have a more varied diet, of which only a small percentage is contributed by milk. Some adults may have more permeable guts than others (e.g., Crohn's disease, Irritable Bowel Syndrome) and may be more affected by β-CMs than adults with normal functioning gastrointestinal tracts.

Bovine Milk Consumption and Risk of Heart Disease

A link between the development of cardiovascular disease and milk protein intake has been suggested (McLachlan, 2001; Laugesen and Elliott, 2003; Tailford et al., 2003). However, some populations that consume a diet rich in milk from a breed carrying the A^2 allele exclusively have virtually no heart disease (McLachlan, 2001).

Rabbits and rhesus monkeys fed a diet containing cholesterol-free casein developed hypercholesterolemia and atheroma (Meeker and Kesten, 1941Kritchevsky et al., 1978; Terpstra et al., 1981). In the absence of cholesterol, β-casein A^1 fed rabbits demonstated elevated serum levels of cholesterol, LDL-cholesterol, HDL-cholesterol, and triglyceride compared with wheyfed rabbits. β-casein A^2 fed rabbits had the lowest levels.Rabbits fed β-casein A^1 exhibited a more extensive surface area of the aorta covered by fatty streaks and thicker fatty streak lesions in the aortic arch than rabbitsfed β-casein A^2 ($P < 0.05$) (Taliford et al., 2003). The thickness increase in the aortic arch lesion in the β-casein A-fed ^{1} groups became more pronounced after addition of cholesterol to the diet. Most western diets contain a great deal of bovine milk and milk products producing the A^1 variant and dietary cholesterol.Both may contribute to a heightened risk ofcardiovascular disease.

β-CM-7, formed by the digestion of A^1 and related variants of β-casein, has an exposed amino terminal tyrosine which can form tyrosyl radicals, and thus catalyze the oxidation of LDL-cholesterol (Torreiles and Guerin, 1995). Other effects on the pulmonary system exerted by bioactive digestion products of β-caseinmay also play a role(Shah, 2000; Meisel and FitzGerald, 2000; Fiat et al., 1989).However, LDL oxidation byβ-CM-7 constitutes an important link between the risk of heart disease and milk protein consumption.

Effects on Blood Pressure, Vascular Function, and Inflammatory Markers Overweight Individuals

Milk proteins contain peptides that inhibit the activity of angiotensin converting enzyme (ACE),a key enzyme in the regulation of peripheral blood pressure. (Takano,1998:). Casokinins and lactokinins are casein-derived and whey-derived inhibitors, respectively (FitzGerald and Meisel,1999;Nurminen et al., 2000). Casokinins and lactokinins can significantly reduce blood pressure in spontaneously hypertensive rats (Abubakar et al.,1998;Nurminen et al.,2000). Sekiya *et al.* (1992)showed that a daily intake of 20g casein brought about a drop in both DBP and SBP in hypertensive human volunteers. Similar effects of casein and fermented milk on SBP and DBP have been reported(Hata et al.,1996;Seppo et al., 2003). SBP was significantly lowered by the intake of fermented milk enriched with whey proteins after 8 weeks.. Recent studies have shown that administration of ACE-inhibitors exerts beneficial effects on inflammatory markers (Andrzejzak et al.,2007). Chronic inflammation is important to the pathogenesis of atherosclerosis (Humbert et al.,2005). The levels of inflammatory markers IL-6, nor TNF-α were not affected by whey protein or casein supplementation. (Pal and Ellis,2010).

Pal and Ellis (2010) conducted a randomized, single blind, parallel design study in Perth, Australia,to investigate the action of whey protein isolate on blood pressure, vascular function and inflammatory markers in comparison with the effects of casein and glucose supplementation.Tf Overweight and obese individuals aged 18–65 with a body mass index between 25–40 kg/m^2 were recruited for the study. Exclusion criteria comprised regular consumption of medications like lipid lowering drugs and antihypertensives likely to affect the study outcomes, smoking, major illnesses,cancer within the last 5 years, cardiovascular events in the last 6 months respiratory disease, type 1 or type 2 diabetes, eating disorders, over two standard alcoholic drinks daily, and. pregnancy or lactation.This study was carried out over a 12-week period with a 4-week washout period before commencement. The subjects were blinded to the type of supplement taken..The dietary intake and physical activity levels of all subjects for the duration of the study were similar to those prior to the investigation. Instructions from a dietitian were followed to prevent weight gain during the study period.

The central augmentation index (AI),an indicator of arterial stiffness ,is elevated in hypercholesterolemia (Wilkinson et al.,2008) and is an independent risk factor for c.ardiovascular disease(Qureshi et al.,2007). As people age, the central arteries gradually stiffen (Greenwald,2007;McEniery et al.,2007), with the rate of progression influenced by hypertension, diabetes, and atherosclerosis (Benetos et al.,2002;Wilkinson et al;2002;Tomiyama et al.,2006) . It has been suggested that the increase in arterial stiffening may be associated with an increase in SBP and DBP (Safar,2003;Blacher and Safar,2005).AI was reduced by circa 7 % from baseline at 6 weeks in the whey protein-treated group. In the whey protein-treated group AI was significantly diminished by 14% from baseline at 12 weeks. By contrast,in the casein-treated group AI remained unaltered at 6 and 12 weeks from baseline. AI was significantly lower in the whey protein-treated group at week 12 compared to control and casein-treated groups There were no significant changes in IL-6, CRP, and TNF-α with time (baseline vs. 12 weeks) or between groups at 12 week (Pal and Ellis 2010).

The study of Pal and Ellis (2010)demonstrated that supplementation with whey protein can significantly improve blood pressure and vascular function in overweight and obese individuals.Hence, whey protein supplementation potential could be used as a component in dietary plans and in functional foods for the management of the risk factors of metabolic syndrome.

Effect on Cardioprotective Effects of Black Tea

In healthy female volunteers consumption of freshly brewed black tea produced a significant improvement in flow-mediated dilation asmeasured by high-resolution vascular ultrasound .. Tea brought about vasorelaxation in rat aortic rings and enhanced endothelial nitric oxide synthase activity in endothelial cells by phosphorylation. Addition of milk abrogated the effects of tea.in both studies. Caseins producedthese inhibiting effects probably by complex formation with catechins. present in tea(Lorenz et al., 2007).

Bovine Milk Consumption and the Risk of Type 1 Diabetes

Antibodies to bovine β-casein were detected in patients recently diagnosed with type 1 diabetes and in adults with latent autoimmune diabetes . Patients with celiac disease had levels of antibodies similar to type 1 diabetics. Both conditions are characterized by abnormal HLA-haplotypes.T he enhanced immune responses to dietary antigens from bovine milk may trigger the two diseases. It appears there is a relationship between type 1 diabetes and antibody response to β-casein from bovine milk(Monetini et al., 2002).

Although type 1 diabetics have the highest antibody titer compared to their siblings ,parents, and healthy controls, in all groups there is an inverse relationship between antibodies raised specifically to the A^1 and A^2 proteins and age. The highest titers of anti-casein A^1 antibodies are observed in diabetics. The observations infer that type 1 diabetics may have a defective immuno-tolerance to bovine milk antigens(Padberg et al., 1999).

Consumption of cows' milk of known genetic variant composition (A^1 or A^2) was compared with the incidence of type 1 diabetes in children up to the age of 14 years (Elliott et al., 1999). A significant relationship was observed between consumption of β-casein A^1 and the incidence of type 1 diabetes (McLachlan, 2001). Finland, Sweden, Demark, and Norway, countries with the highest consumption of milk with this allele, had the highest incidence of type 1 diabetes. The incidence is reduced in countries like Iceland and Germany with a lower consumption of β-casein $A^{1'}$ Icelanders consume more bovine serum albumin and immunoglobulins but imilar amounts of lactoferrin compared with people in other Nordic ountries.None of these milk components correlate with the incidence of type1 diabetes .The results indicate that consumption of β-casein A^1 from bovine milk was the predominant risk factor for type 1 diabetes(Birgisdottir et al., 2002).

Bovine Milk Consumption and Neurological Symptoms

It has been postulated that neurological conditions such as autism, schizophrenia, post-partum depression, and and sudden infant death syndrome may be related to β-casein intake since β-CM-7 reaches the brain cells and cause symptoms associated with the diseases (Sun et al., 1999; Sun and Cade, 1999; Sun et al., 2003). Antibodies to β-CM-7 are present in the majority of patients with schizophrenia and patients with autism. Elevated concentrations of β-CM-7are detected in the blood and cerebral spinal fluid of these patients due to deficient or defective enzymes that cleave β-CM-7

Areas of the brain including the occipital and temporal cortex, which are speech centers,are affected by β-CM-7. Both schizophrenia and autism patients have abnormal speech patterns and a disruption in normal dopamine,serotonin, and γ-aminobutyric acid (GABA) activity associated with diminished social interactions. Thus it is advisable for patients with these neurological disorders to consume a casein-free diet (Sun et al., 1999; Sun and Cade, 1999; Sun et al., 2003).

After receiving intraperitoneal injections of β-CM-7. rats became restless and respiratory rate was increased within the first minute. After seven minutes, the rats appeared to be moving around less and had less social interactions. No behavioral changes were detected in rats infused with saline or in those who were co-infused with β-CM and the opiate receptor antagonist naloxone. These observations indicate an involvement of β-CMs in the pathogenesis of schizophrenia and autism (Sun and Cade, 1999).

β-CM-5 and β-CM-7 are strong opioid agonists. β-CM-5 is much more potent than morphine while β-CM-7 is approximately equipotent.S ome animal studies have shown a depression in ventilation, aSIDS symptom, only when the β-CMs are injected intracerebroventricularly. β-CMs are present in cerebrospinal fluid of infants with impaired respiratory function but not in normal infants. In order to establish if a significant correlation exists between the opioid effect of β-CMs from bovine milk and depression of respiration and behavioral changes, endogenous and exogenous opioid peptides in the plasma and cerebrospinal fluid of neonates with respiratory disturbance in comparison with healthy controls should be determined (Sun et al., 2003).

β-CM-7 with opioid activity is preferentially derived from β-casein A^1 but not from A^2 during digestion. Whether bovine milk with the β-casein A^2 allele should be provided to high-risk individuals is not known and more work is required.

It was suspected that milk -derived peptides may increase the risk of insulin-dependent diabetes (T1D) based on the relationship of type 1 diabetes and consumption of variants A1 and B β-casein from bovine milk.β-casomorphin-7 (BCM-7) from β-casein may, by virtue of its immunosuppressant activity,reduce tolerance to dietary antigen in the gut immune system,and thus contribute to the onset of T1D. The genetic variants A1, A2, and B of bovine β-casein are also found in human milk. The β-casomorphins among these bovine variants resemble their human counterparts in amino acid sequence. In vitro studies reveal that BCM-7 can be formed from A1 and B during typical digestive processes but BCM-7 is not a product of A2 digestion.Results of epidemiological research and studies employing animal models do not associate milk proteins with the development of T1D. Ecological databased on A1/ A2 variations among livestock breeds do not demonstrate a causal relationship, even in countries with a large milk consumption (Clemens,2011).

Populations consuming milk with elevated levels of ss-casein A2 variant exhibit a reduced incidence of cardiovascular disease and type 1 diabetes. with the A2 variant and have less severe symptoms of autism and schizophrenia. ss-casein A1 and related forms produce a bioactive opioid peptide, ss-casomorphin-7 (ss-CM-7), during digestion. The immature gastrointestinal tract of infants may absorb ss-CM-7. Adults, on the other hand, appear to reap the biological activity locally on the intestinal brush border. ss-CM-7 can potentially affect numerous opioid receptors in the nervous, endocrine, and immune systems. Whether there is a definite health benefit to milk containing the A2 genetic variant requires further investigation.(Bell et al., 2005).

Milk and Oral Health

Bovine and human caseins and lactoferrin prevent attachment of cariogenic mutans streptococci to hydroxyapatite which has been coated with saliva or purified saliva host ligands.However, human and bovine milk coated on hydroxyapatite favors attachment of commensal *Actinomyces naeslundii* and other streptococci, and β-casein-derived caseino-phosphate peptide enhances maintenance of tooth minerals, , but randomized clinical trials are needed to reveal if dairy products could be a complementary treatment for oral health(Johansson Holgerson ,2011).

Anticancer and Immunological Effects

Caseinophosphopeptides are derived from casein They can form soluble organophosphate saltsand play a role as calcium carriers.T hey have anticancer effects and enhance the activity of immunocompetent cells and neonatal intestinal cells. Caseinophosphopeptides are produced on an industrial-scale and have been considered for application as components in 'functional foods' and pharmaceutical products (Meisel and FitzGerald ,2003).

Conclusion

Populations consuming milk with an abundance of β-casein A^2 variant are less prone to cardiovascular disease and type 1 diabetes. Consumption of milk with the A^2 variant may help attenuate the symptoms of autism and schizophrenia. The mechanism of action focuses on β-casein A^1 and related forms preferentially that are able to produce a bioactive opioid peptide, β-casomorphin-7 (β-CM-7) during digestion. Infants may absorb β-CM-7 due to an immature gastrointestinal tract. β-CM-7 can potentially affect numerous opioid receptors in the nervous, endocrine, and immune systems(Bell and Stacey,2006). Whether there is a definite health benefit to milk containing the A^2 genetic variant is unknown and further investigations are warranted.A voluminous amount of information pertaining to the biological effects of the milk protein is available but much less is known about casein. More research is needed.

References

Abubakar A, Tadao S, Kitazawa H, Kawai Y, Itoh T. (1998)Structural analysis of new antihypertensive peptides derived from cheese whey protein by proteinase K digestion.*J Dairy Sci* 81:3131–3138

Andiran, F. , Dayi, S. and Mete, E. (2003) Cows milk consumption in constipation and anal fistula in infants and young children. *J. Paediatr. Child Health* 39 , 329-331.

Andrzejczak D, Górska D, Czarnecka E. (2007)Influence of enalapril, quinapril and losartan on lipopolysaccharide (LPS)-induced serum concentrations of TNF-alp IL-1 beta, IL-6 in spontaneously hypertensive rats (SHR).*Pharmacol Rep* 59:437–446.

Bell, SJ. ,Grochoski, G T. andClarke, A J. (2006) 'Health Implications of Milk Containing &b.beta;-Casein with the A^2 Genetic Variant', Critical Reviews in Food Science and Nutrition, 46:93 - 100

Benetos A, Waeber B, Izzo *J et al.* (2002) Influence of age, risk factors, and cardiovascular and renal disease on arterial stiffness: clinical applications.*Am J Hypertens* 15:1101–1108.

Birgisdottir, B. E. ,Hill, J. P. , Harris, D. P. and Thorsdottir, I. (2002) Variation in consumption of cow milk proteins and lower incidence of type 1 diabetes in Iceland vs. the other 4 Nordic countries. *Diabetes Nutr. Metab.* 15 , 240-245.

Clemens RA.(2011)Milk A1 and A2 peptides and diabetes. Nestle Nutr Workshop Ser Pediatr Program. 67,187-195.

Elliott, R. B. , Harris, D. P. , Hill, J. P. , Bibby, N. J. and Wasmuth, H. E. (1999) Type 1 (insulin-dependent) diabetes mellitus and cow milk: Casein variant consumption. *Diabetologia* 42 , 292-296.

Fiat, A.-M. , Levy-Tolediano, S. , Caen, J. P. and Jolles, P. (1989) Biologically active peptides of casein and lactoferrin implicated in platelet function. *J. Dairy Res.* 56 , 351-355.

FitzGerald R, Meisel H. (1998)Lactokinins: whey protein derived ACE inhibitory peptides.*Nahrung*1999;43:165–167. Takano T. Milk derived peptides and hypertension reduction.*Int Dairy J* 8:375–381.

FitzGerald RJ, Meisel H. `(1999)Lactokinins: whey protein-derived ACE inhibitory peptides.*Nahrung* 43:165–167.

FitzGerald RJ, Meisel H. (2000)Milk protein-derived peptide inhibitors of angiotensin-I-converting enzyme. *Br J Nutr* 84(Suppl 1):S33–S37.

Greenwald SE.(2007)Ageing of the conduit arteries. *J Pathol* 211:157–172.

Hata Y, Yamamoto M, Ohni M *et al.* (1996) A placebo-controlled study of the effect of sour milk on blood pressure in hypertensive subjects. *Am J Clin Nutr*64:767–771.

Humbert M, Monti G, Brenot F *et al.* (1995)Increased interleukin-1 and interleukin-6 serum concentrations in severe primary pulmonary hypertension.*Am J Respir Crit Care Med* 151:1628–1631.

Jinsmas, Y. and Yoshikawa, M. (1999) Enzymatic release of neocasomorphin and β-casomorphin from bovine β-casein. *Peptides* 20 , 957-962.

Johansson I,Lif Holgerson P. (2011)Milk and oral health. Nestle Nutr Workshop Ser Pediatr Program.;67,55-66.

Kritchevsky, D. , Tepper, S. A. and Story, J. A. (1978) Influence of soy protein and casein on atherosclerosis in rabbits. *Fed Proc.* 37 — abstract 2801 17.

Laugesen, M. and Elliott, R. (2003) Ischaemic heart disease, type 2 diabetes, and cow milk A1 β-casein. *NZ Med. J.* 116 , 1-19. 18. McLachlan, C NS (2001) β-casein A1, ischaemic heart disease mortality, and other illnesses. *Medical Hypoth.* 56 , 262-272.

Lorenz M,Jochmann N,von Krosigk A,Martus P,Baumann G,Stangl K,Stangl V. (2007)Addition of milk prevents vascular protective effects of tea. Eur Heart J. 28, 219-223.

McEniery CM, Wilkinson IB, Avolio AP. (2007)Age, hypertension and arterial function. *Clin Exp Pharmacol Physiol* 34:665–671.

Meeker, D. R. and Kesten, H. D. (1941) Effect of high protein diets on experimental atherosclerosis of rabbits. *Arch. Pathol.* 31 , 147-162.

Meisel H,FitzGerald RJ.(2003) Biofunctional peptides from milk proteins: mineral binding and cytomodulatory effects. Curr Pharm Des. 9,1289-1295.

Meisel, H. and FitzGerald, R. J. (2000) Opioid peptides encrypted in intact milk protein sequences. *Br. J. Nutr.* 84 , S27-S31. .

Monetini, L., Cavallo, M. G., Manfrini, S., Stefannini, L., Picarelli, A., DiTola, M., Petrone, A., Bianci, M., LaPresa, M., DiGiulio, C., Baroni, M. G., Thrope, R., Walker, B. K.,

IMDIAB group and Pozzilli, P. (2002) Antibodies to bovine beta-casein in diabetes and other autoimmune diseases. *Horm. Metabl. Res.* 34 , 455-459.

Nurminen ML, Sipola M, Kaarto H *et al.*(2000)Alpha-lactorphin lowers blood pressure measured by radiotelemetry in normotensive and spontaneously hypertensive rats. *Life Sci*66:1535–1543.

Padberg, S. , Schumm-Draeger, P. M. , Petzoldt, R. , Becker, F. and Federlin, K. (1999) The significance of A1 and A2 antibodies against beta-casein in type-1 diabetes mellitus. *Dtsch. Med. Wochenschr.* 124 , 1518-1521.

Pal, S.S and Ellis ,V.–(2010)The Chronic Effects of Whey Proteins on Blood Pressure, Vascular Function, and Inflammatory Markers in Overweight Individuals *Obesity*18, 1354–1359.

Qureshi G, Brown R, Salciccioli *L et al.* (2007)Relationship between aortic atherosclerosis and non-invasive measures of arterial stiffness.*Atherosclerosis* 195:e190–e194.\|

Safar ME (2001) ystolic blood pressure, pulse pressure and arterial stiffness as cardiovascular risk factors.*Curr Opin Nephrol Hyperten*10:257–261.

Blacher J, Safar ME.(2005)Large-artery stiffness, hypertension and cardiovascular risk in older patients.*Nat Clin Pract Cardiovasc Med*2:450–455.

Sekiya S, Kobayashi Y, Kita E, Imamura Y, Toyama S. (1992)Antihypertensive effects of tryptic hydrolysate of casein on normotensive and hypertensive volunteers (in Japanese). *J. Jap. Soc. Nutr. Food Sci* 45:513–517

Seppo L, Jauhiainen T, Poussa T, Korpela R. (2003)A fermented milk high in bioactive peptides has a blood pressure-lowering effect in hypertensive subjects.*Am J Clin Nutr* 77:326–330.

Shah, N. (2000) Effects of milk-derived bioactives: An overview. *Br. J. Nutr* 84 , S3-S10.

Singh, M. , Rosen, C. L. , Chang, K. J. and Haddad, G. G. (1989) Plasma beta-casomorphin-7 immunoreactive peptide increases after milk intake in newborn but not in adult dogs. *Pediat. Res* 26 , 34-38.

Sun, Z. , Cade, J. R. , Fregly, M. and Privette, R. M. (1999) β-casomorphin induces Fos-like immunoreactivity in discrete brain regions relevant to schizophrenia and autism. *Autism* 3, 67-83.]

Sun, Z. , Zhang, Z. , Wang, X. , Cade, R. , Elmir, Z. and Fregly, M. (2003) Relation of β-casomorphin to apnea in sudden infant death syndrome. *Peptides* 24 , 937-943.

Sun, Z. and Cade, J. R. (1999) A peptide found in schizophrenia and autism causes behavior changes in rats. *Autism* 3 , 85-95.

Svedberg, J. , de Hass, J. , Leimenstoll, G. , Paul, F. and Teschemacher, H. (1985) Demonstration of beta-casomorphin immunoreactive materials in in vitro digests of bovine milk and in small intestine contents after bovine milk ingestion in adult humans. *Peptides.* 6 , 825-839.

Tailford, K. A. , Berry, C. L. , Thomas, A. C. and Campbell, J. H. (2003) A casein variant in cow's milk is atherogenic. *Atherosclerosis* 170 , 13-19.

Terpstra, A. H.M. , Harkes, L. and van der Veen, F. H. (1981) The effect of different proportions of casein in semipurified diets on the concentration of serum cholesterol and the lipoprotein composition in rabbits. *Lipids* 16 , 114-119. 39.

Tomiyama H, Hashimoto H, Hirayama *Y et al.* (2006)Synergistic acceleration of arterial stiffening in the presence of raised blood pressure and raised plasma glucose. *Hypertension*47:180–188.

Torreiles, J. and Guerin, M.-C. (1995) Casein-derived peptides promote peroxidase-dependent oxidation of human blood low-density lipoproteins. *C. R. Soc. Biol.* 1789 , 933-942.

Umbach, M. , Teschemacher, H. , Praetorius, K. , Hirschhauser, R. and Bostedt, H. (1985) Demonstration of a beta-casomorphin immunoreactive material in the plasma of newborn calves after milk intake. *Regul. Pept* 12 , 223-230.

USDA (2001) State and national standardization lactation averages by breed for Cows calving in 1999 — http://iapreview.ars.usda.gov/research/ publications/publications.htm? SEQ_NO_115 = 102557,

USDA Dairy (2002) *Part II: Changes in the United States dairy industry* pp. 1991-2002. — http://www.aphis.usda.gov/vs/ceah/cahm/Dairy_Cattle/ Dairy02/dairy2002-ptII.pdf

USDA's Food Guide (September 1993) *Human Nutrition Information Services* — Miscellaneous Publication Number 1514

Virtanen, S. M. , Laara, E. , Hypponen, E. , Reijonen, H. , Rasanen, L. , Aro, A. , Knip, M. , Honen, J. ,Akerblom, H. K. and the Childhood Diabetes in Finland Study Group (2000) Cow's milk consumption, HLA-DQB1 genotype, and type 1 diabetes. *Diabetes* 49 , 912-917.

Virtanen, S. M. and Knip, M. (2003) Nutritional risk predictors of β cell autoimmunity and type 1 diabetes at a young age. *Am. J. Clin. Nutr* 78, 1053-1067.

Wilkinson IB, Prasad K, Hall *I et al.* (2002). Increased central pulse pressure and augmentation index in subjects with hypercholesterolemi*a.J Am Coll Cardiol* 39:1005–1011.

In: Casein: Production, Uses and Health Effects
Editors: A. M. Ventimiglia, et. al, pp. 233-241
ISBN: 978-1-62100-129-4

Chapter 13

ANTIMICROBIAL ACTIVITIES OF CASEIN AND OTHER MILK PROTEINS

T.B. Ng[1]*, Jack Ho Wong[1], Osama Almahdy[2], Esmail M. EL-Fakharany[2], Ehab EL-Dabaa[3] and EL-Rashdy M. Redwan[2]*

[1] School of Biomedical Sciences, Faculty of Medicine, The Chinese University of Hong Kong, Shatin, N.T., Hong Kong, China.

[2] Antibody laboratory, Protein Research Department, Genetic Engineering and Biotechnology Research Institute, Mubarak City for Scientific Research and Technology Applications, New Borg EL-Arab post-code 21394, Alexandria, Egypt.

[3] Department of Biochemistry, Theodore Bilharz Institute, Cairo, Egypt.

ABSTRACT

Milk contains a variety of proteins with important biological activities. Many of them including casein, lactoferrin,alpha-lactalbumin and beta-lactoglobulin possess antimicrobial activities either in the native state or after chemical modification. Alpha(S2)-casein and alpha-lactalbumin acquire anti-HIV activity after modification with3-hydroxyphthalic anhydride Casein, alpha-lactalbumin, and beta-lactoglobulin inhibit HIV-1 protease and integrase to an appreciable extent but do not affect HIV-1 reverse transcriptase. Milk proteins (casein and beta-lactoglobulin) other than bovine lactoferrin do not prevent hepatitis C infection, Methylated or ethylated beta-casein alpha-lactalbumin and beta-lactoglobulin exert an antiviral action against bacteriophage M13, proportional to the extent of esterification and, hence, to the increased basicity of the modified proteins The tri-, tetra-, and pentaphosphorylated forms of human beta-casein inhibit *H. influenzae* ,a pathogen implicated in middle ear infection in infants. Kappa-casein inhibits *Helicobacter pylori* adhesion to gastric mucosa.Lactoferrin and its fragments lactoferricin and lactoferrampin manifest inhibitory activity against a large number of viruses, bacteria and fungi.

INTRODUCTION

Milk and colostrum is an abundant source of proteins/peptides with pivotal roles in both neonates and adults. Milk proteins and peptides are potential health-enhancing nutraceuticals. They may be utilized as nutraceuticals in the treatment of asthma, cancer, dental diseases, diarrhea, hypertension, immunodeficiency, mineral malabsorption, and thrombosis,. Casein lactoferrin, alpha lactalbumin, beta-globulin,lysozyme- and lactoperoxidase are some of the proteins present in milk. Casein is protective in experimental bacteremia by eliciting myelopoiesis. Casein hydrolyzate has a protective action in diabetic animals, suppresses tumor growth and reduces colicky symptoms in infants. Casein-derived peptides have antihypertensive effects. Glycomacropeptide (GMP)--a peptide derived from kappa casein,display antibacterial and antithrombotic activities (Zimecki , and Kruzel ,2007)The other milk proteins also manifest important biological activities.The intent of the present article is to review the antimicrobial effects including antiviral, antibacterial and antifungal activities of milk proteins with an emphasis on casein. Casein has been protective in experimental bacteremia by eliciting myelopoiesis.

1. ANTIVIRAL ACTIVITY

1.1. Anti- Human Immunodeficiency Virus (Anti-HIV) Activity

Chemical modification of lysines by aromatic acid anhydrides, preferentially by 3-hydroxyphthalic and trimellitic anhydrides and trimellitic anhydride chloride, converted casein into potent inhibitors of (i) binding between the HIV-1 gp 120 envelope glycoprotein and the CD4 cell receptor,due to binding to CD4, and (ii) infection by HIV-1. Modified bovine milk proteins are also potent HIV-1 inhibitors and may have potential for anti-HIV-1 prophylaxis (Neurath et al. 1995).

Several polyanionic reagents such as dextran sulfates, heparin sulfates, and negatively charged proteins demonstrate anti-HIV activity in vitro. Beta-lactoglobulin (betaLG), upon modification by 3-hydroxyphthalic anhydride (3HP),acquires highly potent inhibitory activity. The introduction of multiple negatively charged carboxyl groups along the polypeptide backbone causes repulsion within the protein molecule thereby affecting the secondary and tertiary structures, causing the protein to become unstructured. Similarly,alpha(S2)-casein and alpha-lactalbumin acquires anti-HIV activity after modification with 3HP Most potent inhibition of HIV-1 replication was obtained with 3HP-modified alpha-lactalbumin, which was the least cytotoxic. (Berkhout et al.,.1997).

Wang and Ng(2000) tested various milk proteins including casein, lactoferrin, angiogenin-1, alpha-lactalbumin, beta-lactoglobulin, lactoperoxidase, lactogenin and glycolactin for inhibitory activity toward human immunodeficiency virus-1 reverse transcriptase (HIV-1 RT), alpha-glucosidase, beta-glucosidase and beta-glucuronidase. Casein, alpha-lactalbumin and beta-lactoglobulin had little or no inhibitory effect. Lactoferrin exhibited the most potent inhibitory action. Lactoperoxidase, lactogenin, angiogenin-1 and glycolactin inhibited HIV-1 RT activity with reduced potencies. Succinylation with succinic anhydride elevated the inhibitory effect of casein glycolactin, alpha-lactalbumin,beta-lactoglobulin, and human lactoferrin. The inhibitory effect of the various milk proteins on the activities of alpha-glucosidase, beta-glucosidase and beta-glucuronidase was meager. Succinylation tended to increase the alpha-glucosidase-inhibitory effect of milk proteins but their beta-glucosidase-inhibitory and beta-glucuronidase-inhibitory activities remained unaffected(Wang and Ng,2000).. A bovine milk protein with an N-terminal sequence resembling polymeric immunoglobulin receptor inhibited HIV-1 reverse transcriptase with an IC(50) of 4.8 microM..(Ng and Ye,2004).

The aforementioned milk proteins inhibited HIV enzymes to different extents. Cascin, alpha-lactalbumin, and beta-lactoglobulin inhibited HIV-1 protease and integrase to an appreciable extent but did not affect HIV-1 reverse transcriptase. Lactoferrin strongly inhibited HIV-1 reverse transcriptase but only weakly inhibited HIV-1 protease and integrase. Glycolactin and angiogenin-1 mildly inhibited HIV-1 reverse transcriptase but more powerfully suppressed HIV-1 protease and integrase. In comparison with the other milk proteins glycolactin strongly inhibited HIV-1 protease and integrase and mildly inhibited HIV-1 reverse transcriptase. Lactogenin strongly inhibited HIV-1 integrase, moderately inhibited HIV-1 reverse transcriptase and weakly inhibited HIV-1 protease.(Ng et al.,2001).

Unlike some chemically modified (strongly negative) milk proteins, unmodified alpha(s2)-, beta- and kappa-casein, as well as several negatively and positively charged fragments, did not inhibit IV replication. Bovine lactoferrin (bLF)inhibited V-1 with an IC50 of 0.4 microM. Lactoferricin, a positively charged domain of bLF, was modestly inhibitory, inferring that other domains of the lactoferrin molecule are also involved in inhibition o9f HIV. bLF blocked HIV-1 variants that employ either the CXCR4 or the CCR5 coreceptor. In order to elucidate the mechanism of action of, bLF a bLF-resistant HIV-1 variant.was employed The bLF-resistance phenotype is mediated by the viral envelope protein, which contains two interesting mutations associated with an altered virus-host interaction and a modified receptor-coreceptor interaction. The data show that bLF targets HIV-1 entry (Berkhout et al., 2002).

1.2. Anti- Human Cytomegalovirus (Anti-HCMV) Activity

Unmodified milk proteins except lactoferrin are generally not active against HIV and. HCMV.Several other milk proteins tested showed strong antiviral effects only after chemical modification to acquire a polyanionic (for anti-HIV activity) or polycationic (for anti-HCMV activity) character..(Florisa et al. 2003).

MRC-5 fibroblasts infected with reference strain AD 169 were treated with different concentrations of methylated alpha-lactalbumin (Met-ALA) or methylated beta-lactoglobulin (Met-BLG), as well as with their peptic hydrolysates, and with the highly basic polypeptides such as are L-polylysines (4-15 kDa). The antiviral activity was calculated by comparing the number of infected cells in the presence and absence of the tested substances. Both Met-ALA and Met-BLG, as well as their peptic hydrolysates, decreased the infectious activity of cytomegalovirus in fibroblast cells. As expected, L-polylysines showed the highest antiviral activity. However, the tested basic proteins and polypeptides despite their lower antiviral activities might be potentially quite useful in fight of arising drug resistance activities and the persistence capacities of this virus (Chobert et al.,2007).

1.3. Anti- Hepatitis C Virus (Anti-HCV) Activity

Milk proteins (casein and beta-lactoglobulin) other than bLF did not prevent HCV infection, bLF prevented HCV infection in PH5CH8 human hepatocyte cells which are susceptible to HCV infection. bLF inhibited viral entry to the cells by direct interaction with HCV immediately after mixing of bLF and HCV inoculum. Bovine lactoferricin,which has antibacterial activity, had no anti-HCV activity, indicating that some other region is responsible for the anti-HCV activity of lactoferrin. Prevention of HCV infection by bLF was a general phenomenon. bLF inhibited HCV infection with all five inocula examined, and bLF inhibited HCV infection in human MT-2C T-cells susceptible to HCV infection. In addition, bLF prevented infection with hepatitis G virus, a distant relative of HCV.(Ikeda et al.,2000).

1.4. Anti-Avian Influenza A (H5N1) Activity

Both native and esterified lactoferrin exhibited the highest potency against A/chicken/Egypt/086Q-NLQP/2008 HPAI (H5N1) strain of clade 2.2.1, followed by β-lactoglobulin.The antiviral activity of lactoferrin and β-lactoglobulin was augmented following esterification α-Lactalbumin displayed the lowest antiviral activity than even after esterification.Casein was not tested. (Taha, et al., 2010).

1.5. Anti-Influenza Virus A Subtype H1N1 Activity

Addition of methylated β-lactoglobulin (Met-BLG) to MDCK cells infected with influenza virus subtype H1N1 resulted in a concentration-dependent decrement in hemagglutination activity. The antiviral activity of Met-BLG was dependent on the mlik protein concentration, viral load, and duration of infection. Viral inhibition was accompanied by a protection of MDCK cells When viral stock solution was mixed with Met-BLG in absence of MDCK cells, the morphology and viability of viral particles were adversely affected as revealed by electron microscopy, and the number of intact virus particlesunderwent a decline (Sitohy et al.,2010).

1.6. Anti-Herpes Simplex Virus Type 1(Anti-HSV-1) Activity

The antiviral activity of methylated alpha-lactalbumin (Met-ALA), methylated and ethylated beta-lactoglobulins (Met- and Et-BLG) was tested against acyclovir (ACV)-sensitive and -resistant strains of HSV-1 using fixed or suspended Vero cell lines. Esterified whey proteins and their peptic hydrolyzates demonstrated a protective action against HSV-1, which was less potent than those of ACV and L-polylysines. Real-time PCR disclosed that Met-ALA inhibited HSV-1 replication. (Sitohy et al., 2007).

1.7. Anti-Tomato Yellow Leaf Curl Virus (Anti-TYLCV) Activity

The inhibitory effect of native and esterified alpha-lactalbumin, beta-lactoglobulin, and lactoferrin and their esterified derivatives on tomato yellow leaf curl virus (TYLCV) infection in tomato plants was examined by Abdelbacki et al.,(2010) by spraying the proteins on TYLCV-infected plants. Alpha-lactoferrin and alpha-lactalbumin produced the most pronounced inhibition of viral replication after 7 and 15 days, respectively. Native and esterified lactoferrin brought about total inhibition after 7 days. Native beta-lactoglobulin showed inhibition after 7 and 15 days. Esterified beta-lactoglobulin was more effective after 7 days. The relative amount of viral DNA was less affected by esterified alpha-lactalbumin whereas native alpha-lactalbumin inhibited virus replication completely after 15 days. These results indicate that native or modified whey proteins fractions can be used for controlling TYLCV infections in plants. (Abdelbacki et al., 2010).

1.8. Anti-Bacteriophage M13 Activity

Esterified milk proteins [methylated (Met) or ethylated (Et) beta-casein (BCN)alpha-lactalbumin (ALA), and beta-lactoglobulin (BLG)] exerted an antiviral action against bacteriophage M13, proportional to the extent of esterification and, hence, to the increased basicity of the modified proteins. The antiviral activity of Met-BLG was abrogated after pepsinolysis but not following trypsinolysis. The antiviral activity of Met-BLG was much more potent than that of the native basic proteins histone and lysozyme. (Sitohy et al.; 2006).

1.9. Anti-Lactococcal Bacteriophage (bIL66, bIL67, and bIL170) Activity

Esterified milk proteins [methylated (Met) or ethylated (Et) beta-lactoglobulin (BLG) and alpha-lactalbumin (ALA)], drastically curtailed the lytic activities and replication of lactococcal bacteriophages (bIL66, bIL67, and bIL170). Met-BLG was more active toward bIL66 and bIL67, Lysozyme manifested weaker antiviral activity. Antiviral activity of Et-BLG was somewhat weaker than that of the Met derivative. Esterified ALA also showed considerable but slightly lower antiviral activity compared with other proteins. Replication of bIL67 was inhibited by the presence of esterified ALA or BLG. The presence of Met-BLG in the incubation medium protected L. lactis against lysis when it was subjected to an infection by bIL67 (Sitohy et al.,2005) is a potent inhibitor for several enveloped and naked viruses, such as rotavirus, enterovirus and adenovirus. It is resistant to tryptic digestion and breast-fed infants excrete high levels of faecal lactoferrin so that its effect on viruses replicating in the gastrointestinal tract is of great interest. (Neurath et al..1995). Lactoferricin, a peptide formed from lactoferrin by enzymatic cleavage, exhibits antiviral activity (Gifford et al.,2005).

2. Antibacterial Activity

Casein has been protective in experimental bacteremia by eliciting myelopoiesis (Kanwar et al.,2009).. Lactoperoxidase shows antibacterial properties(Neurath et al..1995).. Lysozyme is effective in treatment of periodentitis and prevention of tooth decay(Neurath et al.,..1995) Antimicrobial peptides are hidden in the casein molecules.(van Hooijdonk et al.,2000).

2.1. Anti-Haemophilus Influenza Activity

The average beta-casein concentration in 151 human milk samples was 5.37+/-2.26 mg/ml. The phosphoforms in untreated milk are present in the following order ranked by concentration: tetra- > di- > non- > mono- > tri- > pentaphosphorylated beta-casein. The tri-, tetra-, and pentaphosphorylated forms of human beta-casein at a concentration of 0.6 to 0.9 mg/ml inhibited *H. influenzae* ,a pathogen implicated in middle ear infection in infants,,.in the antiadhesion by more than 60% The phosphoform distribution of beta-casein in individual donors varies widely. *Anti-H. influenzae* activity was detected in vitro among human beta-casein molecules with three or more phosphate groups.(Kroning et al.,1999).

2.2. Anti-Helicobacter Pylori Activity

Human milk protects the infant against infectious diseases protection through several mechanisms: specific antibody targeted protection against pathogens in the infant's environment (through milk IgA, IgG, and IgM) and broad-spectrum, nonspecific protection provided through several distinct mechanisms. These are: bactericidal effects (lactoferrin), bacteriostatic action (lactoferrin, lysozyme), lysis of microorganisms (lysozyme), antiviral effects (lactoferrin,), and ligand action (inhibition of *Helicobacter pylori* adhesion to gastric mucosa by kappa-casein). (antiproteases).(Hamosh, 1998).The iron sequestering property of lactoferrin is the basis of its bacteriostatic effect,. In addition to its bacteriostatic action, lactoferrin can exert a direct bactericidal effect owing to its binding to the lipid A part of bacterial lipopolysaccharide,accompanied by an increase in membrane permeability. This action is due to lactoferricin, a peptide formed from lactoferrin by enzymatic cleavage (Orsi, 2004).Another lactoferrin-derived peptide, lactoferrampin, also manifests antibacterial activity(Haney et al., 2009).

3. Antifungal Activity

Jenssen and Hancock (2009) have written an excellent review on the antifungal effects of lactoferrin. The antifungal activity of lactoferrin against *Candida* species follows the order: *C. tropicalis* > *C. krusei* > *C. albicans* > *C. guilliermondii* > *C. parapsilosis* > *C. glabrata*, with the first species being the most susceptible and the last the most resistance.The proposed mode of antifungal action of lactoferrin was perturbation of fungal cell wall which was corroborated by cryo-scanning electron microscopy demonstrating cell wall changes leading to surface blebs, swelling and cell collapse.Similar cell wall damage has been noted, after treatment with human and bovine lactoferrin, The candidacidal activity of human lactoferrin is attributed to direct interaction of the milk protein with the fungal cell surface, rather than iron sequestration.On the contrary, iron sequestration by lactoferrin is crucial for host defense against *Aspergillus fumigatus* . The combination of lactoferrin with clotrimazole,fluconazole, amphotericin B and 5-fluorocytosine produces an additive or a synergistic effect.. Administration of recombinant human lactoferrin improved survival rate in an *in vivo* rat model of co-infection with *C. albicans* and *S. epidermis* . It has also been indicated that lactoferrin can mediated its antifungal activity through the stimulation of host cell immune mechanisms both *in vitro* and*in vivo.* The antifungal activity of lactoferrin and lactoferricin (Gifford et al., 2005) towards Candida has been attributed to a direct action on Candida cell membranes. (Orsi, 2004).Lactoferrampin also exerts antifungal activity(Haney et al., 2009).

CONCLUSION

From the foregoing account it can be seen that casein exhibits antiviral and antibacterial activities, sometimes only after chemical modification. Its spectrum of antimicrobial activities is less impressive than that of lactoferrin and its fragment lactoferricin. The presence of a multitude of antimicrobial proteins in milk endows milk with health promoting and antipathogenic properties.Regular consumption of milk is recommended by nutritionists and physicians.

ACKNOWLEDGMENTS

We are grateful to Food and Health Bureau, Government of Hong Kong Special Administrative Region for award of a research grant (Research Fund for the Control of Infectious Diseases, project number 10090812).

REFERENCES

Abdelbacki AM, Taha SH, Sitohy MZ, Abou Dawood AI, Abd-El Hamid MM, Rezk AA. Inhibition of tomato yellow leaf curl virus (TYLCV) using whey proteins. *Virol J.* 2010 ;7:26

Berkhout B, Derksen GC, Back NK, Klaver B, de Kruif CG, Visser S. Structural and functional analysis of negatively charged milk proteins with anti-HIV activity. *AIDS Res Hum Retroviruses.* 1997 ;13:1101-7.

Berkhout B, van Wamel JL, Beljaars L, Meijer DK, Visser S, Floris R. Characterization of the anti-HIV effects of native lactoferrin and other milk proteins and protein-derived peptides. *Antiviral Res.* 2002 ;55:341-55.

ChobertJM, Sitohy M, Billaudel S, Dalgalarrondo M, Haertlé T. Anticytomegaloviral activity of esterified milk proteins and L-polylysines. *J Mol Microbiol Biotechnol.* 2007;13:255-8.

Florisa R, Recio I, Berkhout B, Visser S. Antibacterial and antiviral effects of milk proteins and derivatives thereof. *Curr Pharm Des.* 2003;9:1257-75.

Gifford JL, Hunter HN, Vogel HJ. Lactoferricin: a lactoferrin-derived peptide with antimicrobial, antiviral, antitumor and immunological properties. *Cell Mol Life Sci.* 2005 ;62:2588-98.

Hamosh M. Protective function of proteins and lipids in human milkBiol Neonate. 1998;74:163-76.

Haney EF, Nazmi K, Lau F, Bolscher JG, Vogel HJ Novel lactoferrampin antimicrobial peptides derived from human lactoferrin. *Biochimie.* 2009 ;91:141-54.

Ikeda M, Nozaki A, Sugiyama K, Tanaka T, Naganuma A, Tanaka K, Sekihara H, Shimotohno K, Saito M, Kato N. Characterization of antiviral activity of lactoferrin against hepatitis C virus infection in human cultured cells. *Virus Res.* 2000 ;66:51-63.

Jenssen H and Hancock R.E.W. Antimicrobial properties of lactoferrin Advances in Lactoferrin Research2009, , 19-29.

Kanwar JR, Kanwar RK, Sun X, Punj V, Matta H, Morley SM, Parratt A, Puri M, Sehgal R. Molecular and biotechnological advances in milk proteins in relation to human health. *Curr Protein Pept Sci.* 2009 ;10:308-38.

Kroening TA, Baxter JH, Anderson SA, Hards RG, Harvey L, Mukerji P. Concentrations and anti-Haemophilus influenzae activities of beta-casein phosphoforms in human milk. *J Pediatr Gastroenterol Nutr.* 1999 ;28:486-91..

Neurath AR, Debnath AK, Strick N, Li YY, Lin K, Jiang S. Blocking of CD4 cell receptors for the human immunodeficiency virus type 1 (HIV-1) by chemically modified bovine milk proteins: potential for AIDS prophylaxis. *J Mol Recognit.* 1995;8:304-16.

Ng TB, Lam TL, Au TK, Ye XY, Wan CC. Inhibition of human immunodeficiency virus type 1 reverse transcriptase, protease and integrase by bovine milk proteins. *.Life Sci.* 2001;69):2217-23.

Ng TB, Ye XY. A polymeric immunoglobulin receptor-like milk protein with inhibitory activity on human immunodeficiency virus type 1 reverse transcriptase. *Int J Biochem Cell Biol.* 2004 36:2242-9.

Orsi N. The antimicrobial activity of lactoferrin: current status and perspectives. *Biometals.* 2004 ;17:189-96.

Sitohy M, Chobert JM, Haertlé T. Esterified whey proteins can protect Lactococcus lactis against bacteriophage infection. Comparison with the effect of native basic proteins and L-polylysines. *J Agric Food Chem.* 2005 53:3727-34.

Sitohy M, Billaudel S, Haertlé T, Chobert JM. Antiviral activity of esterified alpha-lactalbumin and beta-lactoglobulin against herpes simplex virus type 1. Comparison with the effect of acyclovir and L-polylysines. *J Agric Food Chem.* 2007;55:10214-20.

Sitohy M, Chobert JM, Karwowska U, Gozdzicka-Jozefiak A, Haertlé T. Inhibition of bacteriophage m13 replication with esterified milk proteins. *J Agric Food Chem.* 2006 ;54:3800-6.

Taha SH, Mehrez MA, Sitohy MZ, Abou Dawood AG, Abd-El Hamid MM, Kilany WH. Effectiveness of esterified whey proteins fractions against Egyptian Lethal Avian Influenza A (H5N1). *Virol J.* 2010;7:330

van Hooijdonk AC, Kussendrager KD, Steijns JM. In vivo antimicrobial and antiviral activity of components in bovine milk and colostrum involved in non-specific defence. *Br J Nutr.* 2000 ;84 Suppl 1:S127-34.

Wang H, Ye X, Ng TB. First demonstration of an inhibitory activity of milk proteins against human immunodeficiency virus-1 reverse transcriptase and the effect of succinylation. *Life Sci.* 2000 ;67:2745-52.

Zimecki M, Kruzel MLMilk-derived proteins and peptides of potential therapeutic and nutritive value. *J Exp Ther Oncol.* 2007;6:89-106.

In: Casein: Production, Uses and Health Effects
Editors: A. M. Ventimiglia, et. al, pp. 243-258
ISBN: 978-1-62100-129-4

Chapter 14

DENTAL HEALTH EFFECTS OF CASEIN DERIVED PRODUCTS

Gianmaria Fabrizio Ferrazzano and Aniello Ingenito
University of Naples "Federico II", Naples, Italy.

1. THE ORAL HEALTH

Oral health is a part of general health and hence affects the total well-being of individuals. Dental and oral diseases affect various aspects of quality of life. Dental caries has been associated with numerous adverse effects on children's health including pain, restricted dietary intake, impaired growth, and reduced body weight [Hamissi et al., 2008].

Even though the onset of dental caries has decreased through the use of preventive systems (improved oral hygiene, usage of fluoride-containing toothpaste, fluoride content in drinking water, sealing), tooth decay is still one of the most prevalent chronic worldwide diseases [Al-Malik and Rehbini, 2006; de Almeida et al., 2003; Ferrazzano et al., 2006; Ferro et al., 2007; Jacobsen and Young, 2003; Selwitz et al., 2007; Simonsen, 2002].

In fact, a decline in DMFT (decayed, missing, filled teeth) indices from 5.0 to 0.9 in the period from 1980 to 2002 was observed among 12-year-olds in Denmark; in Sweden from 3.1 to 0.9 during 1985–2001. A decline in the DMFT index from 5.1 to 3.8 was found in Poland during 1991–2000. National epidemiological surveys conducted by the Brazilian Ministry of Health in 1996 and 2003 showed a DMFT reduction of 9.15% (3.06 to 2.78) in 12-year-olds. [Cypriano et al., 2008; Milciuviene et al., 2009].

Furthermore, significant gains in oral health levels have been observed over the last few years in countries like Germany, England, USA, Scandinavia, Scotland, Norway and Australia [Cypriano et al., 2008].

However, with the decline of dental caries in several countries the chance of extending oral health care to other groups of the population has increased significantly [Petersen, 2005].

In fact, there are profound disparities in oral health. The indigenous poor populations, immigrants, racial and ethnic minority groups, and medically compromised patients are those who suffer the worst oral health conditions [Skeie et al., 2006].

Recent evidence indicates that immigrants and minority ethnic groups should be regarded as "whole populations at risk" on the verge of oral health deterioration. People crossing national and cultural frontiers originally are characterized by disease patterns, health behaviours and health care modalities different from those at their destinations [Vered et al., 2008].

Recent reports from developed countries indicate inequalities in caries prevalence among preschoolers, with a higher caries prevalence and severity in certain ethnic and immigrant groups [Hallett and O' Rourke, 2006; Wyne, 2008].

For example, in both Denmark [Sundby and Petersen, 2003] and the UK [Gray et al., 2000], children from different ethnic groups present higher levels of caries than is seen in the native population. In The Hague (Netherlands) [Truin et al., 2005], a study has described higher caries prevalence among low-social-class children of Turkish and Moroccan origin when compared with Dutch children from the same social class.

This indicates that there are still shortcomings in the areas of both preventive and curative dental care. More oral health education programs must be deployed in an attempt to control oral diseases and school-based approaches should be combined with family- and community-directed preventive programs. Also, specific strategies should be developed for the different situations concerning oral health status [Hamissi et al., 2008].

There is a need to identify areas with high caries levels, even if the DMFT =3 goal had already been achieved, and develop oral health strategies for the groups with the highest caries prevalence. Finally, individualized approaches should be applied, promoting reasonable, effective and lasting health actions. In this view, because caries affects individuals disproportionately, it is essential that those at the highest risk are identified early so that preventive therapies can be targeted toward those who are most likely to benefit.

A significant improvement in dental caries worldwide levels might be reached by implementation of caries preventive strategies, paying special attention to high-risk groups, represented by people with low economic-social status in developed countries and people who lives in less developed countries.

For this reason, the identification and the characterization of natural active principles (derived from animals) [Aimutis, 2004], having anti-caries activity, appear to be helpful in order to elaborate a new preventive methodology.

In this way, the severity of this illness should be reduced, first of all, only by changing the dietary behaviours of the population, then, by adding these natural active principles to food, toothpaste, mouthwash, tablets. These anticariogenic agents would be particularly useful, being non toxic and relatively cheap.

2. Dental Enamel Demineralization and Remineralization Processes

During dental caries, enamel and dentine are in contact with an acid H+ releasing system. This may be plaque during an in situ or in vivo experiment or an acidic solution or acidic gel in vitro. The H+ ions cause local dissolution of the tissue, and the resulting dissolution products (calcium and phosphate ions) diffuse out of the tooth [Arends et al., 1997].

Over the course of human life, enamel and dentin undergo unlimited cycles of demineralization and remineralization. The crux of this argument lies in the balance of these two competing phenomena. A tip in the balance one way or the other will either lead to stronger healthier teeth or greater susceptibility to dental caries and other oral complications. This mineral loss compromises the mechanical structure of the tooth and could lead to cavitation over a long period of time. Saliva alone has the capability to increase plaque pH with bicarbonates, although typically this process may take up to 2 hours. The susceptibility of apatite in enamel surface layers makes it critical to control the acidity of the plaque fluid and the calcium and phosphate ion concentrations in saliva [Featherstone, 2000].

The subsequent remineralization process is nearly the reverse. When oral pH returns to near neutral, calcium and phosphate ions in saliva incorporate themselves into the depleted mineral layers of enamel as new apatite. The demineralized zones in the crystal lattice act as nucleation sites for new mineral deposition. Essentially, the sudden drop in pH following meals produces an under-saturation of those essential ions in the plaque fluid with respect to tooth mineral. This promotes the dissolution of the enamel. Instead, at elevated pH, the ionic supersaturation of plaque shifts the equilibrium the other way, causing a mineral deposition in the tooth.

Remineralization is the natural repair process for caries lesions. However, when the bacterial challenge is high or the salivary components are lacking, remineralization is insufficient to halt or reverse the caries process [Featherstone, 2009]. There is a need to find ways to enhance the remineralization process and to transfer such knowledge into clinical therapy. Fluoride delivered topically in various vehicles has produced major reductions in the prevalence and incidence of dental caries [DePaola et al., 1993]. None, however, has shown that fluoride can eliminate decayed surfaces and teeth in all individuals. For this reason, both more effective means of remineralization and more effective antibacterial treatment in conjunction with remineralization are needed.

Remineralization requires calcium and phosphate, which generally come from saliva. Remineralization is enhanced by fluoride. The primary modes of action of fluoride are enhancement of remineralization, inhibition of demineralization, and inhibition of plaque bacteria, with remineralization being the most important [ten Cate and Featherstone, 1991]. Enhancement of remineralization can occur only with calcium and phosphate being present, from whatever source [ten Cate, 2009]. Effective ways to enhance remineralization and to make it more effective are to supply calcium and phosphate, and to deliver fluoride more effectively over longer periods of time. Numerous studies have shown the potential benefits of some milk derivates in enhancing enamel remineralization [Reynolds, 2009].

3. Casein Phosphopeptide-Amorphous Calcium Phosphate (CPP-ACP)

As observed from literature review, several mechanisms by which dairy products may reduce enamel demineralization have been proposed.

Levine supposed that milk proteins are adsorbed onto the enamel surface and may impede enamel demineralization; milk fat is adsorbed onto the enamel surface and may have a protective role; milk enzymes may have a role in reducing the growth of acidogenic plaque bacteria [Levine, 2001].

Herod, also, reported that milk and cheese could reduce the effects of metabolic acids, and could help restore the enamel that is lost during eating. Postulated mechanisms involve buffering, salivary stimulation, reduction of bacterial adhesion, reduction of enamel demineralization, and/or promotion of remineralization by casein and ionizable Ca and P [Herod, 1991].

Several constituents in dairy products could exert a direct effect on the tooth surface: calcium, phosphorus, casein and casein phosphopeptides.

The use of calcium and phosphate ions for remineralization has not been successful, due to the low solubility of calcium phosphates, particularly in the presence of fluoride ions. Insoluble calcium phosphates are not easily applied, do not localize effectively at the tooth surface, and require acid for solubility to produce ions capable of diffusing into enamel subsurface lesions. In contrast, soluble calcium and phosphate ions can be used only at very low concentrations, due to the intrinsic insolubility of the calcium phosphates, particularly the calcium fluorophosphates. Soluble calcium and phosphate ions do not substantially incorporate into dental plaque or localize at the tooth surface to produce effective concentration gradients to drive diffusion into the enamel subsurface. Instead, a new calcium phosphate remineralization technology has now been developed based on casein phosphopeptide-amorphous calcium phosphate (CPP-ACP), where it is claimed that the CPP stabilizes high concentrations of calcium and phosphate ions, together with fluoride ions, at the tooth surface by binding to pellicle and plaque. Although the calcium, phosphate, and fluoride ions are stabilized by the CPP from promoting dental calculus, the ions are freely bioavailable to diffuse down concentration gradients into enamel subsurface lesions, thereby effectively promoting remineralization *in vivo* [Reynolds, 2009]. CPP, containing the active sequence –Ser(P)-Ser(P)-Ser(P)-Glu-Glu-, has a remarkable ability to stabilize calcium and phosphate as nanoclusters of ions in metastable solution [Cochrane et al., 2008]. Through the active sequence, the CPP binds to forming nanoclusters of calcium and phosphate ions to form nanocomplexes of around 1.5 nm radius, preventing the growth of the nanoclusters to the critical size required for nucleation and phase transformation.

Casein is the predominant phosphor-protein in bovine milk and accounts for almost 80 percent of its total protein, primarily as calcium phosphate stabilized micellular complexes [Aimutis, 2004]. Although several studies have investigated the use of casein [Bavetta and McClure, 1957; Reynolds and del Rio, 1984] as an anticariogenic additive to food, toothpaste or drinking water, its use has not been implemented because of its unpalatability and the large amount required for efficacy [Reynolds, 1998]. In contrast, CPP does not have these limitations. The potential for a specific anticariogenic activity is at least 10 times greater on a weight basis for CPP than it is for casein [Azarpazhooh and Limeback , 2008].

Casein phosphopeptides (CPPs) are phosphorylated casein-derived peptides produced by proteolytic digestion of αs1-, αs2-, and ß-casein during the natural digestive process in vivo, by action of proteolytic enzymes in vitro, and by proteolytic starter cultures during manufacturing of dairy products as fermented milk, yogurt and cheese [Bouhallab and Bouglé, 2004; Cai et al., 2003; Ferrazzano et al., 2007; Walker et al., 2006].

CPPs-ACP localize ACP in dental plaque, which buffers the free calcium and phosphate ion activities, helping to maintain a state of supersaturation with respect to tooth enamel, depressing demineralization and enhancing remineralization [Cross et al., 2004; Cross et al., 2005].

In particular, CPPs stabilize calcium and phosphate ions under neutral and alkaline conditions forming metastable solutions that are supersaturated with respect to the basic calcium phosphate phases. Under these conditions, the CPPs bind their equivalent weights of calcium and phosphate.

The preventive action of CPPs, in vivo, takes place when there are demineralizing agents (acid pH) that can enhance the release of calcium from the CPP-ACP complex, thus increasing the Ca cation concentration and promoting a supersaturation condition that will prevent demineralization and enhance the remineralization of early enamel caries [Reynolds et al., 2003].

On the basis of the generally accepted molecular formula for ACP [Ca3(PO4)2 - nH2O], ACP also may be considered a tricalcium phosphate. There is no conclusive evidence that ACP is an integral mineral component in hard tissues. It likely plays a special role as a precursor to bioapatite and as a transient phase in biomineralization [Azarpazhooh and Limeback, 2008].

In summary, CPP-ACP complexes have a multiple action mechanism: on one hand, providing an oversaturation of calcium and phosphate ions in the dental biofilm and saliva, conferring the potential to be biological delivery vehicles for calcium and phosphate; on the other hand, inhibiting adhesion of cariogenic bacteria to the hydroxyapatite making it possible to modulate the activity of plaque bacteria and favour colonization by less cariogenic bacteria. This can help to reduce acid formation in the biofilm and reduce enamel demineralization. The final effect is that saliva and plaque are kept in calcium and phosphate supersaturation with respect to the enamel which allows a reduction in demineralization and favours remineralization [Azarpazhooh and Limeback, 2008; Llena et al., 2009].

3a. Anticariogenic Activity of Casein Phosphopeptide-Amorphous Calcium Phosphate

Several laboratory, animal and human in situ experiments have demonstrated that both synthetic casein phosphopeptide-amorphous calcium phosphate (CPP-ACP) nanocomplexes contained in mouthrinses and sugar-free chewing gum, and natural CPPs contained in dairy products (such yogurt) are anticariogenic [Ferrazzano et al., 2007; Ferrazzano et al., 2008; Iijima et al., 2004; Manton et al., 2008; Morgan et al., 2008; Shen et al., 2001].

The CPP-ACP nanocomplexes, in purified form [Reynolds et al., 1995] or in a crude micellar form [Guggenheim et al., 1999], have been shown to reduce caries activity in the rat caries model.

In a study by Reynolds et al. solutions (100 μL) containing different concentrations of CPP-ACP were applied to the rats' molar teeth twice daily. Other groups of animals received 100 μL of either 500-ppm-fluoride ions (positive control) or distilled water (negative control). The animals consumed a highly cariogenic sucrose/gluten diet that did not contain dairy products. The caries activity of the enamel smooth surfaces was reduced significantly in a dose-response fashion (0.1 percent and 1.0 percent weight per volume CPP-ACP, respectively, produced a 14 percent and a 55 percent reduction in smooth-surface caries activity respect to the distilled water control).

The authors found a similar reduction in the caries activity of fissures (0.1 percent and 1.0 percent w/v CPP-ACP, respectively, produced a 15 percent and a 46 percent reduction in fissure caries activity). CPP-ACP at 0.5% w/v produced a reduction in caries activity similar to that of 500 ppm fluoride. A solution containing both 0.5% w/v CPP-ACP and 500 ppm fluoride produced a significantly greater reduction in caries activity than either CPP-ACP or fluoride alone at the same concentrations [Reynolds et al., 1995].

Furthermore, Guggenheim et al. found that CPP-ACP taken with a cariogenic diet in rats significantly reduced the numbers of streptococcus sobrinus by interfering with bacterial adherence and therefore colonization [Guggenheim et al., 1995].

In an in vitro study, Reynolds demonstrated that CPP-stabilized calcium phosphate solutions maintained high concentration gradients of calcium and phosphate ions and ion pairs in subsurface carious lesions in the enamel of human third molars, resulting in high rates of enamel remineralization. In particular, a 1.0% w/v CPP-ACP solution used twice daily produced a 51 ± 19% reduction in enamel mineral loss caused by frequent sugar-solution exposure [Reynolds, 1997]. The twice daily use of the 1.0% CPP-ACP solution resulted in a 144% increase in calcium level and a 160% increase in inorganic phosphate level in the inter enamel plaque recovered from the removable intra-oral appliance [Reynolds, 1997]. These results suggested an anticariogenic mechanism for the CPP-ACP where the CPP stabilize and localize ACP at the tooth surface, thereby buffering plaque pH, depressing enamel demineralization, and enhancing remineralization. The remineralizing capacity was greater for solutions with higher levels of CPP stabilized free calcium and phosphate ions.

Another laboratory experiment, conducted by Rose showed that CPP-ACP binds well to dental plaque, providing a large calcium reservoir that may inhibit demineralization and assist in subsequent remineralization [Rose, 2000a]. In a second experiment, the same researcher also showed that in streptococcal model plaques, 0.1 percent CPP-ACP provides a large number of possible binding sites for calcium and reduces the free calcium diffusion coefficient by about 65 percent at pH 7 and 35 percent at pH 5. During a cariogenic episode, 0.1 percent CPP-ACP prevented mineral loss and provided a potential source of calcium for subsequent remineralization, thus restricting the caries process [Rose, 2000b].

The ability of the CPP-ACP technology added to sugar-free chewing gum to remineralize enamel subsurface lesions has been demonstrated in several randomized, controlled, double-blind *in situ* clinical studies [Shen et al., 2001]. In particular, 30 subjects in randomized, cross-over, double-blind studies wore removable palatal appliances with six human-enamel half-slabs inset containing subsurface demineralized lesions. The appliances were inserted immediately before gum-chewing for 20 min and then retained for another 20 min. This was performed four times per day for 14 days. At the completion of each treatment, the enamel half-slabs were paired with their respective demineralized control half-slabs, embedded, sectioned, and subjected to microradiography and densitometric image analysis, for measurement of the level of remineralization. The addition of CPP-ACP to either sorbitol- or xylitol-based gum resulted in a dose-related increase in enamel remineralization, with 0.19, 10.0, 18.8, and 56.4 mg of CPP-ACP producing an increase in enamel remineralization of 9, 63, 102, and 152%, respectively, relative to the control gum, independent of gum weight or type. Microradiographs of the enamel lesions before and after remineralization showed that the CPP-ACP promoted remineralization throughout the body of the lesion. Electron microprobe analyses of sections of the remineralized enamel indicated that the mineral deposited was hydroxyapatite with a higher Ca:P ratio than normal apatite [Shen et al., 2001].

Furthermore, acid challenge of the enamel remineralized by the CPP-ACP nanocomplexes in situ showed that the remineralized apatite was more resistant to acid challenge than the normal calcium-deficient carbonated tooth enamel [Cai et al., 2007].

It has been reported that CPP-ACP interacts with fluoride ions to produce novel nanoclusters of calcium, fluoride, and phosphate ions [Reynolds et al., 2008]. The identification of this novel CPP-stabilized form of calcium, fluoride, and phosphate ions is consistent with the observed additive anti-cariogenic effects of the CPP-ACP nanocomplexes and fluoride [Reynolds et al., 2008]. The anti-cariogenic mechanism of fluoride is the localization of the fluoride ion at the tooth surface. This localization promotes remineralization of enamel with fluorapatite. It is clear that, for the formation of fluorapatite [$Ca_{10}(PO_4)_6F_2$], calcium and phosphate ions must also be present with the fluoride ions. The reported additive anti-cariogenic effect of CPP-ACP and fluoride may therefore be attributable to the localization of the novel calcium, fluoride, and phosphate ion nanoclusters at the tooth surface by the CPP, which co-localizes calcium, phosphate, and fluoride as bioavailable ions in the correct molar ratio to form fluorapatite [Reynolds et al., 2008].

The application of a CPPs toothpaste and sodium fluoride (Colgate Neutrafluor 9000 ppm) (NaF) can provide significant additional prevention of enamel demineralization when resin-modified glass ionomer cement (RMGIC) is used for bonding molar tubes for orthodontic patient as preventive actions [Sudjalim et al., 2007].

A dentifrice formulation containing 2% CPP-ACP nanocomplexes plus 1100 ppm F has been shown to be superior (2.6 times) to a dentifrice containing only 1100 ppm F in remineralization of enamel subsurface lesions *in situ* with mineral that was more resistant to acid challenge [Reynolds et al., 2008]. The CPP-ACP nanocomplexes plus fluoride dentifrice resulted in significantly greater incorporation of fluoride into the subsurface enamel as fluorapatite, as shown by electron microprobe analysis [Reynolds et al., 2008]. These results indicate that CPPs are an excellent delivery vehicle to co-localize bioavailable calcium, fluoride, and phosphate ions at the tooth surface to remineralize subsurface enamel lesions with fluorapatite.

A randomized, controlled caries clinical trial of CPP-ACP-containing sugar-free chewing gum demonstrated that the CPP-ACP gum significantly slowed progression of caries and enhanced regression of caries compared with the control sugar-free gum [Morgan et al., 2008]. In the 24-month study, 2720 schoolchildren were randomly assigned to either a CPP-ACP or control sugar-free gum. All children received accepted preventive procedures, including fluoridated water, fluoridated dentifrice, and access to professional care. They were instructed to chew their assigned gum for 10 min three times per day, with one session supervised on school days. Standardized digital radiographs were taken at baseline and at the completion of the trial. The radiographs, scored by a single examiner, were assessed for approximal caries at both the enamel and dentin levels. Of the 14589 approximal surfaces that were included in the CPP-ACP gum group data analysis, 13749 (94.2%) remained unchanged, 785 (5.4%) transitioned to a higher radiographic score, indicating caries progression, and 55 (0.4%) transitioned to a lower score, indicating caries regression. In comparison, for the control gum, 13075 (93.3%) of the 14017 baseline approximal surfaces remained unchanged, 906 (6.5%) progressed, and 36 (0.3%) regressed. This study is indeed impressive in that a modest clinical benefit in a population with relatively low caries experience can be ascribed to chewing gum with CPP-ACP for approximal surfaces.

CPP-ACP in paste form has recently been the subject of limited clinical investigation [Andersson et al., 2007]. The study involved 26 adolescents with 152 white spot lesions on 60 teeth immediately following orthodontic treatment. The children were randomized to either a group receiving treatment with CPP-ACP paste (Topacal C-5) application for 3 months, followed by 3 months of fluoride dentifrice use, or a group receiving treatment with a daily application of a 0.05% NaF mouthrinse in combination with fluoride dentifrice for 6 months. After 12 months, the group assigned to the CPP-ACP regimen had significantly more white spot lesions disappear than did the control group. The authors commented that the visual assessment indicated a more favorable outcome with CPP-ACP treatment.

The ability of natural CPPs (contained in the yogurt) to prevent demineralization and promote remineralization of dental enamel was also in vitro assessed [Ferrazzano et al., 2008] . Eighty human molars were used. After standardizing an in vitro demineralization procedure for producing artificial caries at two different pH values (Group 1: pH 4.8; Group 2: pH 3.97), this procedure was used on teeth, but with the addition of natural CPPs (derived from yogurt) (Group 3: pH 4.8; Group 4: pH 3.97). The effects of these procedures were evaluated by quantitative and qualitative analysis (SEM). Statistical analysis showed significant differences in weight changes between the groups with and without natural CPPs. The SEM observation showed the protective effects of natural CPPs. The authors concluded that CPPs contained in the yogurt have an inhibitory effect on demineralization and promote the remineralization of dental enamel. This result reflected what was found in literature, where the association between dairy products consumption and reduction of dental caries emerged [Merritt et al., 2006]. In particular, milk and milk products could release calcium and phosphorus and, increasing their concentrations in the dental plaque, inhibit demineralization and favour remineralization by a common-ion effect [Silva et al., 1987]. However, as suggested by Walker, in liquid milk the majority of the casein and calcium and phosphate ions are bound in micelles and then upon consumption would not necessarily be available to promote enamel remineralization [Walker at al., 2006]. In fact, bovine milk contains around 30 mM-calcium, where only approximately 10 mM is not bound in casein micelles and of that only approximately 2 mM is free calcium ions [Neville et al., 1994]. The calcium in casein micelles is unlikely to be available to diffuse across a relatively intact enamel surface layer into a subsurface lesion and therefore milk may have limited ability to remineralize enamel lesions in situ. Therefore, milk couldn't be used as a natural product against dental caries, because it needs the addition of 2.0–5.0 g CPP-ACP to substantially increases its ability to remineralize enamel subsurface lesions [Walker et al., 2006]. Instead, regarding yogurt, its protein content is generally higher than that of milk because of the addition of non-fat dry milk during processing and concentration, which increases the protein content of the final product. In addition, yogurt is an excellent source of calcium and phosphorus. In fact, fermentation has little effect on the mineral content of milk and therefore the total mineral content remains unaltered in the yogurt. Furthermore, because of the lower pH of yogurt compared with that of milk, calcium is present in yogurt mostly in ionic form [Adolfsson et al., 2004]. Additionally, CPPs content in yogurt is higher than that in milk due to proteolytic activity of microorganisms contained in the yogurt [Rasić and Kurmann, 1983].

4. Hard Tissues Dental Erosion

Dental erosion is the progressive, irreversible loss of dental hard tissue, chemically etched away from the tooth surface by extrinsic and/or intrinsic acids [Petersen et al., 2005], without the involvement of bacteria.

Dental erosion appears to be a growing problem in several countries, affecting 8–13% of adults. Epidemiological studies have shown that the prevalence of dental erosion in children varies widely between 2 and 57% [Linnett and Seow, 2001].

Lussi and colleagues reported that tooth erosion resulting from chemical agents occurred in 30 percent of patients aged 26 to 30 years and in 42.6 percent of patients aged 45 to 50 years [Lussi et al., 1991]. The disease appears particularly in industrialized countries, reflecting the wide availability and the frequent consumption of acidic beverages, fruit juices, carbonated beverages, wines, and sports drinks.

There are several behavioral factors influencing erosion. Acids which cause dental erosion may be extrinsic or intrinsic in origin [Lussi et al., 1993; Lussi and Jaeggi, 2008]. Typical acid sources come from the diet, medications, occupational exposure and lifestyle activities, but dietary acids are the most extensively studied aetiological agents and the most important extrinsic factors [Lussi et al., 2004b]. Gastrointestinal disease is an example of the intrinsic factors that result in regurgitation of acid into the mouth [Correr et al., 2009; de Carvalho Sales-Peres et al., 2007; Lussi and Jaeggi, 2008; Young, 2005].

Frequent and excessive consumption of specific dietary drinks as well as soft drinks and alcohol abuse may damage tooth structure. Unusual drinking and swallowing habits—for example, retaining an acid beverage in the mouth before swallowing—increase the contact time of an acidic substance with the teeth, and thus increase the risk of erosion. Bedtime consumption of acidic beverages is also considered a risk factor. Health-conscious individuals tend to ingest acidic drinks and juices frequently. In addition, they have better-than-average oral hygiene. While good oral hygiene is of proven value in the prevention of periodontal disease and dental caries, frequent tooth brushing with abrasive oral hygiene products may enhance tooth damage.

The erosive potential of sports drinks depends on the acid type or food acid incorporated in the formulation, such as citric acid, maleic acid and phosphoric acid which decrease the pH of the oral environment, but also depends on the high titratable acidity, viscosity, concentrations of calcium, phosphate, and fluoride [Cochrane et al., 2009; Lussi and Jaeggi, 2008]. Preventive methods to reduce the degree of tooth demineralization and to enhance the acid resistance to tooth structure and the remineralization process have been investigated [Azarpazhooh and Limeback, 2008; Cochrane et al., 2008; Rios et al., 2008; Wiegand and Attin, 2003].

Several strategies have been used, such as the topical application of fluoride or calcium-phosphate formulations [Ganss et al., 2001; Ramalingam et al., 2005].

Another option is the addition of calcium, phosphate, iron, ferrous sulphate, to mouth-rinses or toothpastes. The addition of different ions or complexes to erosive agents, rendering the agent supersaturated with respect to tooth mineral, shows protection against erosion. Similarly promising appears to be the formation of a protective layer on dental hard tissue, composed of different ions or molecules and attached directly to the tooth surface or to the pellicle [Jaeggi and Lussi, 2006].

In recent years, research groups have investigated the preventive effects of different fluoride formulations on dental erosion, or their effects on abrasion resistance after an erosive attack. Differences among protective capabilities were found. Using different approaches, several groups demonstrated increased wear resistance after the application of a highly fluoridated gel with a pH of 4.75 on eroded enamel [Attin et al., 1999; Ganss et al., 2001; Lussi et al., 2004a]. It was suggested that, after topical application, some mineral dissolved from the enamel surface, increasing the local pH and leading to precipitation of CaF2-like material, which leads to less softening than without this layer [Ganss et al., 2001]. It has been shown in vitro that fluoride treatment was unlikely to provide a preventive effect against erosion, because an acidic drink will rapidly dissolve accessible CaF2 and remove traces of a previous topical fluoride treatment [Larsen and Richards, 2002]. In a cycling model of erosion (citric acid, 5 min, pH 3.5) and abrasion, less wear was shown with fluoride toothpaste than with non-fluoride toothpaste [Bartlett et al., 1994]. Another group found, in an in situ study, a dose response of F present in toothpastes after 25 min of in vitro exposure to grapefruit juice [Zero et al., 2006]. The two dentifrices with F contents of 1100 ppm and 1150 ppm significantly enhanced remineralization of the previously softened enamel, and they improved resistance against erosion when compared with the placebo toothpaste (0 ppm F).

It must be noted that dental erosion appears to be on the increase, despite the fact that most toothpastes are fluoridated [Lussi et al., 2008]. Thus, the effectiveness of fluoride against dental erosion, in concentrations usual for toothpastes, appears to be limited to specific research models, and may be dependent on the active fluoride and the acidic attack.

4a. The Role of Casein Phosphopeptide-Amorphous Calcium Phosphate Against Dental Erosion

Agents based on milk products (CPP- ACP) have been researched for contrasting dental erosion caused by acidic drinks [Yamaguchi et al., 2007; Ramalingam et al., 2005].

Several studies have shown that capability against erosion: Ramalingam et al. conducted an in vitro study to determine a minimal concentration of casein phosphopeptide-stabilized amorphous calcium phosphate (CPP-ACP) which when added to a sports drink would eliminate such erosion in vitro.

Human enamel specimens were immersed in: (1) the sports drink Powerade; (2) Powerade plus 4 concentrations of CPP-ACP (0.063%, 0.09%, 0.125%, 0.25%); or (3) double deionized water. Windows of test and control enamel were profiled, and the enamel surface characteristics were examined under scanning electron microscopy (SEM). Erosive step lesions occurred in specimens immersed in Powerade (mean depth=38.70kA±5.60), which were eliminated by the addition of CPP-ACP to Powerade at all test concentrations except 0.063% CPP-ACP. Microscopic surface irregularities on test enamel were observed, apparent as adherent granules or globules. These may represent re-deposited mineral phases following mobilization of calcium and phosphate from CPP-ACP. The authors concluded that adding casein phosphopeptide-stabilized amorphous calcium phosphate to the sports drink Powerade significantly reduced the beverage's erosivity without affecting the product's taste [Ramalingam et al., 2005].

The effect of CPP-ACP paste (Tooth Mousse) on the demineralization process in dentin was evaluated by ultrasonic velocity [Yamaguchi et al., 2007]. Demineralization was accomplished with an undersaturated 0.1 M lactic acid buffer (pH 4.75). Tooth Mousse or placebo paste was used for 10 min twice a day before demineralization. The control group was stored in artificial saliva. The sonic velocities in the demineralization and placebo groups were significantly lower than in the control and the experimental groups after 21 and 28 days. It was concluded that Tooth Mousse was effective in preventing demineralization.

Another in vitro study compared the hardness of normal enamel with enamel eroded by a cola soft drink and enamel remineralized by casein phosphopeptide–amorphous calcium phosphate (CPP-ACP) or artificial saliva. 40 extracted sound central and lateral incisors were alternately immersed in a cola soft drink or artificial saliva for 10 cycles of five seconds each, for two times at six-hour intervals. Then, samples were randomly divided into four groups and (CPP-ACP, artificial saliva, CPP-ACP plus artificial saliva, deionized water). The hardness was measured at baseline, after erosion and after remineralization. The cola soft drink significantly decreased enamel hardness. CPP-ACP and CPP-ACP plus artificial saliva significantly increased the hardness of eroded enamel. CPP-ACP and CPP-ACP plus artificial saliva increased the hardness of eroded enamel significantly more than artificial saliva did [Panich and Poolthong, 2009].

Rees and colleagues reported that a single topical application of proenamel or tooth mousse would prevent enamel erosion. Enamel samples were treated with either proenamel or tooth mousse applied for 15 min. The control group was placed in distilled water for 15 min. Then, all specimens were then exposed to an erosive challenge of 0.2% citric acid for 1h. Enamel loss was determined using surfometry. The mean amount of enamel removed in the control group was 5.02 microm (S.D. 1.16). The mean enamel loss in the proenamel group was 2.60 microm (S.D. 0.90) and the mean enamel loss in the tooth mousse group was 3.28 microm (S.D. 1.22). The authors concluded that both tooth mousse and proenamel may offer a degree of protection from enamel erosion [Rees et al, 2007].

Another study evaluated the use of a CPP–ACP paste (Tooth Mousse) as a protective factor on in vitro enamel demineralization produced by a soft drink (Coca Cola), using Atomic Force Microscopy (AFM). Results shown that use of a CPP–ACP paste had a protective effect on in vitro demineralization model of dental enamel [Poggio et al., 2009].

5. Conclusions

Dental caries continues to be a major health problem in adults and children. Erosion of dental hard tissue is becoming increasingly important to the long-term health of dentition. There is a great need for improved and novel remineralization methods both to alter the caries balance for the better, especially in high risk individuals, and to protect dental enamel against erosion.

The CPP-ACP nano- complexes have been shown to remineralize enamel subsurface lesions *in vivo,* to slow the progression of caries significantly, to promote the regression of early caries in randomized, controlled, clinical trials and to give protection against soft drinks erosion. Therefore, evidence exists to support the clinical use as agent for the control of dental caries, and it can be used as an adjunct preventive therapy to reduce caries in high risk patients, to reduce dental erosion in patients with gastric reflux or other disorders, to reduce decalcification in orthodontic patients, to repair enamel in cases involving white-spot lesions, orthodontic decalcification or fluorosis.

REFERENCES

Adolfsson O, Meydani SN, Russell RM. Yogurt and gut function. *Am J Clin Nutr* 2004;80(2):245-56. Review.

Aimutis WR. Bioactive properties of milk proteins with particular focus on anticariogenesis. *J Nutr* 2004;134(4):989S-95S.

Al-Malik MI, Rehbini YA. Prevalence of dental caries, severity, and pattern in age 6 to 7-year-old children in a selected community in Saudi Arabia. *J Contemp Dent Pract* 2006;7(2):46-54.

Andersson A, Sköld-Larsson K, Hallgren A, Petersson LG, Twetman S. Effect of a dental cream containing amorphous cream phosphate complexes on white spot lesion regression assessed by laser fluorescence. *Oral Health Prev Dent* 2007;5(3):229-33.

Arends J, Ruben JL, Inaba D. Major topics in quantitative microradiography of enamel and dentin: R parameter, mineral distribution visualization, and hyper-remineralization. *Adv Dent Res* 1997;11(4):403-14.

Attin T, Deifuss H, Hellwig E. Influence of acidified fluoride gel on abrasion resistance of eroded enamel. *Caries Res* 1999;33(2):135-9.

Azarpazhooh A, Limeback H. Clinical efficacy of casein derivatives: a systematic review of the literature. *J Am Dent Assoc* 2008;139(7):915-24.

Bartlett DW, Smith BG, Wilson RF. Comparison of the effect of fluoride and non-fluoride toothpaste on tooth wear in vitro and the influence of enamel fluoride concentration and hardness of enamel. *Br Dent J* 1994;176(9):346-8.

Bavetta LA, McClure FJ. Protein factors and experimental rat caries. *J Nutr* 1957;63(1):107-117.

Bouhallab S, Bouglé D. Biopeptides of milk: caseinophosphopeptides and mineral bioavailability. *Reprod Nutr Dev* 2004;44(5):493-8.

Cai F, Manton DJ, Shen P, Walker GD, Cross KJ, Yuan Y, Reynolds C, Reynolds EC. Effect of addition of citric acid and casein phosphopeptide-amorphous calcium phosphate to a sugar-free chewing gum on enamel remineralization in situ. *Caries Res* 2007;41(5):377-83.

Cai F, Shen P, Morgan MV, Reynolds EC. Remineralization of enamel subsurface lesions in situ by sugar-free lozenges containing casein phosphopeptide-amorphous calcium phosphate. *Aust Dent J* 2003;48(4):240-3.

Cochrane NJ, Cai F, Yuan Y, Reynolds EC. Erosive potential of beverages sold in Australian schools. *Aust Dent J* 2009;54(3):238-44.

Cochrane NJ, Saranathan S, Cai F, Cross KJ, Reynolds EC. Enamel subsurface lesion remineralisation with casein phosphopeptide stabilised solutions of calcium, phosphate and fluoride. *Caries Res* 2008;42(2):88-97.

Correr GM, Alonso RC, Correa MA, Campos EA, Baratto-Filho F, Puppin-Rontani RM. Influence of diet and salivary characteristics on the prevalence of dental erosion among 12-year-old schoolchildren. *J Dent Child* (Chic) 2009;76(3):181-7.

Cross KJ, Huq NL, Palamara JE, Perich JW, Reynolds EC. Physicochemical characterization of casein phosphopeptide-amorphous calcium phosphate nanocomplexes. *J Biol Chem* 2005;280(15): 15362-9.

Cross KJ, Huq NL, Stanton DP, Reynolds EC. NMR studies of a novel calcium, phosphate and fluoride delivery vehicle-alpha(S1)-casein(59-79) by stabilized amorphous calcium fluoride phosphate nanocomplexes. *Biomaterials* 2004;25(20):5061-9.

Cypriano S, Hoffmann RH, de Sousa Mda L, Wada RS. Dental caries experience in 12-year-old schoolchildren in southeastern Brazil. *J Appl Oral Sci* 2008;16(4):286-92.

de Almeida CM, Petersen PE, Andre SJ, Toscano A. Changing oral health status of 6- and 12-year-old schoolchildren in Portugal. *Community Dent Health* 2003;20(4):211-6.

de Carvalho Sales-Peres SH, Magalhães AC, de Andrade Moreira Machado MA, Buzalaf MA. Evaluation of the erosive potential of soft drinks. *Eur J Dent* 2007;1(1):10-3.

DePaola PF, Soparkar PM, Triol C, Volpe AR, Garcia L, Duffy J, Vaughan B. The relative anticaries effectiveness of sodium monofluorophosphate and sodium fluoride as contained in currently available dentifrice formulations. *Am J Dent* 1993;6 Spec No:S7-12.

Featherstone JD. The science and practice of caries prevention. *J Am Dent Assoc* 2000;131(7):887-99.

Featherstone JD. Remineralization, the natural caries repair process--the need for new approaches. *Adv Dent Res* 2009;21(1):4-7.

Ferrazzano GF, Cantile T, Ingenito A, Chianese L, Quarto M. New strategies in dental caries prevention: experimental study on casein phosphopeptides. *Eur J Paediatr Dent* 2007;8(4):183-7.

Ferrazzano GF, Cantile T, Quarto M, Ingenito A, Chianese L, Addeo F. Protective effect of yogurt extract on dental enamel demineralization in vitro. *Aust Dent J* 2008;53(4):314-9.

Ferrazzano GF, Scaravilli MS, Ingenito A. Dental and periodontal health status in Campanian children and relation between caries experience and socio-economic behavioural factors. *Eur J Paediatr Dent* 2006;7(4):174-8.

Ferrazzano GF, Aamto I, Cantile T, Sangianantoni G, Ingenito A. In vivo remineralizing sffect of GC Tooth Mousse on early dental esame lesions: Sem analysis. *Int Dent J* 2011 Aug; 61(4): 2120-6

Ferro R, Besostri A, Olivieri A, Stellini E, Mazzoleni S. Preschoolers' dental caries experience and its trend over 20 years in a North-East Italian Health District. *Eur J Paediatr Dent* 2007;8(4):199-204.

Ganss C, Klimek J, Schäffer U, Spall T. Effectiveness of two fluoridation measures on erosion progression in human enamel and dentine in vitro. *Caries Res* 2001;35(5):325-30.

Gray M, Morris AJ, Davies J. The oral health of South Asian five-year-old children in deprivated areas of Dudley compared with White children of equal deprivation and fluoridation status. *Community Dent Health* 2000;17(4):243-5.

Guggenheim B, Neeser JR, Golliard M, Schüpbach P. Salivary pellicle modified by milk components mediated caries prevention. *1995.41st ORCA Congress Abstracts*, pag 182.

Guggenheim B, Schmid R, Aeschlimann JM, Berrocal R, Neeser JR. Powdered milk micellar casein prevents oral colonization by Streptococcus sobrinus and dental caries in rats: a basis for the caries-protective effect of dairy products. *Caries Res* 1999;33(6):446-54.

Hallett KB, O'Rourke PK. Caries experience in preschool children referred for specialist dental care in hospital. *Aust Dent J* 2006;51(2):124-129.

Hamissi J, Ramezani GH, Ghodousi A. Prevalence of dental caries among high school attendees in Qazvin, Iran. *J Indian Soc Pedod Prev Dent* 2008;26 Suppl 2:S53-5.

Herod EL. The effect of cheese on dental caries: a review of the literature. *Aust Dent J* 1991;36(2):120-5.

Iijima Y, Cai F, Shen P, Walker G, Reynolds C, Reynolds EC. Acid resistance of enamel subsurface lesions remineralized by a sugar-free chewing gum containing casein phosphopeptide-amorphous calcium phosphate. *Caries Res* 2004;38(6):551-6.

Jacobsen P, Young D. The use of topical fluoride to prevent or reverse dental caries. *Spec Care Dentist* 2003;23(5):177-9.

Jaeggi T, Lussi A. Prevalence, incidence and distribution of erosion. *Monogr Oral Sci* 2006;20:44-65.

Larsen MJ, Richards A. Fluoride is unable to reduce dental erosion from soft drinks. *Caries Res* 2002;36(1):75-80.

Levine RS. Milk, flavoured milk products and caries. *Br Dent J* 2001;191(1):20.

Linnett V, Seow WK. Dental erosion in children: a literature review. *Pediatr Dent* 2001;23(1):37-43.

Llena C, Forner L, Baca P. Anticariogenicity of casein phosphopeptide-amorphous calcium phosphate: a review of the literature. *J Contemp Dent Pract* 2009;10(3):1-9.

Lussi A, Jaeggi T. Erosion-diagnosis and risk factors. *Clin Oral Investig* 2008;12 Suppl 1:S5-13.

Lussi A, Jaeggi T, Schaffner M. Prevention and minimally invasive treatment of erosions. *Oral Health Prev Dent* 2004a;2 Suppl 1:321-5.

Lussi A, Jaeggi T, Schärer S. The influence of different factors on in vitro enamel erosion. *Caries Res* 1993;27(5):387-93.

Lussi A, Jaeggi T, Zero D. The role of diet in the aetiology of dental erosion. *Caries Res* 2004b;38 Suppl 1:34-44.

Lussi A, Megert B, Eggenberger D, Jaeggi T. Impact of different toothpastes on the prevention of erosion. *Caries Res* 2008;42(1):62-7.

Lussi A, Schaffner M, Hotz P, Suter P. Dental erosion in a population of Swiss adults. *Community Dent Oral Epidemiol* 1991;19(5):286-290.

Manton DJ, Walker GD, Cai F, Cochrane NJ, Shen P, Reynolds EC. Remineralization of enamel subsurface lesions in situ by the use of three commercially available sugar-free gums. *Int J Paediatr Dent* 2008;18(4):284-90.

Merritt J, Qi F, Shi W. Milk helps build strong teeth and promotes oral health. *J Calif Dent Assoc* 2006;34(5):361-6.

Milciuviene S, Bendoraitiene E, Andruskeviciene V, Narbutaite J, Sakalauskiene J, Vasiliauskiene I, Slabsinskiene E. Dental caries prevalence among 12-15-year-olds in Lithuania between 1983 and 2005. *Medicina* (Kaunas) 2009;45(1):68-76.

Morgan MV, Adams GG, Bailey DL, Tsao CE, Fischman SL, Reynolds EC. The anticariogenic effect of sugar-free gum containing CPP-ACP nanocomplexes on approximal caries determined using digital bitewing radiography. *Caries Res* 2008;42(3):171-84.

Neville MC, Keller RP, Casey C, Allen JC. Calcium partitioning in human and bovine milk. *J Dairy Sci* 1994;77(7):1964-75.

Panich M, Poolthong S. The effect of casein phosphopeptide-amorphous calcium phosphate and a cola soft drink on in vitro enamel hardness. *J Am Dent Assoc* 2009;140(4):455-60.

Petesen PE. Sociobehavioral risk factors in dental caries –international perspectives. *Community Dent Oral Epidemiol* 2005;33(4):274-279.

Petersen PE, Bourgeois D, Ogawa H, Estupinan-Day S, Ndiaye C. The global burden of oral diseases and risks to oral health. *Bull World Health Organ* 2005;83(9):661-9.

Poggio C, Lombardini M, Dagna A, Chiesa M, Bianchi S. Protective effect on enamel demineralization of a CPP-ACP paste: an AFM in vitro study. *J Dent* 2009;37(12):949-54.

Ramalingam L, Messer LB, Reynolds EC. Adding casein phosphopeptide-amorphous calcium phosphate to sports drinks to eliminate in vitro erosion. *Pediatr Dent* 2005;27(1):61-7.

Rasić JL, Kurmann JA. Bifidobacteria and their role. Microbiological, nutritional-physiological, medical and technological aspects and bibliography. *Experientia Suppl* 1983;39:1-295.

Rees J, Loyn T, Chadwick B. Pronamel and tooth mousse: an initial assessment of erosion prevention in vitro. *J Dent* 2007;35(4): 355-357.

Reynolds EC. Remineralization of enamel subsurface lesions by casein phosphopeptide-stabilized calcium phosphate solutions. *J Dent Res* 1997;76(9):1587-95.

Reynolds EC. Anticariogenic complexes of amorphous calcium phosphate stabilized by casein phosphopeptides: a review. *Spec Care Dentist* 1998;18(1):8-16. Review.

Reynolds EC. Casein phosphopeptide-amorphous calcium phosphate: the scientific evidence. *Adv Dent Res* 2009;21(1):25-9.

Reynolds EC, Cai F, Cochrane NJ, Shen P, Walker GD, Morgan MV, Reynolds C. Fluoride and casein phosphopeptide-amorphous calcium phosphate. *J Dent Res* 2008;87(4):344-8.

Reynolds EC, Cai F, Shen P, Walker GD. Retention in plaque and remineralization of enamel lesions by various forms of calcium in a mouthrinse or sugar-free chewing gum. *J Dent Res* 2003;82(3):206-11.

Reynolds EC, Cain CJ, Webber FL, Black CL, Riley PF, Johnson IH, Perich JW. Anticariogenicity of calcium phosphate complexes of tryptic casein phosphopeptides in the rat. *J Dent Res* 1995;74(6):1272-9.

Reynolds EC, del Rio A. Effect of casein and whey-protein solutions on caries experience and feeding patterns of the rat. *Arch Oral Biol* 1984;29(11):927-933.

Rios D, Magalhães AC, Polo RO, Wiegand A, Attin T, Buzalaf MA. The efficacy of a highly concentrated fluoride dentifrice on bovine enamel subjected to erosion and abrasion. *J Am Dent Assoc* 2008;139(12):1652-6.

Rose RK. Binding characteristics of *Streptococcus mutans* for calcium and casein phosphopeptide. *Caries Res* 2000a;34(5):427-431.

Rose RK. Effects of an anticariogenic casein phosphopeptide on calcium diffusion in streptococcal model dental plaques. *Arch Oral Biol* 2000b;45(7):569-575.

Selwitz RH, Ismail AI, Pitts NB. Dental caries. *Lancet* 2007;369(9555):51-9. Review.

Shen P, Cai F, Nowicki A, Vincent J, Reynolds EC. Remineralization of enamel subsurface lesions by sugar-free chewing gum containing casein phosphopeptide-amorphous calcium phosphate. *J Dent Res* 2001;80(12):2066-70.

Silva MF, Burgess RC, Sandham HJ, Jenkins GN. Effects of water-soluble components of cheese on experimental caries in humans. *J Dent Res* 1987;66(1):38-41.

Simonsen RJ. Pit and fissure sealant: review of the literature. *Pediatr Dent* 2002;24(5):393-414.

Skeie MS, Riordan PJ, Klock KS, Espelid I. Parental risk attitudes and caries related behaviors among immigrant and western native children in Oslo. *Community Dent Oral Epidemiol* 2006;34(2):103-113.

Sudjalim TR, Woods MG, Manton DJ, Reynolds EC. Prevention of demineralization around orthodontic brackets in vitro. *Am J Orthod Dentofacial Orthop* 2007;131(6):705.e1-9.

Sundby A, Petersen PE. Oral health status in relation to ethnicity of children in the municipality of Copenhagen, Denmark. *Int J Paediatr Dent* 2003;13(3):150-7.

ten Cate JM. The need for antibacterial approaches to improve caries control. *Adv Dent Res* 2009;21(1):8-12.

ten Cate JM, Featherstone JD. Mechanistic aspects of the interactions between fluoride and dental enamel. *Crit Rev Oral Biol Med* 1991;2(3):283-96.

Truin GJ, Van Rijkom HM, Mulder J, Van't Hof MA. Caries trends 1996-2002 among 6- and 12-year-old children and erosive wear prevalence among 12-year-old children in the Hague. *Caries Res* 2005;39(1):2-8.

Vered Y, Zini A, Livny A, Mann J, Sgan-Cohen HD. Changing dental caries and periodontal disease patterns among a cohort of Ethiopian immigrants to Israel: 1999-2005. *BMC Public Health* 2008;8:345.

Walker G, Cai F, Shen P, Reynolds C, Ward B, Fone C, Honda S, Koganei M, Oda M, Reynolds E. Increased remineralization of tooth enamel by milk containing added casein phosphopeptide-amorphous calcium phosphate. *J Dairy Res* 2006;73(1):74-8.

Wiegand A, Attin T. Influence of fluoride on the prevention of erosive lesions - a review. *Oral Health Prev Dent* 2003;1(4):245-253.

Wyne AH .Caries prevalence, severity, and pattern in preschool children. *J Contemp Dent Pract* 2008;9(3):24-31.

Yamaguchi K, Miyazaki M, Takamizawa T, Inage H, Kurokawa H. Ultrasonic determination of the effect of casein phosphopeptide-amorphous calcium phosphate paste on the demineralization of bovine dentin. *Caries Res* 2007;41(3):204-7.

Young WG. Tooth wear: diet analysis and advice. *Int Dent J* 2005;55(2):68-72.

Zero DT, Hara AT, Kelly SA, González-Cabezas C, Eckert GJ, Barlow AP, Mason SC. Evaluation of a desensitizing test dentifrice using an in situ erosion remineralization model. *J Clin Dent* 2006;17(4):112-6.

INDEX

A

B

C

D

I

J

K

L

M

N

O

P

Q

R

S

T

U

V

W

X

Y

Z